Radivoj V. Krstić

General Histology of the Mammal

An Atlas for Students of Medicine and Biology

With a Foreword by R. J. Reiter

Translated by S. Forster

With 190 Plates Drawn by the Author

Springer-Verlag
Berlin Heidelberg New York Tokyo

Dr. R. V. KRSTIĆ, Professeur associé, Université de Lausanne,
Institut d'Histologie et d'Embryologie, 1005 Lausanne/Switzerland

Translator:
STEVEN FORSTER, 2 School Lane, Stannington,
Sheffield S6 6DD, South Yorkshire/Great Britain

Title of the original edition: Die Gewebe des Menschen und der Säugetiere. Ein Atlas zum Studium für Mediziner und Biologen.
© Springer-Verlag Berlin Heidelberg 1978, 1982, 1984

ISBN 3-540-15346-2 Springer-Verlag Berlin Heidelberg New York Tokyo
ISBN 0-387-15346-2 Springer-Verlag New York Heidelberg Berlin Tokyo

Library of Congress Cataloging in Publication Data. Krstić, Radivoj V., 1935. General histology of the mammal. Translation of: Die Gewebe des Menschen und der Säugetiere. Bibliography. Includes index. 1. Histology – Atlases. 2. Mammals – Histology – Atlases. I. Title. [DNLM: 1. Histology – Atlases. QS 517 K93g] QM557.K7713 1985 599′.08′24 85-12650

This work is subject to copyright. All rights reserved, whether the whole or part of the material is concerned, specifically those of translation, reprinting, re-use of illustrations, broadcasting, reproduction by photocopying machine or similar means, and storage in data banks. Under § 54 of the German Copyright Law where copies are made for other than private use, a fee is payable to "Verwertungsgesellschaft Wort", Munich.

© Springer-Verlag Berlin Heidelberg 1985
Printed in Germany

The use of registered names, trademarks, etc. in this publication does not imply, even in the absence of a specific statement, that such names are exempt from the relevant protective laws and regulations and therefore free for general use.

Product liability: The publisher can give no guarantee for information about drug dosage and application thereof contained in this book. In every individual case the respective user must check its accuracy by consulting other pharmaceutical literature.

Typesetting, printing and bookbinding: Brühlsche Universitätsdruckerei, Giessen
2121/3130-543210

*Dedicated to the illustrious memory
of Professor* WOLFGANG BARGMANN

Foreword

In this book, concerned with the spatial and structural representation of the cell and its particular specializations, the author has deviated considerably from the usual plan for other books related to this subject and has presented the illustrative material in the form of detailed and accurate drawings. The layout of the book provides the reader with a brief narrative account of the particular organelle accompanied by a full-plate illustration on the facing page. Most of the narrative accounts are accompanied by a short bibliography of germane references in the event the reader desires to pursue the subject matter in greater depth. In my estimation there is no other presentation currently available which utilizes this approach to demonstrate the cellular components and their associated morphophysiology with such elegance.

The text is clearly written and, although the individual accounts are brief, they are highly informative with all of the important details being provided. The accuracy of the textural presentations and the continuity of expression provide strong evidence that the author has spent an enormous amount of time in preparing the text and in painstakingly drawing the illustrations. Certainly, an obvious strength of this volume is the high quality of the illustrative renditions, all of which were drawn by Dr. KRSTIĆ. These attest to the profound and comprehensive nature of the author's knowledge of the field of cellular and structural biology.

This book is truly a work of love and art by one who is gifted both didactically and artistically. Dr. KRSTIĆ should be congratulated for providing us with an extremely accurate and detailed account of the 2-dimensional, and in many cases the 3-dimensional, view of the cell and its component organelles. It has no equal.

Whereas the volume was not designed as a text book, it will be extremely valuable as a supplement to the more classical books that cover this subject matter. Furthermore, it will provide many with a concise and easy to follow review of a wide variety of biomorphological elements. Dr. KRSTIĆ has a talent and a perspective which is refreshing and allows a holographic view of the cell and its specializations. The book should have widespread appeal among histologists, electron microscopists, cell biologists and cytophysiologists. Every student and faculty member should have a copy of this important volume within easy reach.

San Antonio, Texas, Summer, 1985   RUSSELL J. REITER

Preface to the English Translation

Following two reprints of the German original and publication of a Japanese edition, the English translation of this atlas is now presented.

The translation of the book amounts to a new edition. Some textual changes compared with the German original have been required by the progress made in histology over the past few years. Some drawings have been altered in the light of new discoveries, or to facilitate comprehension. Recent references have also been added. The terminology in the present work is in keeping with that in my book *Illustrated Encyclopedia of Human Histology* (Springer 1984).

As stated in the preface to the German edition, this atlas is not intended to replace a textbook of histology; its goal is simply to complement other histological texts with three-dimensional drawings, illustrating general histology subjects and making them easier for the student to digest.

Aside from my cordial thanks to Springer-Verlag for its initiative concerning the present translation, I am most obliged to Mr. STEPHEN FORSTER for his very competent translation and to Mrs. D. NICOLAS for help in the preparation of this book.

Fully conscious of the existence of deficiencies in such an atlas, I remain open and thankful for any constructive criticism from the benevolent reader.

Le Mont-sur-Lausanne, Summer 1985   R. V. KRSTIĆ

Preface to the German Edition

The spatial interpretation of histological structures usually presents great difficulties to the less experienced. The images produced by transmission electron microscopy lack depth, and scanning electron micrographs, though often impressive, provide virtually no information about the inner structure of cells and tissues. Drawings can remedy these deficiencies by simultaneously depicting the internal and external morphology. For this reason, I decided to illustrate the subject matter of general histology largely by means of three-dimensional line drawings, a procedure I employed with cell structure in *Ultrastructure of the Mammalian Cell* (Springer, 1979).

The material in the present work has been arranged in accordance with the usual systematic approach adopted by textbooks of histology. This atlas is intended to supplement, not replace, a textbook and as such serves to clarify and expound the tissue structure of humans and animals for those studying or engaged in medicine, veterinary science, biology, and electron microscopy, for the physician wishing to further his knowledge, as well as for the interested layman.

The didactic concept of this book is based on a gradual introduction to the complexities of tissue structures and this inevitably entails a certain degree of repitition. In many drawings, it was impossible to present all details and faithfully reproduce all proportions. Here, I beg the reader's indulgence.

The references I have cited for further reading do not necessarily relate to the plate in question, but deal generally with the tissue under discussion. In many cases, reference is made to plates in my earlier atlas *Ultrastructure of the Mammalian Cell*, where the reader will find additional information.

I am most grateful to Prof. G. AUMÜLLER, Marburg on Lahn, for his kindness in critically reading the manuscript. I extend very warm thanks to Prof. OTTO BUCHER, the head of my institute, for his valuable comments and suggestions for improvement. I also thank Ms. CHRISTA THOMMEN and Mr. PIERRE-ANDRÉ MILLIQUET for technical assistance with the transmission and scanning electron-microscopic preparations that served as the basis for my drawings, Ms. HELMTRUD BRUDER-GERVERDINCK for language editing, Ms. SIMONE EVANS and Mrs. MARLIS VAN HOOYDONCK for typing the manuscript, and, not least, the Springer-Verlag for respecting my wishes with regard to the layout of the book.

I am always open to suggestions for improvement and constructive criticisms on the part of the reader.

Le Mont-sur-Lausanne   R. V. KRSTIĆ

Contents

Numbers refer to Plates

I. INTRODUCTION

Plate

Origin of Tissues 1
Differentiation. Closed and Wide-Meshed Cell Unions . . 2
Four Tissue Groups 3
Growth, Regeneration, Hyperplasia, Hypertrophy, Atrophy,
    Involution, Degeneration, and Necrosis 4
Metaplasia 5
Transplantation, Implantation, and Explantation 6

II. EPITHELIAL TISSUE

Surface Epithelia
Classification of Surface Epithelia 7
Localization of Various Surface Epithelia 8
    Simple Squamous Epithelium 9
    Simple Cuboidal Epithelium 11
    Simple Columnar Epithelium 14
    Pseudostratified Columnar Epithelium 17
    Transitional Epithelium 19
    Stratified Columnar Epithelium 21
    Nonkeratinized Stratified Squamous Epithelium 22
    Keratinized Stratified Squamous Epithelium (Epidermis) 23

    Secretory Surface Epithelia
        Amniotic Epithelium 25
        Vascularized Secretory Epithelium 26
    Atypical Epithelia 27
        Enamel Organ 28
        Thymus 29
        Changes in Form of Epithelial Cells 30
    Functions of Surface Epithelia 31

Glandular Epithelia
    Exocrine Glands
        Development of Exocrine and Endocrine Glands . . . 32
        Classification of Exocrine Glands 33
        Form of Exocrine Glands 34
            Unicellular Glands 35
            Endoepithelial Multicellular Gland 36
            Tubular Gland 37

	Plate
Tubuloacinar Gland	38
Tubuloalveolar Gland	39
Apocrine Alveolar Gland	40
Holocrine Alveolar Gland	41
Scheme of a Compound Tubuloacinar Gland	42
Endocrine Glands	43
Regeneration and Transplantation of Epithelial Tissue	47

III. CONNECTIVE AND SUPPORTING TISSUES

Origin of Connecting and Supporting Tissues	48
Classification of Connective and Supporting Tissues	49
Mesenchyme	50
Gelatinous or Mucous Connective Tissue	52
Reticular Connective Tissue	53
White Adipose Tissue	56
White Adipose Tissue Cell	58
Brown Adipose Tissue	59
Brown Adipose Tissue Cell	61
Loose Connective Tissue	62
Fixed Cells: Fibroblast and Fibrocyte	64
Formed and Amorphous Components of the Intercellular Substance	65
Wandering Cells	
Histiocyte	66
Reticulohistiocytic or Reticuloendothelial System	68
Mast Cell	69
Lymphocytes	70
Monocyte	71
Plasma Cell	72
Eosinophilic Granulocyte	73
Special Forms	
Pigment Connective Tissue	74
Pigment Cell	75
"Cellular" Connective Tissue	76
Retiform Connective Tissue	77
Irregular Dense Connective Tissue	78
Regular Dense Connective Tissue	
Tendon	79
Aponeurosis	81
Elastic Connective Tissue	82
Cartilaginous Tissue	
Hyaline Cartilage	83
Elastic Cartilage	86
Chondrocyte	88
Fibrous Cartilage	89
Notochordal Tissue	90

	Plate
Bony Tissue	
Direct Bone Formation	91
Osteoblasts	94
Indirect Bone Formation	95
Secondary Bone Formation	105
Structure of Bone	106
Osteon	107
Osteocyte	110
Osteoclast	111
Regeneration of Bone	112
Dentin and Cementum	113
Possibilities of Transplanting Connective and Supporting Tissues	114
Graft Rejection. Role of Free Connective Tissue Cells	115

IV. MUSCULAR TISSUE

	Plate
Three Types of Muscular Tissue	116
Occurrence of Smooth Musculature	117
Smooth Muscle Cells	118
Hypothetical Model of Contraction Mechanism of Smooth Muscle Cells	122
Myoepithelial Cells	123
Skeletal Musculature	
Histogenesis of Skeletal Musculature	124
Structure of a Muscle	125
Structure of a Muscle Fascicle	126
White Muscle Fiber and Satellite Cell	127
Red Muscle Fiber	128
Structure of a Muscle Fiber	129
Scheme of a Relaxed and a Contracted Myofibril	130
Interaction Between Actin and Myosin Myofilaments	131
Myotendinal Junction	132
Regeneration of Skeletal Muscle Fibers	133
Cardiac Musculature	
Light-Microscopic Appearance	134
Cardiac Muscle Fibers	135
Electron-Microscopic Appearance	136
Cardiac Muscle Cell	137
Vascularization, Necrosis, Transplantation	139
Impulse-Conducting System	140
Purkinje Fibers	141
Cell of the Impulse-Conducting System	142

V. NERVOUS TISSUE

	Plate
Central Nervous System	
Histogenesis of Nervous Tissue	143

	Plate
Cells Originating from Neural Tube and Neural Crest	144
Central and Peripheral Nervous System	145
Relationship Between Neurons and Glia Cells in Central and Peripheral Nervous System	146
Glia of Central Nervous System	
Ependymal Cells	147
Astrocytes	148
Membrana Limitans Gliae Superficialis and Perivascularis	153
Blood-Brain Barrier	154
Oligodendrocytes	155
Microglia	156
Nerve Cells	
Form of Nerve Cells	157
Types of Nerve Cell	158
Arrangement of Nerve Cells	159
Nerve Cells and Neuropil	160
Neurosecretory Cells	162
Mode of Action of Polypeptide Hormones	163
Nerve Fibers and Schwann's Cells	164
Node of Ranvier	165
Schmidt-Lanterman Incisure	166
Regeneration of Nerve Fibers	167
Synapses	168
Components of Peripheral Nervous System	170
Spinal Ganglion	171
Pseudounipolar Nerve Cell	172
Bipolar Nerve Cell	173
Peripheral or Spinal Nerve	174
Perineurium	176
Endings of Efferent Nerve Fibers	
Motor End Plate	178
Endings of Autonomic Nerves in the Smooth Musculature	180
Endings of Afferent Nerve Fibers	
Neuromuscular Spindle	181
Golgi Tendon Organ	184
Endings of Afferent Nerve Fibers in Epithelial and Connective Tissues	185
Endings of Afferent Nerve Fibers Around Hair Follicle	186
Meissner's Corpuscle	187
Vater-Pacini Corpuscle	188
Sympathetic Trunk	189
Multipolar Autonomic Nerve Cell	190
GENERAL READING	391
INDEX	393

I Introduction

INTRODUCTION

Plate 1. Origin of Tissues

The development of tissues begins at the moment of fertilization, i.e., upon the union of a spermatozoon (Fig. A 1) with an ovum (Fig. A 2). The nucleus of the ovum and the head of the spermatozoon, both of which possess only half the full chromosomal complement, swell and form the female and male pronuclei. The two pronuclei double their deoxyribonucleic acid content, fuse, and thus form the nucleus of the first cell (zygote) of the new organism.

The zygote (Fig. B) divides mitotically along a meridional cleavage plane (Fig. B 1) into two daughter cells – blastomeres (Fig. C). Further meridional and equatorial divisions, following in rapid succession, produce 4, 8, 16, etc. blastomeres (Fig. D), which develop into a mulberry-shaped mass of cells, the morula (Fig. E). The inner cells, which give rise to the embryo, are referred to as the embryoblast (Fig. E 1), and the outer cells, which later form the nutritive organ of the embryo, the trophoblast (Fig. E 2). The blastocyst (Fig. F) or blastula develops by the gradual accumulation of fluid between the embryoblast (Fig. F 1) and trophoblast (Fig. F 2). Initially, small spaces form which soon coalesce, creating the large cavity of the blastocyst, the blastocoele (Fig. F 3).

About 1 week after fertilization (days 7–8), a small cleft can be observed in the embryoblast (Fig. G 1), which widens into the amniotic cavity (Fig. G 2). The base of this cavity, toward the blastocoele (Fig. G 3), consists of a layer of columnar cells, which represents the embryonic ectoderm (Fig. G 4). The extraembryonic cells opposite the ectoderm transform into the amniotic epithelium (Fig. G 5).

Beneath the ectodermal cells (Fig. H 1) a layer of cuboidal cells, the endoderm (Fig. H 2), develops by delamination. The free ends of this layer approach one another (arrows) and fuse. In this manner, the bilaminar embryonic disc and yolk sac (Fig. H 3) are established.

Seven days later, ectodermal cells migrate along the primitive streak between the layers of ectoderm and endoderm (Fig. I 1, 2), forming the middle germ layer, the mesoderm (Fig. I 3). The embryonic disc is now trilaminar.

A section through this structure is presented three-dimensionally in Fig. J. Between the ectodermal and endodermal cells (Fig. J 1, 2), the insertion of the mesodermal layer (Fig. J 3) can be seen.

All groups of tissues (Figs. K–N) develop from the trilaminar germ layer. Each germ layer has specific functions, though it does not possess histogenetic specificity, since the same type of tissue can emerge from different germ layers. Thus, all three germ layers give rise to epithelial tissue (**K**). Nervous tissue (**L**), except for microglial cells (see Plate 156) which develop from the middle layer, originates from the ectoderm. Connective, supporting (**M**), and muscular (**N**) tissues derive from the mesoderm; exceptions are the iris muscles and the myoepithelial cells of some glands, which are of ectodermal origin.

Magnifications: Figs. **A, F,** × 300;
Figs. **G, I,** × 150; Fig. **J,** × 100;
Fig. **K,** × 700; Fig. **L,** × 450;
Fig. **M,** × 550; Fig. **N,** × 450

(Note: all magnifications cited in this atlas are orientational values)

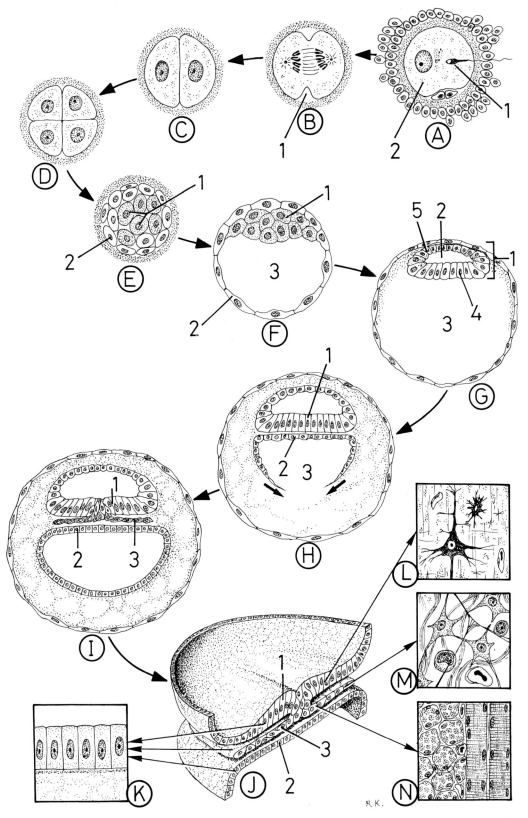

INTRODUCTION

Plate 2. Some Basic Terms: Differentiation, Tissue, Histology, Histophysiology, Closed and Wide-Meshed Cell Unions

It was stated in Plate 1 that all tissues derive from one zygote. Cell multiplication alone, however, does not give rise to a mature tissue; the latter also requires differentiation.

Differentiation signifies a gradual specialization of cells toward their subsequent functions by means of the appearance of characteristic structures (gliofibrils, myofibrils, neurofibrils, tonofibrils).

The development of two tissues and an organ is schematically presented in Fig. **A**.

The histogenetic process commences with the formation of a zygote, i.e., four to eight blastomeres (Fig. **A1**). By the process of division and differentiation, the two types of cell, **A2** and **A3**, are formed, each of which gives rise to a tissue. It is possible, therefore, to define a tissue as a union of identically differentiated cells which may also be complemented by specific noncellular structures. Histology, with histophysiology, is the branch of science dealing with the structure and functions of the various tissues of the body. Two or more tissues together with the nervous and vascular systems comprise an organ (Fig. **A4**).

Two basic types of cell union can be distinguished according to morphology: In closed cell unions, e.g., epithelial tissue (Fig. **B1**), the intercellular spaces are usually very narrow (from 20 nm to several micrometers); in a wide-meshed cell union, e.g., connective tissue (Fig. **B2**), the intercellular spaces are significantly wider and contain intercellular substance.

The central nervous system (CNS) may be regarded as a closed cell union. Here, the intercellular spaces are almost of the same width as in many epithelial tissues. Figure **C** shows an electron-microscopic magnification of part of the body of a nerve cell (Fig. **C1**), around which are a great many closely packed cell processes of unmyelinated (Fig. **C2**) and myelinated (Fig. **C3**) nerve fibers.

The circular inset illustrates the similarity in width of the intercellular spaces in epithelial tissue (Fig. **B1**) and the CNS: A gap (arrow), about 25 nm wide, is lined by two trilaminar unit membranes, approximately 12 nm across.

Magnifications: Fig. **B**, × 700;
Fig. **C**, × 3,500; circular inset, × 85,000

REFERENCES
Bellairs R (1974) Cell differentiation. In: Beck F, Lloyd JB (eds) The cell in medical science, vol 2. Academic, London
Flickinger RA (1982) Evolutionary aspects of cell differentiation. Int Rev Cytol 75:229–241
Monroy A (ed) (1978) Cell differentiation. Elsevier/North Holland, Amsterdam
Nover L, Luckner M, Parthier B (eds) (1982) Cell differentiation. Springer, Berlin Heidelberg New York

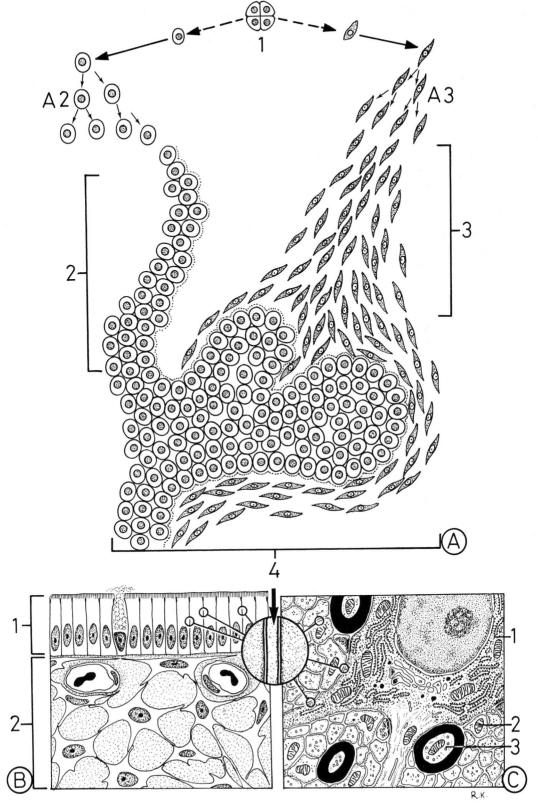

Plate 3. Four Tissue Groups

The bodies of humans and other mammals are made up of four major tissue groups. These are (illustrated by examples):

A) Epithelial tissue or epithelia [section of skin with epidermis and sweat (**1**) and sebaceous (**2**) glands];
B) Connective and supporting tissue (loose connective tissue);
C) Muscular tissue (cardiac muscle);
D) Nervous tissue (section of gray matter of the brain).

Epithelial tissue is subdivided into three major groups – surface epithelia, glandular epithelia, and sensory epithelia. The latter will not be dealt with in this book.
Surface epithelia form a lining over the entire outer and inner surfaces of the body. Glandular epithelia comprise cells that are specialized for external and internal secretion.

Connective tissue fulfills important metabolic and defensive functions in the organism, whereas supporting tissue plays largely a mechanical role.
(Blood is considered as the fluid connective tissue, but it will be not treated in this book.)
Muscular tissue, by virtue of the contractility of its cells or fibers, effects movement of the body and inner organs.
The highly differentiated nervous tissue is primarily responsible for receiving, transmitting, and processing information, originating from the environment or from within the body itself.
The plates in this book provide a systematic description of all the tissues referred to above.

Magnifications: Fig. **A**, $\times 70$;
Fig. **B**, $\times 400$; Fig. **C**, $\times 700$;
Fig. **D**, $\times 450$

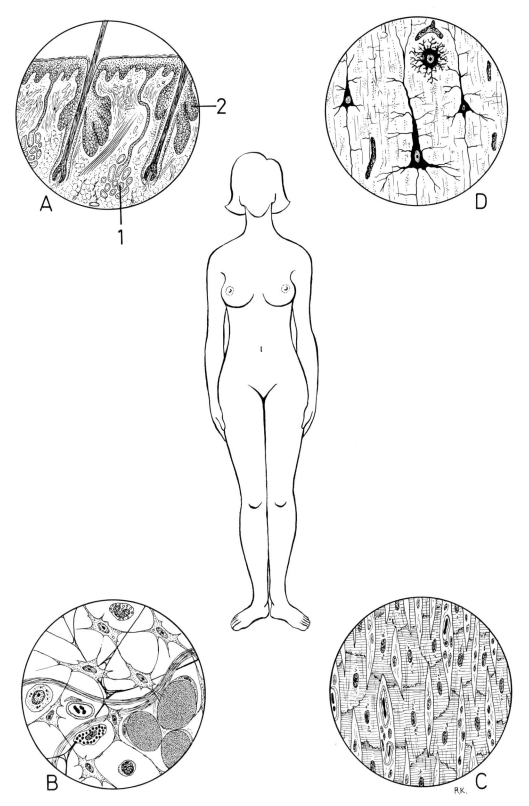

INTRODUCTION

Plate 4. Growth, Regeneration, Hyperplasia, Hypertrophy, Atrophy, Involution, Degeneration, and Necrosis

The basic faculty of cells to utilize high-molecular-weight, organic substances (primarily proteins) to form and replicate their living mass is termed growth. During embryonic development and in young organisms, growth mainly consists of cellular multiplication, achieved by mitotic division.

Physiological regeneration is the ability to replace lost cells in the tissues by cells identical to those lost. It is possible to distinguish two types of regeneration – continuous (blood cells, some epithelia; Plate 47) and cyclic regeneration (mucous coat of uterus; Plate 47). Reparative regeneration occurs following pathological tissue loss or after the operative removal of tissues or organs.

Regenerative power is inversely related to the degree of differentiation: Poorly differentiated tissues retain the ability to regenerate, whereas this ability is largely or completely lost to highly differentiated tissues (muscular and nervous tissue). When increased functional demands are made on tissues with good regenerative power, they adapt by increasing the number of their cells. This process is termed numerical growth or hyperplasia (Fig. **A 1**).

In tissues whose powers of division are limited or no longer existent, adaptation to increased demands simply takes the form of an increase in the size of the cells. This increase in volume is referred to as volumetric growth or hypertrophy (Fig. **A 2**).

Conversely, reduced demands on the output of cells, tissues, and organs lead to a reduction in function and volume termed atrophy (Fig. **B**).

Involution largely signifies a reduction in the number of cells (Fig. **C**). Degeneration (Fig. **D**) is associated with metabolic disorders, which can lead to the death of cells or tissues. Necrosis is the rapid circumscribed death of cells or tissues (Fig. **E**).

REFERENCES

Anderson WA, Sadler W (eds) (1982) Perspective in differentiation and hypertrophy. Elsevier/North Holland, New York

Bellairs R (1974) Cell growth and morphogenesis. In: Beck F, Lloyd JB (eds) The cell in medical science, vol 2. Academic, London

Nicolini C (ed) (1982) Cell growth. Plenum, New York

Wyllie AH, Kerr JF, Currie AR (1980) Cell death: The significance of apoptosis. Int Rev Cytol 68:251–306

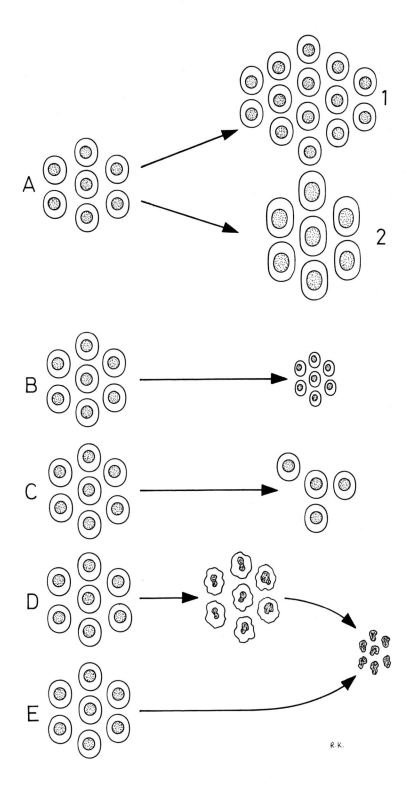

INTRODUCTION

Plate 5. Metaplasia

Metaplasia signifies the transformation of one mature, differentiated tissue into another, related, mature, differentiated tissue. Metaplasia is only possible within the same tissue group and can only occur in the epithelial and connective tissues. Epithelial tissue cannot transform into connective, supporting, or muscular tissue, neither can muscular tissue become nervous tissue. Metaplasia of smooth muscle into striated or cardiac muscle, and vice versa, is not possible. This plate illustrates some examples of metaplasia.

If the ductus deferens of a rat is experimentally converted to function as a ureter, the pseudostratified epithelium of the ductus deferens (Fig. **A1**) transforms into the transitional epithelium of the ureter (Fig. **A2**). This metaplasia may be regarded as functional adaptation.

Long-term irritation causes the ciliated epithelium (Fig. **B1**) of the trachea to be largely replaced by nonkeratinized stratified squamous epithelium (Fig. **B2**). Abnormal chronic mechanical irritations can bring about the transformation of nonkeratinized squamous epithelium (Fig. **C1**) into keratinized squamous epithelium (Fig. **C2**).

Metaplasia is possible, to a limited extent, among the connective tissues; thus, reticular tissue (Fig. **D1**) can transform into adipose tissue (Fig. **D2**). In the class of supporting tissues, however, metaplasia cannot take place: Hyaline cartilage (Fig. **E1**) cannot change into bone (Fig. **E2**).

All the above processes of metaplasia are reversible; if the pathological irritation or functional demand disappear the original tissue returns (broken arrow).

Magnifications: Fig. **A1**, ×800; Fig. **A2**, ×550; Fig. **B1**, ×550; Fig. **B2**, ×150; Fig. **C1, 2**, ×150; Fig. **D1, 2**, ×400; Fig. **E1, 2**, ×150

REFERENCE
Woodworth CD, Mossman BT, Craighead JE (1983) Squamous metaplasia of respiratory tract. Lab Invest 48:578–584

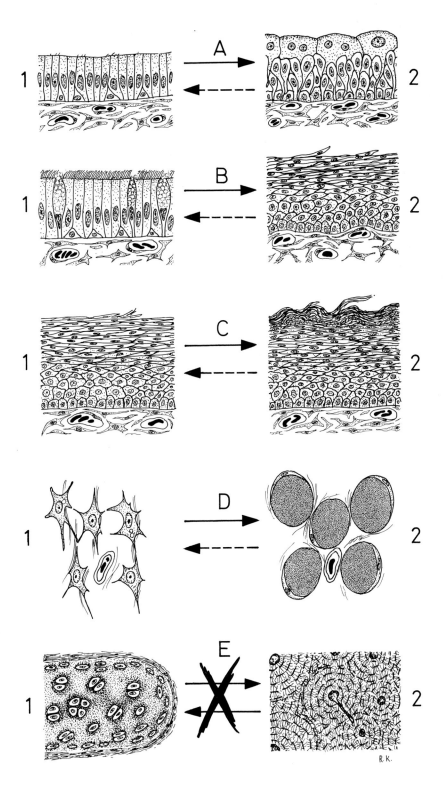

INTRODUCTION

Plate 6. Transplantation, Implantation, and Explantation

The removal of a tissue or organ from its natural environment in the body and its transfer to another position in the body of the same or a different individual is termed transplantation or grafting. Several possibilities of grafting exist.

In autoplastic transplantation (autografting), tissues or organs are transferred within the same individual (Fig. A).

Isoplastic transplantation (isografting) signifies that the grafting occurs between genetically identical individuals of the same species (monozygotic twins, animals of a pure inbred stock; Fig. B).

When an individual receives tissues or organs from a genetically nonidentical member of the same species this is termed homoplastic transplantation (homografting; Fig. C). The best-known example of this is blood transfusion.

The final possibility of grafting, heteroplastic transplantation (heterografting; Fig. D), concerns the transfer of tissues or organs from individuals of one species to individuals of another.

One of the major determining factors in the success of transplantation is the regenerative power of the grafted tissue; tissues with good regenerative abilities are easier to transplant than highly differentiated tissues. Another very important condition for a successful transplantation is the immunological similarity between donor and recipient (histocompatibility). If the differences are too great, an immunological reaction will result, which can lead to rejection of the transplant (graft rejection, see Plate 115).

The insertion of artificial blood vessels (Fig. E1), cardiac valves (Fig. E2), articular surfaces (Fig. E3), etc., made of metal or plastic, as well as electrical devices (e. g., pacemakers; see Plate 142) into an organism is termed implantation. Explantation signifies the transfer of pieces of tissue into a natural (blood plasma and tissue extract) or artificial culture medium (Fig. F). Here, the explants can be taken both from embryos (e. g., incubated eggs; Fig. F1) and from fully developed organisms (Fig. F2).

Auto-, iso-, and homotransplantation serve to replace diseased or lost parts of the body, whereas heterotransplantation and explantation are of interest to biological and medical research.

REFERENCES

Brent L, Fabre JW, Elves MW, Sells RA, Rappaport FT (1983) Transplantation today, vol 7. Grune and Stratton, San Francisco

Gale RP, Fox FC (eds) (1980) Biology of bone marrow transplantation. Academic, New York

Woodruff MFA, Nolan B (1964) Organ and tissue transplantation. In: Rob C, Smith R (eds) Clinical surgery, vol 1. Butterworths, London

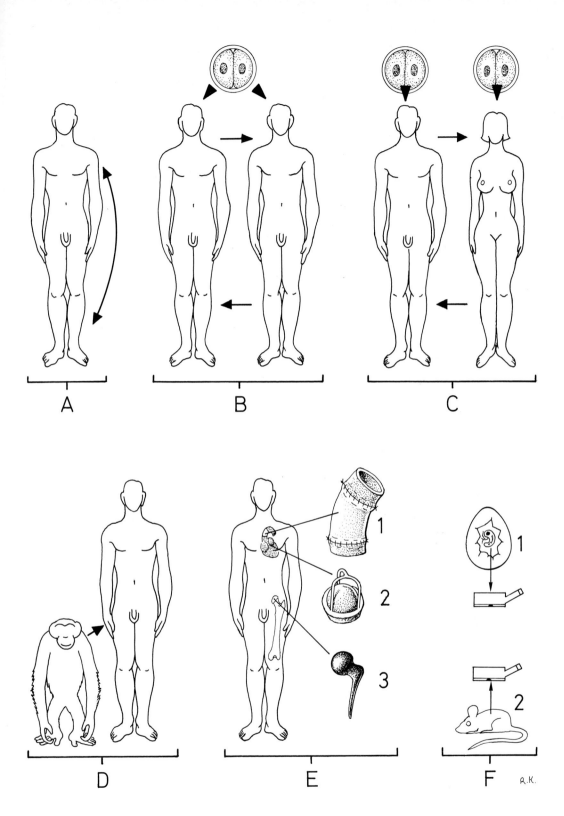

II Epithelial Tissue

Plate 7. Classification of Surface Epithelia. Modified from BUCHER 1980

Surface epithelia are closed, avascular, but innervated cell unions that cover and protect the underlying connective tissue. The various types of surface epithelium are depicted in this plate.

One criterion in differentiating pseudostratified from stratified epithelia is the relationship between the cells and the basal lamina (arrows). In the case of pseudostratified epithelia, all cells are in contact with the basal lamina though they do not necessarily extend to the epithelial surface. In stratified epithelia, on the other hand, only the basal cells are in contact with the basal lamina.

Stratified epithelia are named according to the shape of the cells that form the surface. Thus, the surface cells are prismatic in stratified columnar epithelia and plate-like in stratified squamous epithelia.

REFERENCE
Bucher O (1980) Cytologie, Histologie und mikroskopische Anatomie des Menschen, 10th edn. Huber, Bern

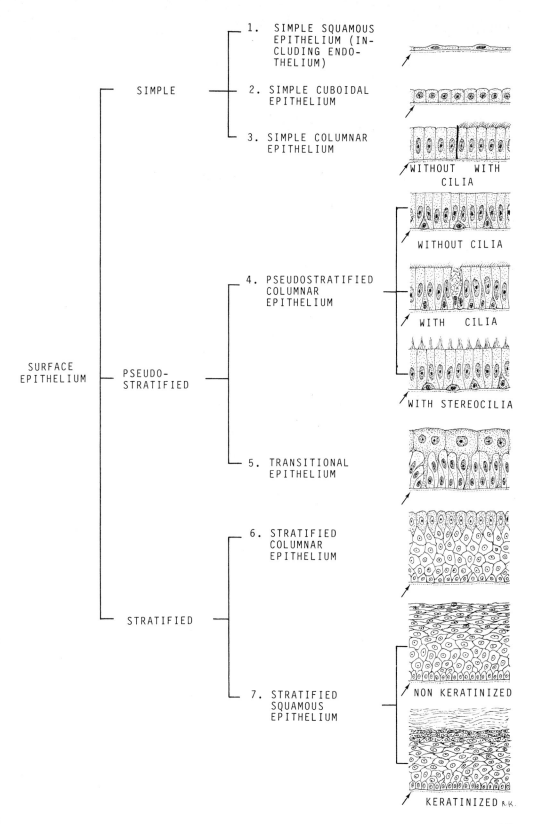

Plate 8. Localization of Various Surface Epithelia

This plate indicates, though without aiming to be complete, the distribution of surface epithelia in an organism.

Simple squamous epithelia (Fig. A) primarily line heart and body cavities (endocardium, pleura, peritoneum), all vessels, etc. Simple cuboidal epithelium (Fig. B) is not so widely distributed; it occurs in various parts of the renal tubules, choroid plexus, pigment epithelium of the retina, etc. Simple columnar epithelium (Fig. C) with cilia (kinocilia) is found in the oviducts and uterus (Fig. C 1), from the cardiac orifice to the anus as a lining of the intestinal tract (Fig. C 2), in the large collecting tubules, in the papillary ducts of the kidney (Fig. C 3), etc.

Pseudostratified columnar epithelium with and without stereocilia (Fig. D 1, 2) is found in the ductus epididymidis and ductus deferens; a pseudostratified epithelium with cilia (Fig. E) is present in the respiratory tract (nasal cavities, trachea, bronchial tree).

Transitional epithelium (Fig. F) is found specifically in the urinary tract (renal pelvis, ureter, bladder, initial section of the urethra).

Stratified columnar epithelium (Fig. G) is uncommon: It is largely found in transitional areas between stratified squamous and pseudostratified columnar epithelia (palate, epiglottis, fornix conjunctivae) and in part of the urethra.

Nonkeratinized stratified squamous epithelium (Fig. H) occurs in areas subject to greater mechanical stress, such as the oral cavity, esophagus, anus, vagina, and anterior epithelium of the cornea. This type of epithelium has to be constantly moistened by glandular secretion.

Keratinized stratified squamous epithelium (Fig. I) as the epidermis forms part of the skin. This epithelium protects the organism from physical and chemical noxae and hinders desiccation (perspiratio insensibilis).

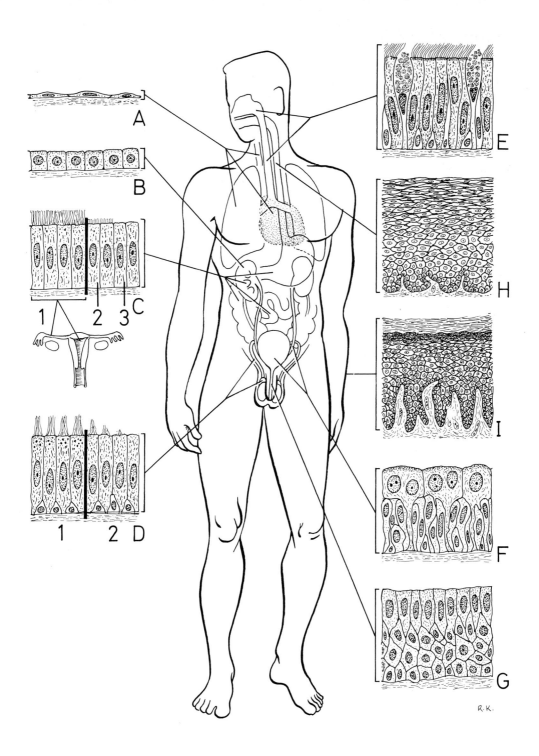

EPITHELIAL TISSUE

Plate 9. Surface Epithelia. Simple Squamous Epithelium.
Example: Vascular Endothelial Cells from a Small Vein of the Rat

This type of epithelium forms an inner coating of the heart and blood and lymph vessels.

On the inner surface of a vein (Fig. A), longitudinal, spindle-shaped, flattened epithelial cells are evident. Some of these cells (inset) are depicted three-dimensionally in Fig. B.

The cytoplasm of the endothelial cells is flattened; the nucleus-containing portion (Fig. B1) of the cell body bulges into the vascular lumen. The cells display elongated interdigitations (Fig. B2); the cell boundaries can be visualized with silver staining of the intercellular matrix substance (Fig. C). The junctions between the cells are strengthened by zonulae occludentes (Fig. B3) and are frequently covered by short marginal folds (Fig. B4). Fairly numerous microvilli (Fig. B5) are found on the free cell surface, especially in the vicinity of the nucleus. They are particularly numerous on resorptive squamous epithelial cells of mesodermal origin (mesothelium), which line the pleural and peritoneal cavities. Mesothelial cells (see Plates 56, 57) have a similar appearance to endothelial cells. The spherical structures on the cell surface (Fig. B6) possibly originate as a result of exocytosis (expulsion) of cell products. The minute invaginations on the free cell surface (Fig. B7) are micropinocytotic vesicles.

The cytoplasm of endothelial cells contains a few mitochondria (Fig. B8), some cisternae of rough endoplasmic reticulum (Fig. B9), and the Golgi apparatus (Fig. B10). Simple squamous epithelium lies, like all other epithelia, upon a basal lamina (Fig. B11). Collagen microfibrils (Fig. B12) separate this basal lamina from the basal lamina (Fig. B13) surrounding smooth muscle cells (Fig. B14).

Within the blood vessel, the endothelial cells provide a smooth surface so as to hinder blood coagulation.

Magnifications: Fig. A, $\times 220$; Fig. B, $\times 14,000$; Fig. C, $\times 700$

REFERENCES

Albert EN, Nayak RK (1976) Surface morphology of human aorta as revealed by the scanning electron microscope. Anat Rec 185:223–234

Andrews PM, Porter KR (1973) The ultrastructural morphology and possible functional significance of mesothelial microvilli. Anat Rec 177:409–426

Baradi AF, Hope J (1964) Observations on ultrastructure of rabbit mesothelium. Exp Cell Res 34:33–44

Edanaga M (1974) A scanning electron microscope study on the endothelium of the vessels: I. Fine structure of the endothelial surface of aorta and some other arteries in normal rabbits. Arch Histol Jpn 37:1–14

Hüttner T, Gabbiani G (1982) Vascular endothelium: Recent advances and unanswered questions. Lab Invest 47:409–411

Messmer K, Hammersen F (eds) (1983) Structure and function of endothelial cells. Progress in applied microcirculation, vol 1. Karger, Basle

Tokunaga J, Osaka M, Fujita T (1973) Endothelial surface of rabbit aorta as observed by scanning electron microscopy. Arch Histol Jpn 36:129–141

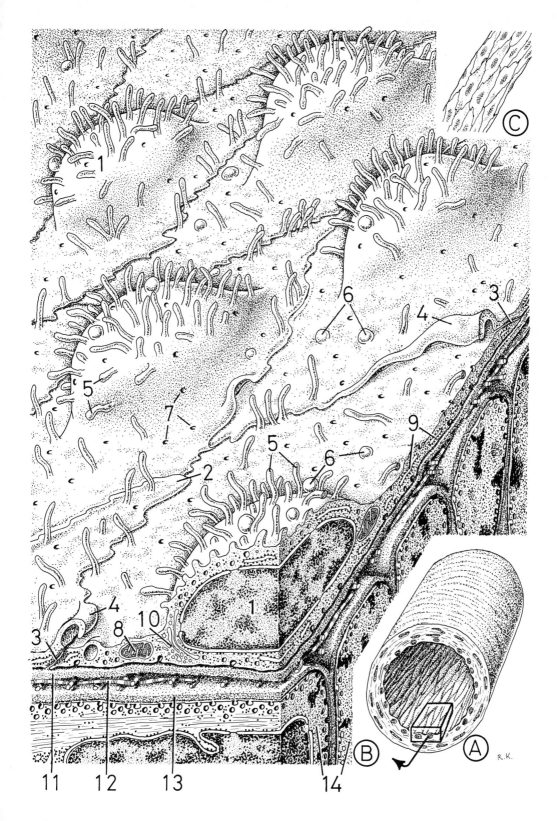

Plate 10. Surface Epithelia. Simple Squamous Epithelium.
Example: Endothelium of the Rat Cornea

Simple squamous epithelium, the so-called corneal endothelium, lines the posterior surface of the cornea (Fig. **A**). The cell boundaries can be visualized distinctly with silver staining, just as in vascular endothelium.

The forms of the cells and their interdigitations can be easily seen with increased magnification. The nuclei (Fig. **B1**) are flattened; unlike vascular endothelial cells, few microvilli (Fig. **B2**) are evident on the cell surface. The corneal endothelial cells lie on a thick Descemet's membrane (Fig. **B3**), which also constitutes their basal lamina. Two sets of collagen microfibrils (Fig. **B4**), perpendicular to one another, with interjacent processes of corneal fibroblasts or keratocytes (Fig. **B5**), can be seen.

Pulmonary alveoli are also lined with simple squamous epithelium, which predominantly consists of type I alveolar cells. Their extremely thin cytoplasm permits gas exchange and the passage of soluble substances between blood and alveolar air (see Plate 31).

Magnifications: Fig. **A**, $\times 65$; Fig. **B**, $\times 8,000$

REFERENCES

Blümcke S, Morgenroth K, JR (1967) The stereo ultrastructure of the external and internal surface of the cornea. J Ultrastruct Res 18:502–518

Miller MM, Revel J-P (1975) Scanning electron microscopy of epithelia prepared by blunt dissection. Anat Rec 183:339–358

Svedbergh B, Bill A (1972) Scanning electron microscopic studies of the corneal endothelium in man and monkey. Acta Ophthalmol (Copenh) 50:321–336

Weibel ER (1971) The mystery of "non-nucleated plates" in the alveolar epithelium explained. Acta Anat (Basel) 78:425–443

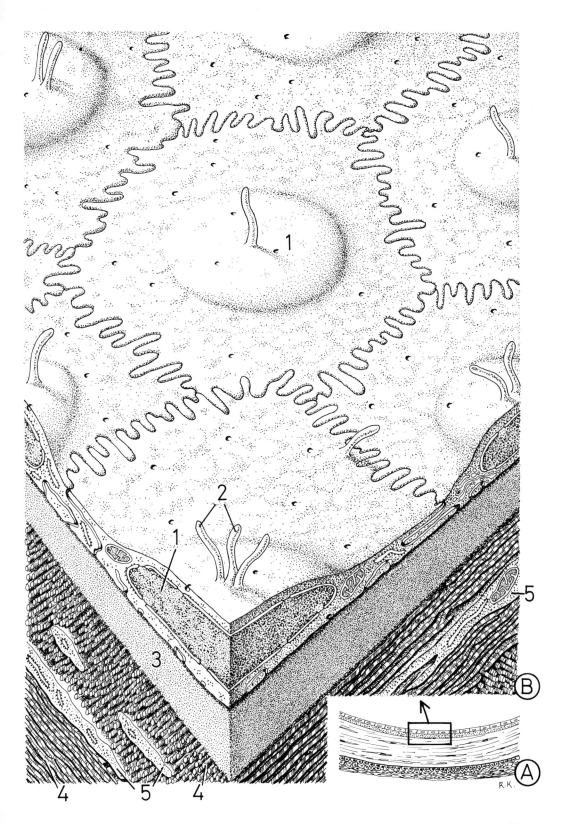

Plate 11. Surface Epithelia. Simple Cuboidal Epithelium.
Example: Distal Tubule of the Nephron

Simple cuboidal epithelium occurs in many glands and glandular ducts (see Plates 37–39), various sections of the renal tubules, choroid plexus (see Plate 12), and as germinal epithelium of the ovary (see Plate 76). Viewed in profile, the cells of this epithelium have a cuboidal form; the cells are, however, small prisms (hence the alternative name "isoprismatic epithelium").

A section from the distal tubule of the nephron (Fig. A) is presented three-dimensionally in Fig. B. The cells of the distal tubule are approximately as wide as they are high. In plan view, they largely appear as hexagons. The round or oval nuclei (Fig. B1) of this epithelium are predominantly located in the apical pole of the cell; invaginations of the plasmalemma penetrate deep into the basal cell pole. The whole of these invaginations form a basal labyrinth (Fig. B2), which permits intensive transport of substances. Part of the basal lamina (Fig. B3) has been folded back so that the polygonal pattern of the basal infoldings may be seen. (See Plate 90 in KRSTIĆ 1979.)

Magnification: Fig. B, ×2,000

REFERENCES

Andrews PM, Porter KR (1974) A scanning electron microscopic study of the nephron. Am J Anat 140:81–116

Bulger RE, Siegel FL, Pendergrass R (1974) Scanning and transmission electron microscopy of the rat kidney. Am J Anat 139:483–502

Hagège I, Richet MG (1970) Etude par microscopie électronique à balayage de la surface apicale des cellules des tubes contournés distaux du rein de rat. CR Seances Acad Sci Ser D 271:331–334

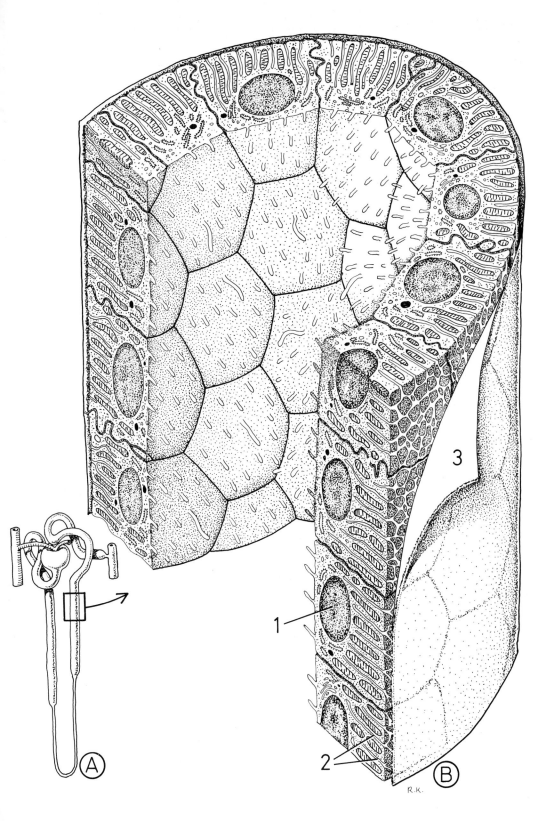

Plate 12. Surface Epithelia. Simple Cuboidal Epithelium from the Rat Choroid Plexus

The villous processes (**1**) of the choroid plexus, which coats the roof of the third and fourth ventricles as well as parts of both lateral ventricles, are lined with cuboidal cells. Their free convex surface is covered with numerous, long microvilli. The round nuclei are primarily located in the middle of the cell body. A basal labyrinth (**2**) is important in the transport of metabolites.

Two blood capillaries (**3**) of the fenestrated type can be observed in the interior of the villous process. The basal laminae between the epithelium and capillaries are not depicted for the sake of clarity.

The epithelium of the choroid plexus is involved in active secretion (production of cerebrospinal fluid).

Magnification: ×3,300

REFERENCES

Agnew WF, Alvarez RB, Yen TGH, Crews AK (1980) Protein synthesis and transport by the rat choroid plexus and ependyma. Cell Tissue Res 208:261–281

Castel M, Sahar A, Erlij D (1974) The movement of lanthanum across diffusion barriers in the choroid plexus of the cat. Brain Res 67:178–184

Cserr HF (1971) Physiology of the choroid plexus. Physiol Rev 51:273–311

Peters A, Swan RC (1979) Choroid plexus of the mature and aging rat. Anat Rec 194:325–354

Segal MB, Burgess AMC (1974) A combined physiological and morphological study of the secretory process in rabbit choroid plexus. J Cell Sci 14:339–350

Shuangshoti S, Netsky MG (1970) Human choroid plexus: morphologic and histochemical alterations with age. Am J Anat 128:73–96

Yamadori T (1972) A scanning electron microscopic observation of the choroid plexus in rats. Arch Histol Jpn 35:89–97

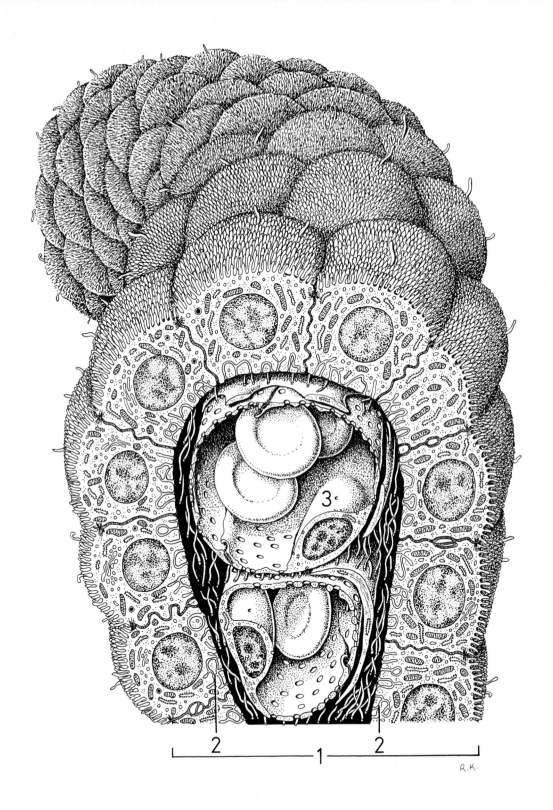

Plate 13. Surface Epithelia. Simple Cuboidal Epithelium.
Example: Pigment Epithelium of Human and Mouse Retina

A further type of simple cuboidal epithelium is the pigment epithelium of the retina. It can be easily lifted from the connective tissue membranes of the eyeball, mounted onto a microscope slide, and observed unstained from above (Fig. A). In this way, the polygonal forms of the cells (Fig. A1) and the large numbers of spindle-shaped melanin granules can be seen.

In vertical section (Fig. B), the brownish pigment epithelial cells (Fig. B1) are seen to be quadrate. With their basal poles they lie on a basal lamina; the apical portions, in which are located many melanin granules (Fig. B2), come into contact with the tips of the rods and cones (Fig. B3). Light travels in the direction of the arrow.

Figure C depicts pigment epithelial cells of the house mouse; this section roughly corresponds to the inset in Fig. B. The cells lie on a basal lamina (Fig. C1), contain a spherical or ellipsoidal nucleus, a moderately developed basal labyrinth (Fig. C2), and more smooth (Fig. C3) than rough (Fig. C4) endoplasmic reticulum. Above the nucleus are located numerous 2- to 4-µm-long spindle-shaped melanin granules (Fig. C5), which continue into the long apical cell processes (Fig. C6). The latter are inserted between the outer segments (Fig. C7) of the rod cells, the membranous discs (Fig. C8) of which can be recognized.

Upon exposure to light, the apical processes, containing the pigment granules, increase in length so that they surround every outer segment. In this way, minute camerae obscurae are formed around every photoreceptor, and this considerably improves the resolving power of the eye.

In darkness the processes shorten, the pigment granules stream back into the cells, and sensitivity increases at the expense of resolving power. These so-called retinomotor movements are much better developed in lower animals than in mammals.

Magnifications: Figs. A, B, × 350; Fig. C, × 5,600

REFERENCES

Dowling JE, Gibbons IR (1962) The fine structure of the pigment epithelium in the albino rat. J Cell Biol 14:459–474

Goldmann Herman K, Steinberg RH (1982) Melanosome metabolism in the retinal pigmented epithelium of the opossum. Cell Tissue Res 227:485–507

Mason TW, Fager RS, Abrahamson EW (1973) Ultrastructure of the receptor and epithelial layers of the bovine retina. J Anat 115:289–308

Novikoff AB, Leuenberger PM, Novikoff PM, Quintana N (1979) Retinal pigment epithelium. Interrelations of endoplasmic reticulum and melanolysosomes in black mouse and its beige mutant. Lab Invest 40:155–165

Samorajski T, Ordy JM, Keefe JR (1966) Structural organisation of the retina in the tree shrew (Tupaia glis). J Cell Biol 28:489–504

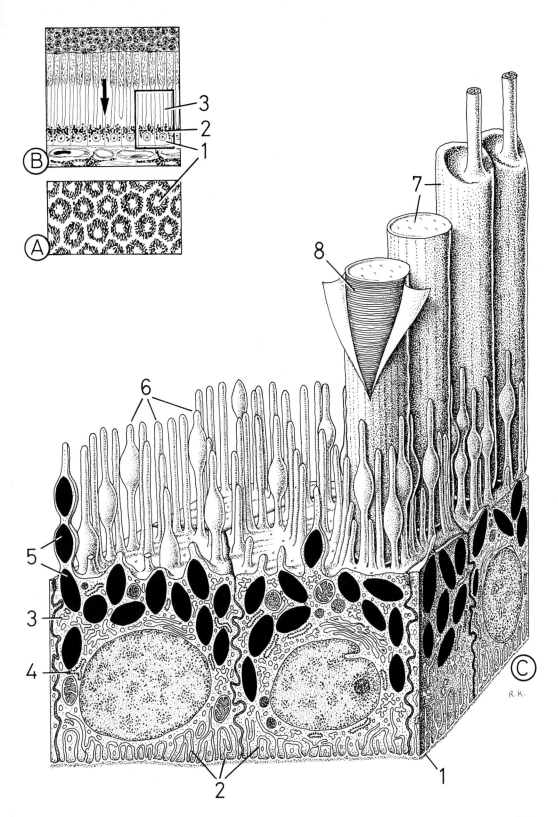

Plate 14. Surface Epithelia. Simple Columnar Epithelium from the Rat Jejunum

This simple epithelium has a large distribution: It occurs in the gastrointestinal tract from the cardiac orifice to the anus. Figure **A** depicts part of an intestinal villus with columnar epithelium. The inset corresponds to the area in Fig. **B**.

The cells of simple columnar epithelium (absorptive cells or enterocytes) are slender; in cross section they have a polygonal form. Neighboring cells are connected by means of terminal bars (Fig. **B1**) and fairly complicated interdigitations. The intercellular spaces can be widened, especially in the lower half of the cell, according to the functional condition of the epithelium. Lateral microvilli (Fig. **B2**) then project into these spaces.

The nuclei of columnar cells are ellipsoidal as a consequence of the shape of the cell. In the cytoplasm are scattered numerous mitochondria, a well-developed Golgi apparatus, and cisternae of smooth and rough endoplasmic reticulum. The microvilli (Fig. **B3**) of the apical pole are very characteristic of these cells and constitute an enormous increase in cell surface area. The microvilli form the striated border, which is also visible under the light microscope.

Goblet cells (Fig. **B4**), which synthesize and secrete mucous granules (Fig. **B5**), are located between the columnar cells. In common with other epithelial cells, absorptive cells lie upon a basal lamina (Fig. **B6**), which separates them from the fenestrated capillaries (Fig. **B7**) and subepithelial loose connective tissue (Fig. **B8**). The kind of epithelium described here is involved in absorption (see Plate 31).

Magnifications: Fig. **A**, × 80; Fig. **B**, × 4,000

REFERENCES

Böck P, Tillmann B, Osterkamp U (1980) Rasterelektronenmikroskopische Untersuchungen an isolierten Darmepithelien. Z Mikrosk Anat Forsch 94:1077–1089

Farquhar MG, Palade GE (1963) Junctional complexes in various epithelia. J Cell Biol 17:375–412

Johnson FR (1975) The absorptive cell. In: Beck F, Lloyd JB (eds) The cell in medical science, vol 3. Academic, London

Oshima Y (1977) An electron microscope study of the intestinal absorption of medium chain and long chain triglycerides in the rat. Arch Histol Jpn 40:153–169

Pavelka M, Ellinger R (1981) Morphological and cytochemical studies on the Golgi apparatus of rat jejunal absorptive cells. J Ultrastruct Res 77:210–222

Stenling R, Helander HR (1981) Stereological studies on the small intestinal epithelium of the rat: I. The absorptive cells of the normal duodenum and jejunum. Cell Tissue Res 217:11–21

Taylor AB, Anderson JM (1972) Scanning electron microscope observations of mammalian intestinal villi, intervillus floor and crypt tubules. Micron 3:430–453

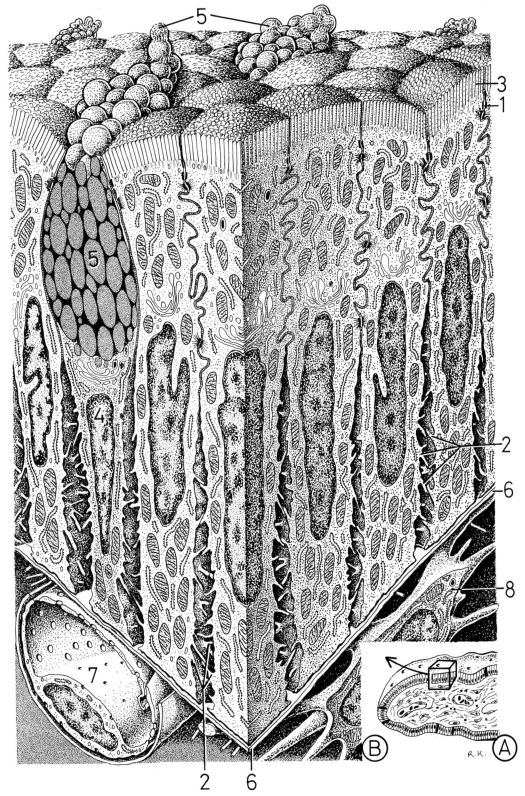

Plate 15. Surface Epithelia. Simple Columnar Epithelium.
Example: Collecting Tubule of the Rat Kidney

In the upper part of the plate, an interlobular artery (**1**) is depicted with an afferent (**2**) and efferent (**3**) arteriole. A nephron, the parts of which are – renal corpuscle (**4**), proximal tubule (**5**), thin segment (**6**), distal tubule (**7**), and connecting portion (**8**) – is joined to a collecting tubule (**9**), which is lined with simple columnar epithelium (**10**).

Part of the basal lamina (**11**) is folded back, revealing the polyhedral basal surfaces of the epithelial cells. A similar pattern can be recognized in the lumen of the collecting tubule, corresponding to the pattern on the cell surface.

The enlarged intercellular spaces (**12**) between the prismatic cells probably assist in reabsorption of water from urine.

Magnification of the sectioned collecting tubule cells: ×900

REFERENCES

Andrews PM (1975) Scanning electron microscopy of human and rhesus monkey kidneys. Lab Invest 32:610–618

Bulger RE, Trump BF (1966) The fine structure of the renal papilla. Am J Anat 118:685–722

Myers CE, Bulger RE, Tisher C, Trump BF (1966) Human renal ultrastructure: IV. Collecting duct of healthy individuals. Lab Invest 15:1921–1950

Pfaller W, Klima J (1976) A critical reevaluation of the structure of rat uriniferous tubule as revealed by scanning electron microscopy. Cell Tissue Res. 166:91–100

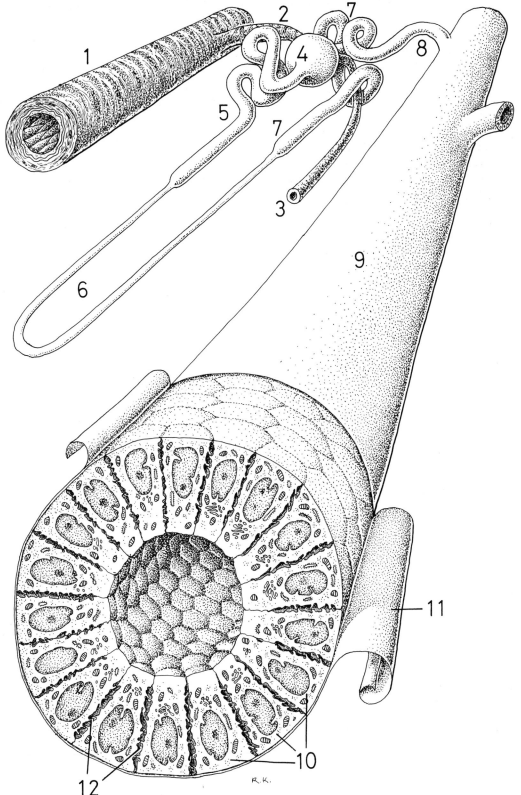

Plate 16. Surface Epithelia. Simple Columnar Epithelium with Cilia from the Human Oviduct

Simple cuboidal epithelium from the mucous coat of the oviduct consists of two morphologically distinct types of cell. One type comprises the ciliated cells (**1**), which contain a large nucleus and relatively well-developed organelles. Numerous long cilia (**2**) project from the apical surface; their rootlets (**3**) extend to the nucleus.

Secretory or nonciliated cells (**4**) constitute the second type of cell. They are also columnar, rather more slender than the ciliated cells, and have a protruding apical surface, upon which only a small number of microvilli are located. Their internal structure corresponds to that of secretory cells; in addition to the well-developed rough endoplasmic reticulum and Golgi apparatus, they contain mitochondria and secretory granules (**5**). During the menstrual cycle, the number and structure of both kinds of cell are altered: In the first half (proliferative or follicular phase) of the cycle, the ciliated cells predominate; in the second half (secretory or luteal phase), the number of ciliated cells decreases and the secretory cells are dominant. In this phase, the latter secrete a hyaluronidase-resistant substance, which moistens the mucosa of the oviduct and promotes the development of the ovum on its way to the uterus. After the secretory cells have yielded their product, they become narrow and the cytoplasm becomes dense.

The movement of the cilia is directed toward the uterus, however in humans this has little to do with transport of the ovum. The current produced by the cilia largely aids the orientation of spermatozoa, which move upstream as a result of their positive rheotaxis.

It is possible that the two kinds of cell may transform into one another. (See Plates 170, 171 in KRSTIĆ 1979.)

Magnification: × 4,000

REFERENCES

Clyman MJ (1966) Electron microscopy of the human fallopian tube. Fertil Steril 17:281–301

Ferenczy A (1974) The surface ultrastructure of the human fallopian tube. A comparative morphophysiologic study. In: Johari O, Corvin I (eds) Scanning electron microscopy. IIT Research Institute, Chicago, pp 614–621

Frowein J, Petersen K-G (1976) Rezeptoren für Steroidhormone. Dtsch Med Wochenschr 101:589–593

Kühnel W, Busch LC (1981) Functional morphology of the oviductal mucosa and the endometrium as viewed by SEM. Biomed Res (Suppl) 2:341–353

Kugler P (1981) Zur Histochemie der Flimmerzellen der menschlichen Endosalpinx. Histochemistry 73:137–150

Motta PM, Andrews PM (1976) Scanning electron microscopy of the endometrium during the secretory phase. J Anat 122:315–322

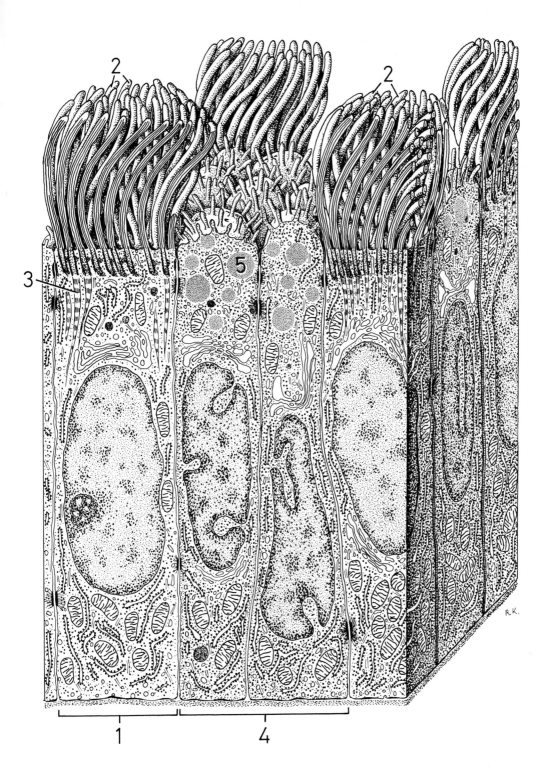

Plate 17. Surface Epithelia. Pseudostratified Columnar Epithelium from the Human Ductus Epididymidis

Without exception, all cells that constitute pseudostratified epithelia lie on a basal lamina, though they do not necessarily extend to the epithelial surface. Their nuclei occur at different levels within the epithelial layer. According to the length of the cells that comprise such an epithelium, it is possible to distinguish two, three, or more rows of nuclei, and this at first sight can resemble stratified epithelium. Pseudostratified columnar epithelium is present in many excretory ducts and in the ductus epididymidis.

Figure **A** shows a transverse section through the ductus epididymidis. Large numbers of spermatozoa (Fig. **A 1**) are located in the lumen of the ductus. The inset corresponds to Fig. **B**.

The cells lying on the basal lamina (Fig. **B 1**) are of two different sizes: The lower, incomplete row consists of basal cells (Fig. **B 2**), whereas the second, uninterrupted row is made up of columnar cells with stereocilia (Fig. **B 3**). The basal cells contain a spherical or ellipsoidal nucleus and poorly developed organelles. Since basal cells are able to differentiate into columnar cells they are seen as "replacement cells." The ellipsoidal, often deeply indented nuclei of the columnar cells (Fig. **B 4**) compose the second row of nuclei in this epithelium. A highly developed Golgi apparatus (Fig. **B 5**) from which arise many secretory granules (Fig. **B 6**) is characteristic of these cells. Rough endoplasmic reticulum (Fig. **B 7**) is usually located near the nucleus, whereas the cisternae of smooth endoplasmic reticulum, lysosomes, and lipofuscin granules are distributed in varying amounts throughout the cytoplasm.

The columnar cells are distinguished by the numerous stereocilia (Fig. **B 8**) on the apical cell surface. These immobile structures, which are probably involved in secretion, increase the free cell surface area by a significant degree.

The secretory product of the columnar cells contains substances that are important in the metabolism of the as yet nonmotile spermatozoa in the ductus epididymidis. (See Plate 105 in KRSTIĆ 1979.)

Magnifications: Fig. **A**, × 150;
Fig. **B**, × 5,000

REFERENCES

Goto K (1981) Surface morphology of the epithelium of human seminiferous tubules, rete testis, ductuli efferentes and ductus epididymidis. Biomed Res (Suppl) 2:361–374

Horstmann E (1962) Elektronenmikroskopie des menschlichen Nebenhodenepithels. Z Zellforsch 57:692–718

Nicander L, Glover TD (1973) Regional histology and fine structure of the epididymal duct in golden hamster (Mesocricetus auratus). J Anat 114:347–364

Shimada T, Huang CT, Obayashi I (1975) Scanning electron microscopy of epididymal ducts in the Japanese monkey (Macacus fuscatus) with special reference to architectural analysis of stereocilia. Arch Histol Jpn 38:101–107

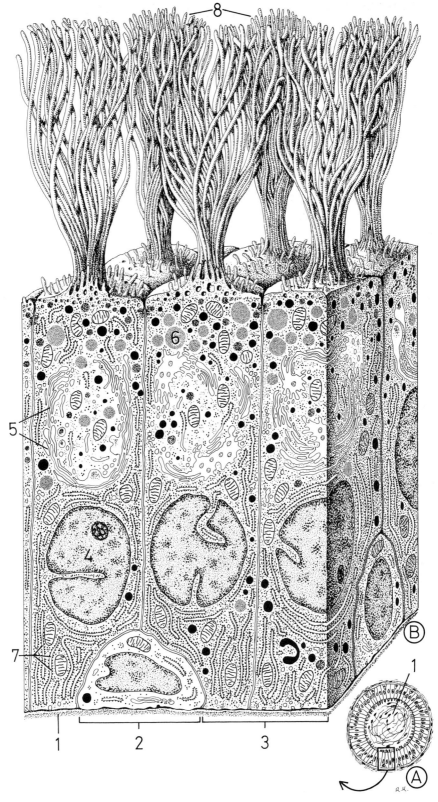

Plate 18. Surface Epithelia. Pseudostratified Columnar Epithelium with Cilia (Ciliated Epithelium) from the Rat Trachea

Pseudostratified ciliated epithelium lines the nasal cavities, trachea, and bronchi. It contains several types of cell. The basal cells (1), the nuclei of which form a basal row, are poor in internal structures. They serve to replace dead ciliated cells. The basal cells daily renew about 2% of the cell population of the epithelium.

The cuneate cells (2) are slender and are located between the ciliated cells; they do not extend to the surface and contain more cytoplasmic structures. These "supporting cells" constitute an intermediate stage between the basal and ciliated cells, i.e., cells in the process of differentiation.

The characteristic ciliated cells (3) of pseudostratified columnar epithelium are rich in organelles and bear numerous cilia (4) on their free surface, corresponding to the kinetosomes (5) in the cell interior. Owing to their high degree of specialization, these cells have lost the capacity to divide and have to be replaced by differentiation of the deeper-lying cells.

The mucus-producing goblet cells (6) are situated between the ciliated cells. The number of mucin granules (7) they contain varies according to functional circumstances. Following secretion from the goblet cells, these granules, together with the product of the seromucous glands of the respiratory tract, form a fine mucous film (not shown), which moistens the free surface of the epithelium. Most particles of dust that are present in the inspired air become attached to this mucous layer. These particles are then transported toward the larynx and pharynx by continuous ciliary motion and eliminated by coughing or swallowing.

In this epithelium are scattered endocrine cells (8), which, like the adjacent epithelial cells, lie on the basal lamina (9).

Magnification: $\times 2{,}900$

REFERENCES

Adams DR (1972) Olfactory and non-olfactory epithelia in the nasal cavity of the mouse. Am J Anat 133:37–50

Andrews PM (1974) A scanning electron microscopic study of the extrapulmonary respiratory tract. Am J Anat 139:399–424

Dalen H (1983) An ultrastructural study of the tracheal epithelium of the guinea-pig with special reference to the ciliary structure. J Anat 136:47–67

Greenwood MF, Holland P (1972) The mammalian respiratory tract surface. A scanning electron microscopic study. Lab Invest 27:296–304

Kawamata S, Fujita H (1983) Fine structural aspects of the development and aging of the tracheal epithelium of mice. Arch Histol Jpn 46:355–372

Rhodin JAG (1966) Ultrastructure and function of the human tracheal mucosa. Am Rev Respir Dis 93:1–15

Stockinger L (1970) Ultrastruktur und Histophysiologie des Respirationstraktes. Mikroskopie 26:83–98

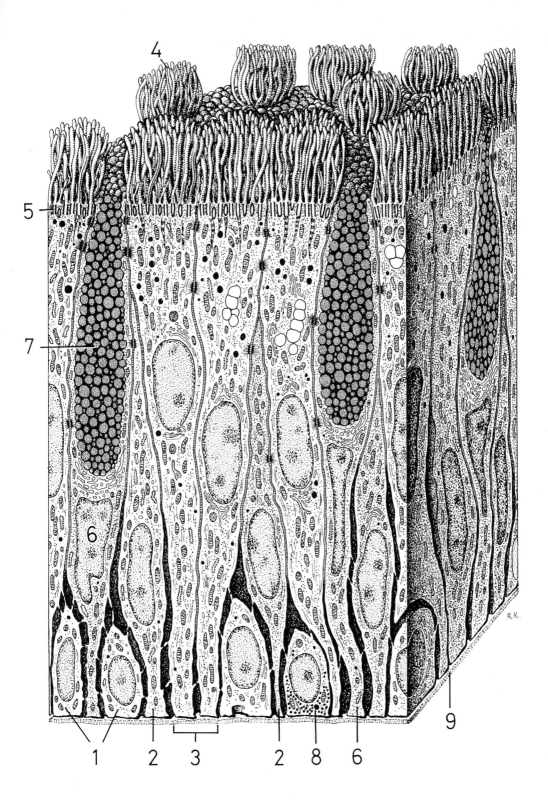

EPITHELIAL TISSUE

Plate 19. Surface Epithelia. Transitional Epithelium from the Empty Bladder of a Monkey

Transitional epithelium is classed as a pseudostratified epithelium since all the cells are most probably in contact with the basal lamina (**1**) by means of, in parts very slender, cytoplasmic feet. The epithelium comprises small basal cells (**2**), then intermediate cells (**3**) in the shape of a tennis racket, and finally the very voluminous facet or superficial cells (**4**).

The polyploid superficial cells contain one large or two smaller nuclei. In the cytoplasm of the apical pole of the cell are found the Golgi apparatus, numerous microfilaments, and discoid vesicles (**5**). This dense cellular zone appears dark under the light microscope and is termed the crusta.

The highly folded apical cell membrane of superficial cells is asymmetrical: The layer of the unit membrane in contact with urine is, in areas called plaques, thicker (~ 4.5 nm) than the cytoplasmic layer (~ 2.5 nm). It is believed that these circumscribed reinforcements of the membrane protect the epithelium against the effects of urine. (See Plate 92 in KRSTIĆ 1979.)

Magnification: ×2,900

REFERENCES

Knutton S, Robertson JD (1976) Regular structures in membranes: the luminal plasma membrane of the cow urinary bladder. J Cell Sci 22:355–370

Monis B, Zambrano D (1968) Transitional epithelium of urinary tract in normal and dehydrated rats. Z Zellforsch 85:165–182

Phillips SJ, Griffin T (1985) Scanning electron microscope evidence that human urothelium is a pseudostratified epithelium. Anat Rec 211:153A–154A

Scheidegger G (1980) Der Aufbau des Übergangsepithels der Harnblase bei Schwein, Schaf, Ratte und Spitzmaus. Acta Anat (Basel) 107:268–275

Severs NJ, Warren RC (1978) Analysis of membrane structure in the transitional epithelium of rat urinary bladder: I. The luminal membrane. J Ultrastruct Res 64:124–140

Staechelin LA, Chlapowski FJ, Boneville MA (1972) Luminal plasma membrane of the urinary bladder: I. Three-dimensional reconstruction from freeze-etch images. J Cell Biol 53:73–91

Walker BE (1960) Electron microscopic observations on transitional epithelium of mouse urinary bladder. J Ultrastruct Res 3:345–361

Warren RC, Hicks RM (1978) Chemical dissection and negative staining of the bladder luminal membrane. J Ultrastruct Res 64:327–340

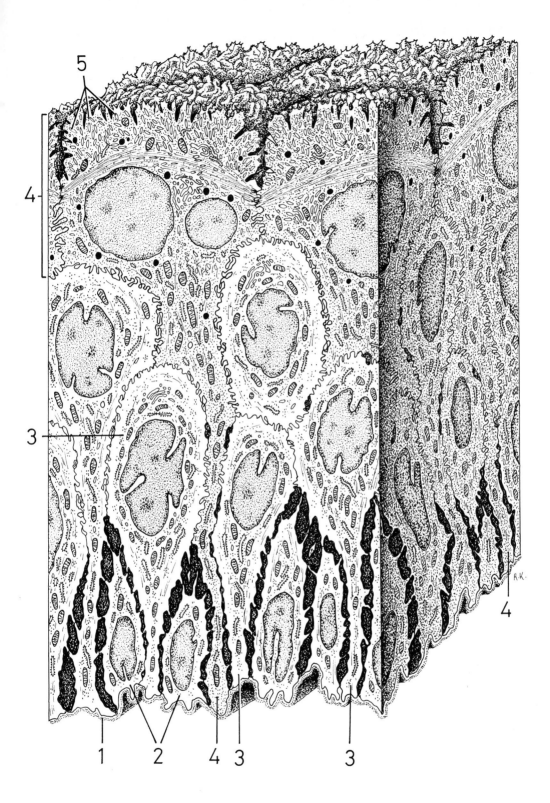

Plate 20. Surface Epithelia. Transitional Epithelium of an Empty and a Filled Rat Bladder

In the empty efferent urinary tract (renal pelvis, ureter, bladder, and initial section of the urethra), the epithelial cells have a columnar form (Fig. **A**). In this state, it is possible to count eight to ten rows of nuclei. The basal epithelial surface is infolded and the apical surfaces of superficial cells are convex.

When the urinary tract is filled, the epithelium becomes thinner. In this state, only two to three rows of nuclei are evident (Fig. **B**). The coherence of the surface is maintained by the reserve folds of the plasmalemma and the discoid vesicles, which fuse with the apical cell membrane upon distension.

The name of this type of epithelium derives from these changes that take place in its structure.

Magnifications: Figs. **A, B,** × 1,900

REFERENCES

Minsky BD, Chlapowski FJ (1978) Morphometric analysis of the translocation of lumenal membrane between cytoplasm and cell surface of transitional epithelial cells during the expansion-contraction cycles of mammalian urinary bladder. J Cell Biol 77:685–697

Noack W (1976) The superficial cells of the transitional epithelium in the expanded and unexpanded rat urinary bladder. Acta Anat (Basel) 93:171–183

Petry G, Amon H (1966) Licht- und elektronenmikroskopische Studien über Struktur und Dynamik des Übergangsepithels. Z Zellforsch 69:587–612

Richter WR, Moize SM (1963) Electron microscopic observations on the collapsed and distended mammalian urinary bladder (transitional epithelium). J Ultrastruct Res 9:1–9

Walton J, Yoshiyama JM, Vanderlaan M (1982) Ultrastructure of the rat urothelium in en face section. J Submicrosc Cytol 14:1–15

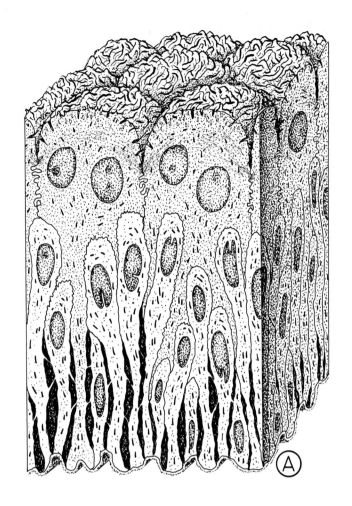

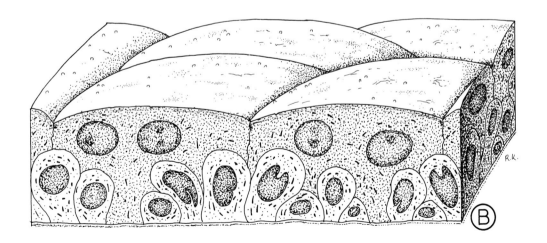

EPITHELIAL TISSUE

Plate 21. Surface Epithelia. Stratified Columnar Epithelium from the Rat Urethra

Stratified columnar epithelium, which has a limited distribution, is found between zones of pseudostratified columnar and nonkeratinized stratified squamous epithelium. As previously mentioned, it occurs in fornix conjunctivae, on the palate, epiglottis, and part of the urethra. It consists of several layers of cells, of which only the lowermost are in contact with the basal lamina (**1**). This cell layer comprises polygonal or cuboidal cells (**2**); the overlying cells are largely spindle-shaped (**3**); those forming the free epithelial surface are columnar (**4**). These columnar cells contain a notable amount of glycogen (**5**) and bear microvilli.

All the epithelial cells interdigitate to a high degree, which increases the elasticity of the epithelium.

The functional properties of this type of epithelium are as yet poorly understood.

Magnification: ×2,900

REFERENCES

Hicks RM (1965) The fine structure of the transitional epithelium of rat ureter. J Cell Biol 26:25–48

Hicks RM (1966) The function of the Golgi complex in transitional epithelium. Synthesis of the thick cell membrane. J Cell Biol 30:623–643

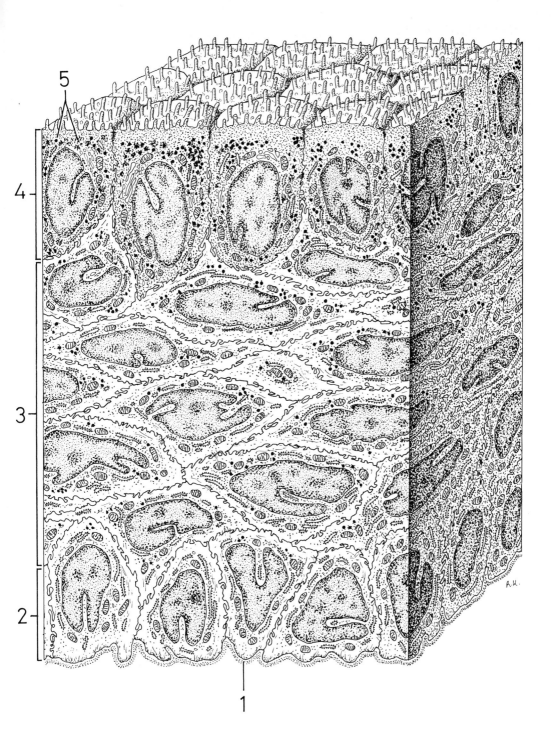

Plate 22. Surface Epithelia. Nonkeratinized Stratified Squamous Epithelium. Example: Epithelium of Rat Cornea

A transverse section through the cornea (Fig. A) reveals its epithelium (Fig. A 1) and endothelium (Fig. A 2; see Plate 10), between which lies the corneal stroma (Fig. A 3). Figure B shows a section from the epithelium. The following layers can be distinguished in nonkeratinized stratified epithelium:

C. The stratum basale (basal cell layer) consists of relatively large and prismatic cells, which are attached to Bowman's membrane (Fig. B 1) by numerous hemidesmosomes.

D. The stratum spinosum (spinous or prickle-cell layer) is made up of large polygonal cells that are joined by extensive interdigitations, studded with desmosomes. Their cytoplasm, like that of the basal cells, contains a significant number of tonofibrils (Fig. B 2). The cells of layers C and D replace, by mitotic division, those that detach from the epithelial surface (stratum germinativum or germinal layer).

E. The stratum pavimentosum, which lies above the spinous layer, consists of cells that adopt an increasingly flattened form; some cells lose their nuclei, though the desmosomes are retained. The cells eventually transform into thin anuclear plates (hence the name, stratum pavimentosum). The free surface of some of these cells is covered with large numbers of microvilli and low ridgelike folds, so-called, microridges or microplicae (arrow).

Like all other nonkeratinized stratified epithelia of similar structure (oral cavity, esophagus, vagina, etc.), the cornea is moistened; in this case, the moistening is performed by the lacrimal glands. The anterior epithelium of the cornea comprises only five to nine cell layers and lies on the avascular corneal stroma; it receives nutrients solely by diffusion.

The diagnostically important corneal reflex is due to the highly developed ramifications of the sensitive nerve endings among epithelial cells.

Magnifications: Fig. A, × 80;
Fig. B, × 2,600

REFERENCES

Allen TD, Potten CS (1975) Desmosomal form, fate, and function in mammalian epidermis. J Ultrastruct Res 51:94–105

Andrews PM (1976) Microplicae: characteristic ridgelike folds of the plasmalemma. J Cell Biol 68:420–429

Parakkal PF (1967) An electron microscopic study of esophageal epithelium in the newborn and adult mouse. Am J Anat 121:175–196

Sperry DG, Wassersug RJ (1976) A proposed function for microridges on epithelial cells. Anat Rec 185:253–258

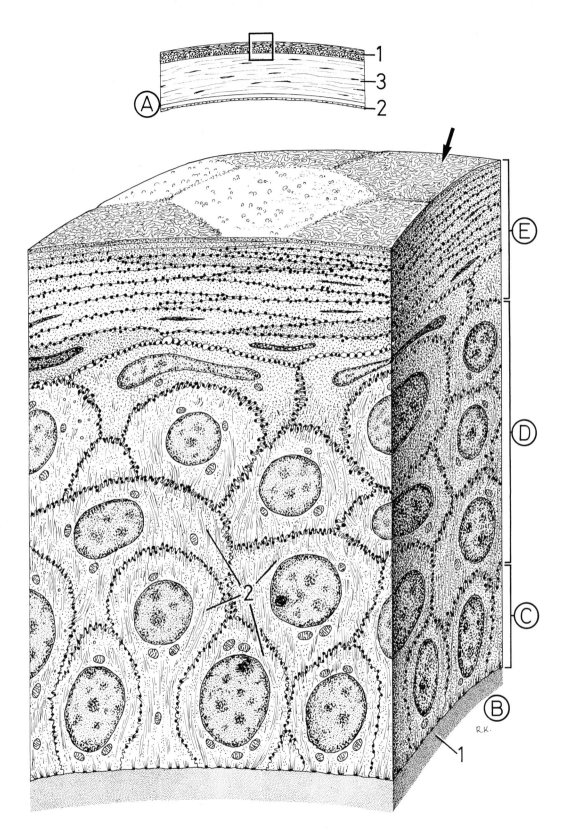

Plate 23. Surface Epithelia. Keratinized Stratified Squamous Epithelium or Epidermis of Human Hairy Skin

Among other functions, keratinized squamous epithelium provides protection for the organism from slight mechanical and chemical damage. The structure of the epithelium and its cells are adapted to this role.

The epidermis consists of the following layers:

- **E.** Stratum corneum
- **D.** Stratum lucidum (only in the glabrous skin of the inner surface of the hand and sole)
- **C.** Stratum granulosum
- **B.** Stratum spinosum ⎫ Stratum
- **A.** Stratum basale ⎬ germinativum
  ⎭ (germinal layer)

A. The small cuboidal basal cells (**1**) are attached by means of basal processes (**2**) – surrounded by the basal lamina – to the feltwork of reticular and collagen microfibrils (**3**) of the subepidermal connective tissue. The supranuclear cytoplasm of the basal cells contains some melanin granules (**4**).

B. One or more layers of large polygonal cells overlie the basal layer. The cells are joined by processes bearing hemidesmosomes (**5**). Numerous tonofibrils (**6**) run through the cytoplasm. After mechanical separation, the cells of this layer appear spinous (hence, prickle cells, stratum spinosum, or spinous layer). The basal and spinous layer together form the stratum germinativum, the germinal layer, in which (largely at night) mitoses take place.

C. In the stratum granulosum or granular layer, where the cells are already flattened, the highly refractile keratohyaline granules (**7**) appear as precursors of keratin. The cells gradually die as a result of an increase in this substance. One or two layers of granular cells are sufficient for light keratinization (as depicted in this plate), whereas three to five layers are required for heavy keratinization. The origin of keratohyalin is as yet unknown.

D. The stratum lucidum or clear layer is very refractile; it only occurs in glabrous, highly keratinized skin of the palms and soles. For didactic reasons, the stratum lucidum, composed of flattened cells, is also presented on this plate. Its cells contain keratin, and it is possible that the clear layer permits movement between the softer germinal and the harder horny layers.

E. The stratum corneum or horny layer can vary in thickness according to mechanical strain. Its cells are dead and their bodies are transformed into keratin scales (**8**). Since horny cells are constantly desquamating, they have to be continually replaced by cells from the germinal layer. In a thick avascular tissue like the epidermis, supplied by unmyelinated nerve fibers (**9**), the broad intercellular spaces (**10**) act as a channel system and are involved in transporting substances. The intercellular spaces in the stratum granulosum and stratum corneum are significantly narrower (occasionally 20 nm) and are interrupted by desmosomes.

Melanocytes (**11**) occur in the epidermis and are illustrated in Plate 24. (See Plate 109, 110, 115, 116, 118, 158, 159 in KRSTIĆ 1979.)

Magnification: ×2,800

REFERENCES

Green H, Fuchs E, Watt F (1982) Differential structural components of the keratinocyte. Cold Spring Harbor Symposia on Qualitative Biology, vol 46, part 1

Menton DN, Eisen AZ (1971) Structure and organisation of mammalian stratum corneum. J Ultrastruct Res 35:247–264

Zelickson AS (ed) (1967) Ultrastructure of normal and abnormal skin. Lea and Febiger, Philadelphia

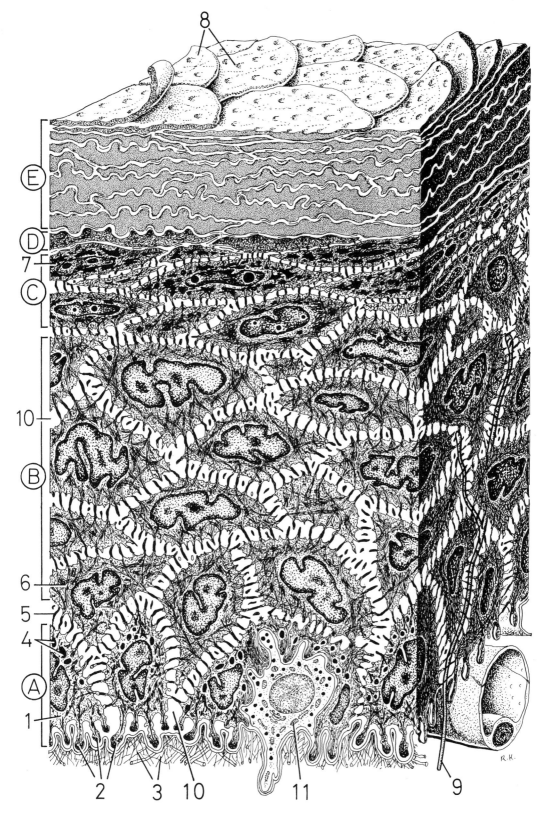

Plate 24. Surface Epithelia. Epidermis. Germinal Layer with a Melanocyte. Same Preparation as in Plate 23

The cells that synthesize the pigment melanin are termed melanocytes (**1**) and are located between the basal cells of the epidermis and in the hair bulb. Melanocytes stem from the neuroectoderm of the neural crest (see Plate 144); during embryonic development, they migrate into the epidermis and contact the basal lamina (**2**).

Melanocytes are very difficult to distinguish in normally stained sections. Following the dopa reaction (in which an enzyme responsible for melanin synthesis is stained) they appear black. Melanocytes are pale, highly branched cells. The cytoplasm contains a round nucleus, well-developed organelles, some free ribosomes, few microfibrils, and a varying number of melanosomes (**3**), precursors of mature melanin granules (**4**).

The melanin granules are about 0.7×0.3 μm in size and are invested with a unit membrane. These granules also occur in the cell processes (arrow), from which many are able to penetrate deep into the intercellular spaces and into the bodies of neighboring cells of the stratum germinativum. It is assumed that melanin granules can be transferred to neighboring germ cells by means of the melanocyte processes. It is also possible that the pigment particles are phagocytized by basal and prickle cells. Melanocytes are not connected to neighboring cells by desmosomes. In white race, only the basal cells contain melanin.

The existence of processes (**5**) oriented toward the subepidermal connective tissue suggests that melanocytes may even be able to migrate out of the corium after embryogenesis has been completed. Melanocytes retain the ability to divide for the duration of their lives. It has been estimated that there are about 1,200–1,500 melanocytes/mm^2 of skin. Pigment synthesis is stimulated by ultraviolet light (increased coloration of the skin following exposure to the sun). Albinos possess melanocytes in their skin, however the melanosomes are unable to turn black and transform into mature melanin granules, owing to the absence of the enzyme tyrosinase.

At the bottom of the plate, the basal processes (**6**) of the basal cells can be seen. Reticular (**7**) and collagen microfibrils (**8**) are connected to the basal lamina (**2**). (See Plates 75, 76 in KRSTIĆ 1979).

Magnification: × 8,000

REFERENCES

Fitzpatrick TB, Szabo G, Seiji M, Quevedo WC Jr (1979) Biology of the melanin pigmentary system. In: Fitzpatrick TB (ed) Dermatology in general medicine, 2nd edn. McGraw Hill, New York

Jimbow K, Takahashi M, Sato S, Kukita A (1971) Ultrastructural and cytochemical studies of melanogenesis in melanocytes of normal human hair matrix. J Electron Microsc (Tokyo) 20:87–92

Petzoldt D (1967) Die morphologische und funktionelle Organisation der Melanocyten in menschlicher Haut. Hautarzt 18:481–488

Plačkova A, Skach M (1975) Feinstrukturelle Untersuchungen zur Keratinisierung der Papilla filiformis der Zunge des Menschen. Z Mikrosk Anat Forsch 89:305–318

Poirier J, Nunez-Dispot CH (1968) Le mélanocyte: I. Structure et ultrastructure. Presse Med 76:1179–1181

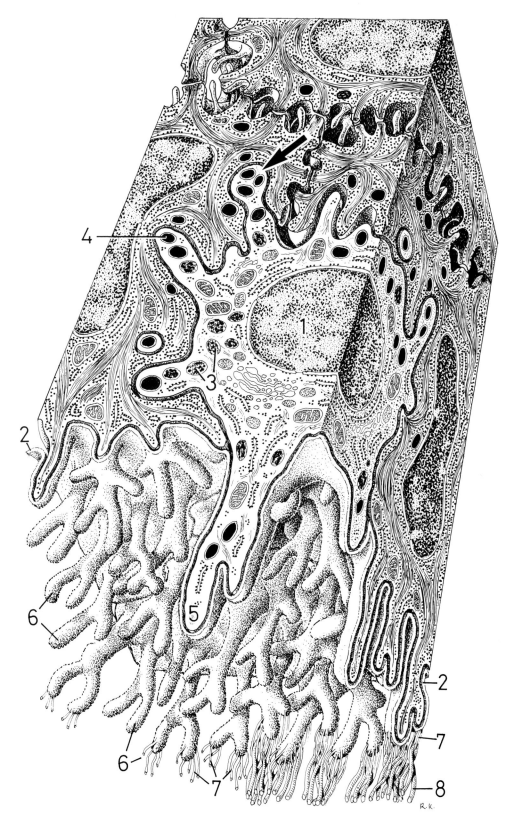

Plate 25. Secretory Surface Epithelia.
Example: Amniotic Epithelium of a Human Fetus

Certain epithelial cells have secretory activities: The surface epithelium of the stomach synthesizes mucus, which serves as a protection against the corrosive effects of HCl and pepsin, and the product of the surface epithelium of the gallbladder protects it against bile acids. The function of the simple cuboidal epithelium of the choroid plexus was mentioned in Plate 12.

Amniotic epithelium develops in the embryoblast from the extraembryonic cells opposite the embryonic ectoderm (Plate 1, Fig. G5). The amniotic cavity becomes wider and provides the necessary conditions for undisturbed development of the embryo.

Figure **A** depicts a 4-month-old human fetus with the amnion (Fig. **A 1**). The chorionic villi (Fig. **A 2**) and umbilical cord (Fig. **A 3**) can be distinguished. The segment in the inset is enlarged in Fig. **B**. In this region, the amniotic epithelium is simple and cuboidal; it lies upon a basal lamina (Fig. **B 1**). The epithelial cells have a well-developed basal labyrinth, spherical nuclei, well-developed organelles, and numerous microvilli on the apical surface. As inclusions in the cytoplasm occur lipid droplets, glycogen (Fig. **B 2**), and so-called amniotic vacuoles (Fig. **B 3**). The latter are mainly located near the broad intercellular spaces (Fig. **B 4**), which are penetrated by numerous microvillous processes. Amniotic epithelium produces about 1–2 l amniotic fluid (liquor amnii), in which the fetus floats freely. This fluid does not only serve to protect the fetus from mechanical damage or desiccation; the turnover of 0.5 l/h at the end of gravidity indicates that amniotic fluid is also of great physiological importance for the development of the fetus, providing it with a uniform hydrostatic support.

Magnification: Fig. **B**, × 4,500

REFERENCES

Hoyes AD (1969) Ultrastructure of the epithelium of the human umbilical cord. J Anat 105:145–162

King BF (1978) A cytological study of plasma membrane modifications, intercellular junctions, and endocytic activity of amniotic epithelium. Anat Rec 190:113–126

Miller MM, Revel J-P (1974) Scanning electron microscopy of the apical, lateral, and basal surfaces of transporting epithelia in mature and embryonic tissues. In: Johari O, Corvin I (eds) Scanning electron microscopy 1974. IIT Research Institute, Chicago

Thomas EC (1965) The ultrastructure of human amnion epithelium. J Ultrastruct Res 13:65–84

Tiedemann K (1979) The amniotic, allantoic and yolk sac epithelia of the cat: SEM and TEM studies. Anat Embryol 158:75–94

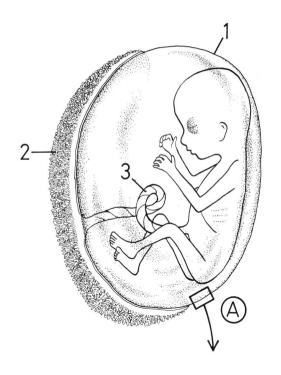
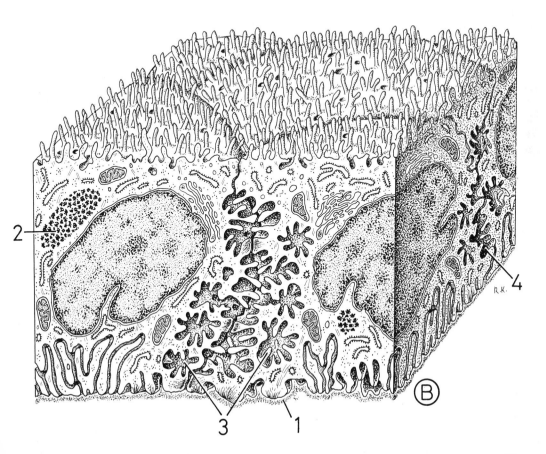

EPITHELIAL TISSUE

Plate 26. Vascularized Secretory Surface Epithelium.
Sole Example: Epithelium of Stria Vascularis of Inner Ear

The surface epithelia hitherto described are all avascular; the stria vascularis of the inner ear is thus an exception. A section from the cochlea (Fig. **A**) is provided for purposes of orientation. On the outer surface is the spiral ligament (Fig. **A 1**), which is lined with the epithelium of the stria vascularis (inset). The cochlear duct (Fig. **A 2**) contains the organ of Corti (Fig. **A 3**). The processes (Fig. **A 4**) of the bipolar cells of the spiral ganglion can also be observed.

The epithelium of the stria vascularis comprises one to three layers and lies on a basal lamina (Fig. **B 1**). In addition to the basal cells (Fig. **B 2**), this epithelium is made up of marginal cells (Fig. **B 3**), which form the surface. These cells contain a large number of mitochondria in small labyrinthine compartments (Fig. **B 4**) and a few microvilli at the hexagonal apical cell pole. A type of intermediate cell (Fig. **B 5**) also occurs in the stria vascularis.

Pericytes (Fig. **B 6**) and a basal lamina (not shown) surround the capillaries (Fig. **B 7**), which run between the epithelial cells. The stria vascularis is the only vascularized epithelium in the body.

The close contact between the capillaries and mitochondria-rich marginal cells is in accordance with the supposition that endolymph, important for the function of the organ of hearing, is produced in the stria vascularis.

Magnifications: Fig. **A**, × 70; Fig. **B**, × 1,500

REFERENCES

Forge A (1982) A tubulo-cisternal endoplasmic reticulum system in the potassium transporting marginal cells of the stria vascularis and effects of the ototoxic diuretic ethacrynic acid. Cell Tissue Res 226:375–387

Fujimoto S, Yamamoto K, Hayabuchi I, Yoshizuka M (1981) Scanning and transmission electron microscope studies on the organ of Corti and stria vascularis in human fetal cochlear ducts. Arch Histol Jpn 44:223–235

Hinojosa R, Rodriguez-Echandia EL (1966) The fine structure of the stria vascularis of the cat inner ear. Am J Anat 118:631–664

Reale E, Luciano L, Franke K, Pannese E, Werbter G, Iurato S (1975) Intercellular junctions in the vascular stria and spiral ligament. J Ultrastruct Res 53:284–297

Rodriguez-Echandia EL, Burgos M (1965) The fine structure of stria vascularis of guinea-pig inner ear. Z Zellforsch 67:600–619

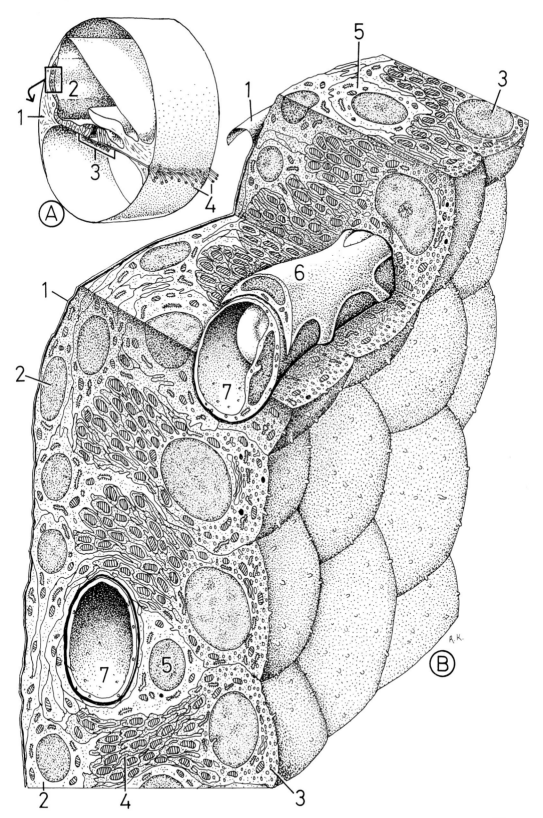

Plate 27. Atypical Epithelia. Examples: Enamel Organ of the Mouse and Thymus of a Human Adolescent

According to definition, epithelia are closed cell unions; the enamel organ and the stroma of the thymus, however, do not conform to this pattern.

On a frontal section through the head of a 21-day-old mouse fetus (Fig. A), it is possible to distinguish parts of the rhinencephalon (Fig. A 1), the eye anlagen with the lenses (Fig. A 2), the nasal cavity (Fig. A 3), the canalis nasopharyngicus (Fig. A; arrow), and the oral cavity (Fig. A 4) with the tongue (Fig. A 5). The epithelium (Fig. A 6) of the oral cavity forms the dental lamina (Fig. A 7), from which the bell-shaped enamel organs (Fig. A 8) develop. In the interior of each enamel organ, an increase in intercellular fluid forces apart the initially closely apposed cells, giving rise to the enamel pulp (Fig. A 9). The mesenchyme condenses in the cup of the enamel organ and forms the dental papilla (Fig. A 10). The area within the inset is illustrated in Plate 28.

The epithelium of the thymus originates from the endoderm of the third pharyngeal pouch; during embryonic development, the cells of this epithelium become infiltrated and separated by migrating lymphocytes. Under the light microscope at low magnification, the capsule (Fig. B 1) and lobules of the thymus are quite distinct. Each lobule is divided into a cortex (Fig. B 2) and medulla (Fig. B 3). Figure C corresponds to the section of medulla in the inset where the Hassall's corpuscles (Fig. B 4) are located.

Reticular-epithelial cells (Fig. C 1) form the epithelial framework of the parenchyma of the thymus. These stellate cells of epithelial origin are connected by cell processes and build up a vascular, sponge-like network. These cells should not be confused with reticular cells of the reticular connective tissue (see Plate 55), which are of mesodermal origin. Many small lymphocytes (Fig. C 2) are located in the interstices of the epithelial network. In Hassall's corpuscles (Fig. C 3), the lamellar, epitheliogenic, reticular-epithelial cells are closely packed in the form of concentric arrays.

The area within the inset in Fig. C is enlarged in Plate 29.

Magnifications: Figs. A, B, × 30; Fig. C, × 750

REFERENCES

Gaudecker BV (1978) Ultrastructure of the age-involuted adult human thymus. Cell Tissue Res 186:507–525

Itoh T, Kasahara S, Aizu S, Kato K, Takeuchi M, Mori T (1982) Formation of Hassall's corpuscles in vitro by the thymic epithelial cell line IT-26R21 of the rat. Cell Tissue Res 226:469–479

Matthiessen ME, Rømert P (1980) Ultrastructure of the human enamel organ: I. External enamel epithelium, stellate reticulum and stratum intermedium. Cell Tissue Res 205:361–370

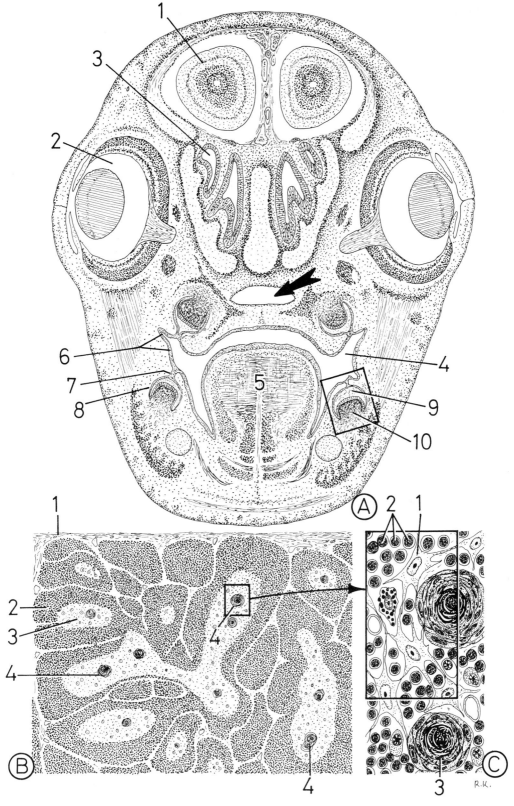
57

Plate 28. Atypical Epithelia: Enamel Organ. Continuation of Plate 27

At the beginning of odontogenesis, a dental lamina (**1**) grows from the stratified squamous epithelium of the oral cavity (**2**) into the subepithelial connective tissue, where it gradually forms the bell-like enamel organs. Every enamel organ (**3**) is seen to be comprised of an outer (**4**) and inner (**5**) epithelium. Between the two epithelia, the stellate epithelial cells (**6**) form a wide-meshed avascular network, the enamel pulp (asterisk), which is atypical for an epithelial tissue.

The outer layer of the enamel organ (**4**) is in contact with cells of the connective tissue of the dental sac (**7**) and numerous blood vessels (**8**).

The cells of the simple columnar, inner epithelium differentiate into ameloblasts (**5**), which synthesize the enamel (substantia adamantina) of the future crown.

As mentioned in Plate 27, mesenchyme enters the concavity of the enamel organ (dental papilla), where it forms the vascular dental pulp (**9**). The mesenchymal cells in contact with the ameloblasts arrange themselves into a vascularized epithelium-like layer and differentiate, under the inductive action of ameloblasts, into dentin-producing odontoblasts (**10**; Plate 113).

Magnification: $\times 500$

REFERENCES

Boyde A, Reith EJ (1976) Scanning electron microscopy of the lateral cell surfaces of incisor ameloblasts. J Anat 122:603–610

Kallenbach E (1976) Fine structure of differentiating ameloblasts in the kitten. Am J Anat 145:283–318

Lavergne J (1975) Etude ultrastructurale de l'émail chez diverses espèces de mammifères. C R Soc Biol (Paris) 169:595–604

Weiss MP, Voegel JC, Frank RM (1981) Enamel crystallite growth: width and thickness study related to the possible presence of octocalcium phosphate during amelogenesis. J Ultrastruct Res 76:286–292

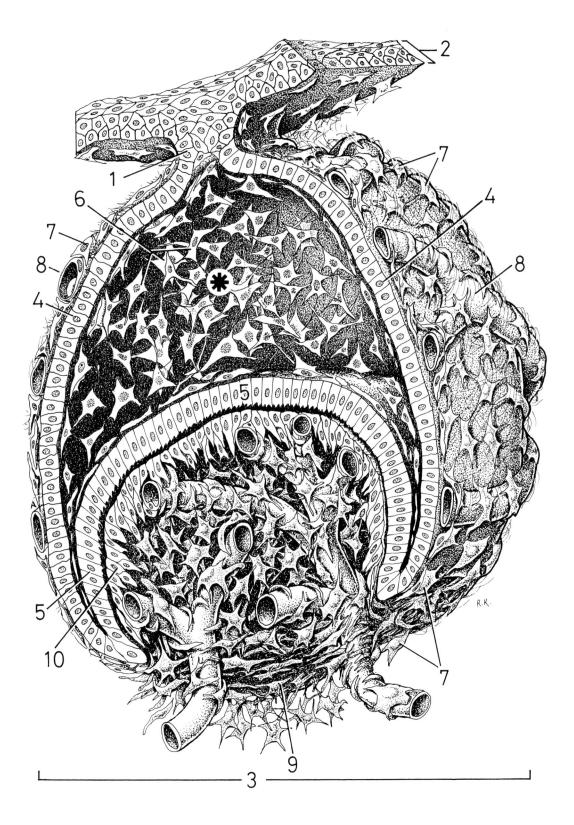

Plate 34. Glandular Epithelia. Form of Exocrine Glands

Simple glands (Fig. A) may be tubular (Fig. A1), mulberrylike or tubuloacinar (Fig. A2), vesicular or tubuloalveolar (Fig. A3), and saclike or alveolar (Fig. A4). Each gland has a terminal secretory portion (unshaded arrow) and a more or less developed excretory duct – termed "neck" in tubular glands (solid arrows). It should be noted here that forms A2 and A3 are, at least in humans, purely of theoretical significance. Simple tubular glands include the glands of the body of the stomach, glands of the small intestine, etc. Simple alveolar glands are represented by small sebaceous glands.

When several simple glands empty into one excretory duct (solid arrows), branched glands (Fig. B) develop. Here, the highly convoluted (Fig. B1) sweat glands or glandulae glomiformes represent a special intermediate form between the simple and branched glands.

Pyloric glands, duodenal glands, etc. belong to the ramified tubular glands (Fig. B2). Purely tubuloacinar and tubuloalveolar glands (Fig. B3, 4) only represent stages in the development toward compound glands. Branched alveolar glands (Fig. B5) include the large sebaceous glands and the tarsal glands of the eyelid. In compound glands (Fig. C), the main excretory duct (double arrows) branches into several smaller excretory ducts (solid arrows), each of which can possess tubular, acinar, tubuloalveolar, and alveolar terminal portions. Compound tubular glands (Fig. C1) are exemplified by the mucous glands of the oral cavity. The main representatives of compound tubuloacinar glands (Fig. C2) are the parotid glands and exocrine pancreas. The submandibular glands are an example of a compound tubuloalveolar gland (Fig. C3), and the secretory mammary gland is an example of a compound alveolar gland (Fig. C4). Large-scale subdivisions of compound glands lead to the development of lobules and lobes.

It should not be supposed that all the glands of the body of a mammal pass neatly into this schematic classification. There always exist individual and, in particular, functional variations which account for the same gland being able to switch rapidly from one category to another.

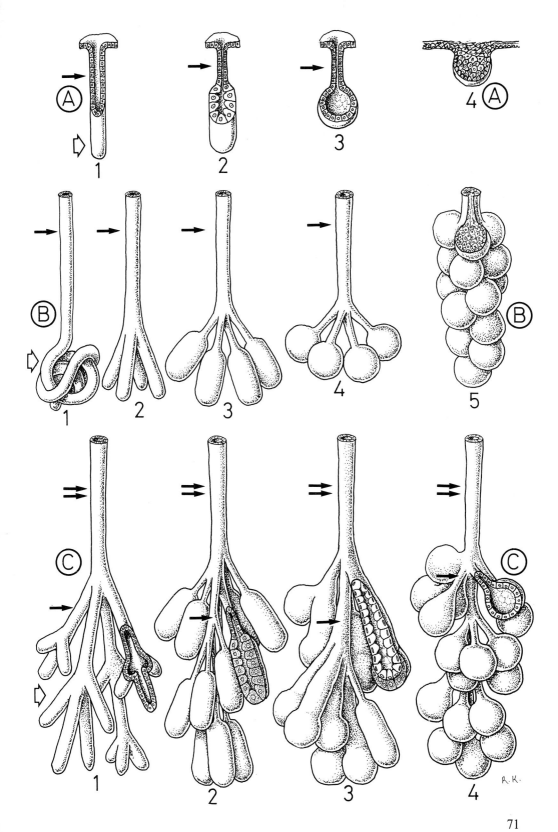

Plate 35. Glandular Epithelia. Unicellular Glands. Example: Goblet Cells from the Simple Columnar Epithelium of the Rat Small Intestine

Goblet cells occur in large numbers in the epithelium of the intestinal and respiratory tracts (see Plates 14, 18). In Fig. **A**, two goblet cells are depicted within insets: The cell on the left corresponds to Fig. **B**, the cell on the right to Fig. **C**.

Fully active cells have a narrow base and a broad apex and hence the shape is that of a goblet. The nucleus is slender and contains condensed chromatin and above it is found a very well-developed Golgi apparatus (Figs. **B1**, **C1**).

Mucigen granules (Figs. **B2**, **C2**) are formed in the Golgi apparatus with the assistance of other organelles and move toward the apical cell pole as mucous droplets (Figs. **B3**, **C3**). The apical plasmalemma temporarily opens and the mucus is expelled from the cell. The cell does not lose cytoplasm in the process. This type of discharge is referred to as eccrine secretion. After expulsion of the mucous droplets the goblet cell becomes slender (so-called slender cell; Fig. **C**). Note the distinct microvilli (Figs. **B4**, **C4**) and zonulae occludentes (Figs. **B5**, **C5**) on the apical pole.

After a pause a new secretory cycle begins; the cell synthesizes its product and discharges it again. A secretory cycle can thus be repeated many times. (See Plates 154, 155 in KRSTIĆ 1979.)

Magnifications: Fig. **A**, × 700; Figs. **B**, **C**, × 5,000

REFERENCES

Cheng H, Bjerknes M (1982) Whole population cell kinetics of mouse duodenal, jejunal, ileal, and colonic epithelia as determined by radioautography and flow cytometry. Anat Rec 203:251–264

Freeman JA (1966) Goblet cell fine structure. Anat Rec 154:121–148

Kurosumi K, Shibuichi I, Tosaka H (1983) Ultrastructural studies on the secretory mechanism of goblet cells in the rat jejunal epithelium. Arch Histol Jpn 44:263–284

Specian RD, Neutra MR (1980) Mechanism of rapid mucus secretion in goblet cells stimulated by acetylcholine. J Cell Biol 85:626–640

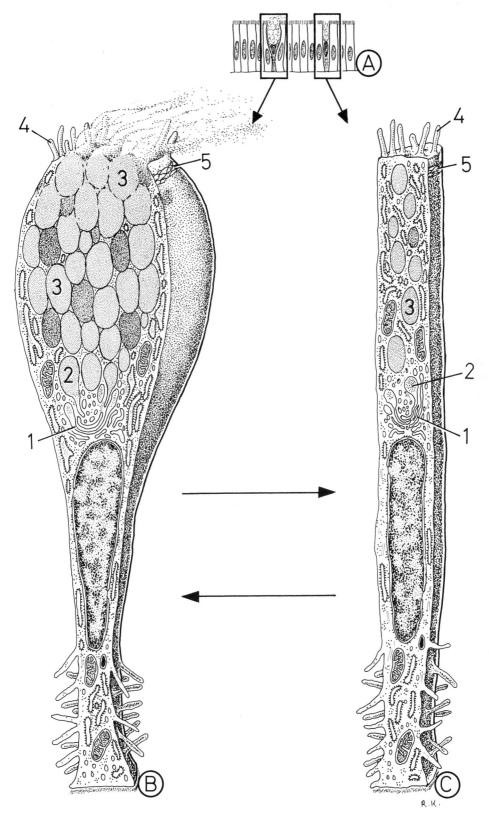

Plate 36. Glandular Epithelia. Endoepithelial Gland of Human Nasal Mucosa

All the mucus-producing cells (**1**) of an endoepithelial gland (**2**) are located in the middle of the pseudostratified columnar epithelium (**3**). They are somewhat lower than the neighboring ciliated cells (**4**), with the result that the glandular lumen is situated in the middle of the epithelium. Endoepithelial glands produce mucus for moistening the nasal mucosa.

Magnification: ×2,300

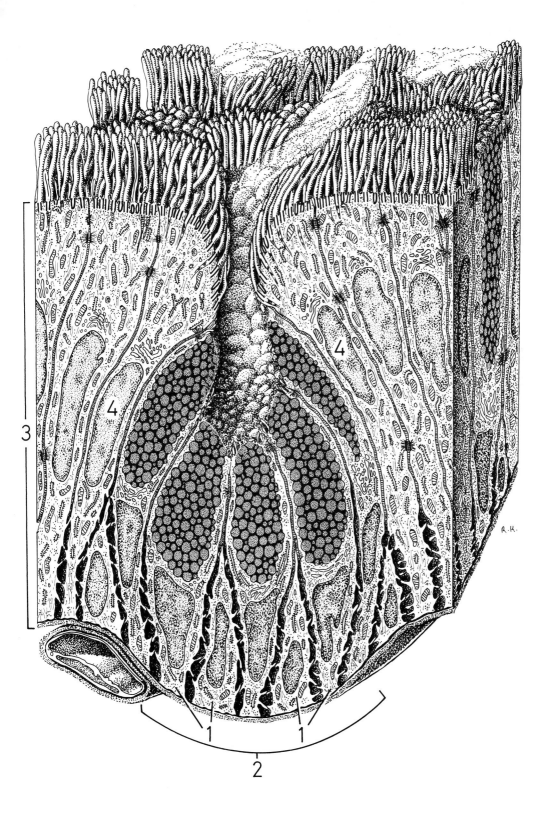

Plate 37. Glandular Epithelia. Terminal Portion of a Simple Tubular Heterocrine Gland from the Rat Stomach

Figure **A** depicts a section from the mucosa of the body of the stomach. Tubular gastric glands proper (Fig. **A 1**) open into the gastric pits (Fig. **A 2**). The area in the inset corresponds to Fig. **B**.

A gastric gland proper is composed of several types of cell: Neck mucous cells (not shown since they are only found in the neck of the gland; Fig. **A 3**); chief cells; parietal cells and isolated endocrine cells. Columnar chief cells (Fig. **B 2**) are distinguished by a very well-developed ergastoplasm and large numbers of free ribosomes. Many pepsinogen granules are found above the nucleus.

Parietal cells (Fig. **B 3**) are appreciably larger than the chief cells and their basal sections frequently protrude toward the exterior. They have large spherical or ellipsoidal nuclei, and mitochondria are very numerous in the cytoplasm. Particularly characteristic are the deep, occasionally branched intracellular canaliculi (Fig. **B 5**), which are penetrated by great numbers of microvilli. Parietal cells provide the hydrogen ions that are essential for the production of hydrochloric acid of gastric juice.

Endocrine cells (Fig. **B 4**) are scattered at the base of the gland. They do not extend to the glandular lumen, but their bases are in contact with the capillaries (Fig. **B 6**). Endocrine cells have spherical nuclei, well-developed organelles, and the infranuclear cytoplasm contains numerous hormone granules.

All these cells secrete their products without loss of cytoplasm, i.e., by an eccrine mechanism. Since the glandular secretion is composed of different products, this type of gland is termed heterocrine. (See Plates 42, 43, 160 in KRSTIĆ 1979.)

Magnifications: Fig. **A**, × 70; Fig. **B**, × 3,000

REFERENCES

Forssmann WG, Orci L, Pictet R, Renold AE, Rouiller C (1969) The endocrine cells in the epithelium of the gastrointestinal mucosa of the rat. An electron microscopic study. J Cell Biol 40:692–715

Gannon B, Browning J, O'Brien P (1983) The microvascular architecture of the glandular mucosa of rat stomach. J Anat 135:667–683

Hage E, Kjaergaard J (1972) Ultrastructure of basal-granulated cells in the rectum of human fetuses and children. Arch Histol Jpn 35:31–36

Helander HF (1981) The cells of the gastric mucosa. Int Rev Cytol 70:217–289

Rubin W, Ross LL, Sleisenger MH, Jeffries GH (1968) The normal human gastric epithelia. Lab Invest 19:598–626

Winckler J (1976) Biogene Amine in Polypeptidhormonbildenden Zellen. Die APUD-Zellen (Pearse). Klin Wochenschr 54:49–58

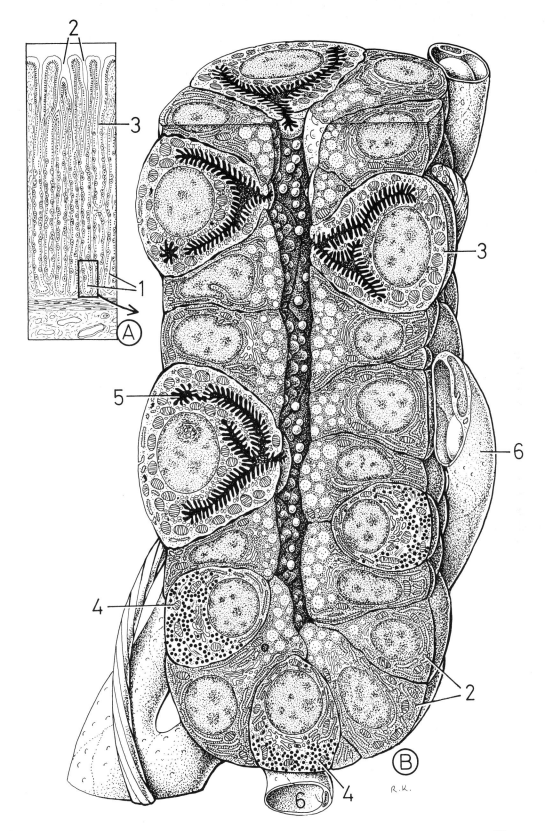

Plate 38. Glandular Epithelia. Acinus of a Tubuloacinar Serous Gland. Example: Rat Pancreas

The light-microscopic image of a section from the exocrine pancreas is presented in Fig. A. An acinus (Fig. A 1) is the exocrine secretory unit that communicates via its intercalated duct (Fig. A 2) with an excretory duct (Fig. A 3). Centroacinar cells (Fig. A 4) are localized in the interior of acini. The inset in Fig. A corresponds to Fig. B.

An acinus is a structure made up of simple epithelium and its shape is similar to that of a mulberry. The acinar cells (Fig. B 1) are cuboidal with globular, central nuclei and large nucleoli. Ergastoplasm and free ribosomes are noticeably widespread, indicative of a high level of protein synthesis. The supranuclear cytoplasm contains large numbers of zymogen granules of various sizes which are discharged in eccrine fashion. For the sake of clarity, the acinar lumen is shown as a broad space, though in reality it is a narrow cleft.

Centroacinar cells (Fig. B 2), characteristic of the exocrine pancreas, also occur in the acinus. They form an incomplete layer, and zymogen granules pass through the gaps between the cells. The centroacinar cells are continuous with the flattened cells of the intercalated duct (Fig. B 3). The lumen of this channel has also been drawn a good deal wider; it is actually narrow, since the pancreatic secretion is thin. For the same reason, no myoepithelial cells are found around the pancreatic acini.

There is a well-developed blood supply to the enzyme-synthesizing glandular units. Thus, on their external surface, which is separated from the secretory cells by a basal lamina (Fig. B 4), many capillaries (Fig. B 5) are found in addition to unmyelinated nerve fibers (Fig. B 6). (See Plate 155 in KRSTIĆ 1979.)

Magnifications: Fig. A, ×400; Fig. B, ×3,000

REFERENCES

Amsterdam A, Ohad I, Schramm M (1969) Dynamic changes in the ultrastructure of the acinar cell of the rat parotid gland during the secretory cycle. J Cell Biol 41:753–773

Bendayan M, Roth J, Perrelet A, Orci L (1980) Quantitative immunocytochemical localization of pancreatic secretory proteins in subcellular compartments of the rat acinar cell. J Histochem Cytochem 28:149–160

Ekholm R, Zelander T, Edlund Y (1962) The ultrastructural organization of the rat exocrine pancreas: I. Acinar cells. J Ultrastruct Res 7:61–72

Ekholm R, Zelander T, Edlund Y (1962) The ultrastructural organization of the rat exocrine pancreas: II. Centroacinar cells, intercalary and intralobular ducts. J Ultrastruct Res 7:73–83

Hand AR (1970) Nerve-acinar relationships in the rat parotid gland. J Cell Biol 47:540–543

Williams DW, Kendall MD (1982) The ultrastructure of the centroacinar cells within the pancreas of the starling (Sturnus vulgaris). J Anat 135:173–181

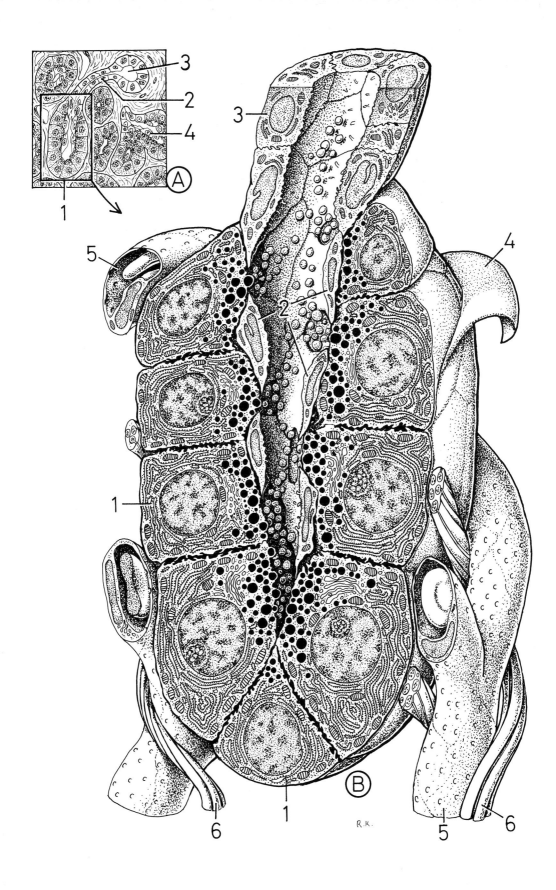

Plate 39. Glandular Epithelia. Seromucous Tubuloalveolar Terminal Portion of a Mixed Salivary Gland. Example: Human Submandibular Gland

In a section through a mixed salivary gland, the mucous secretory portions or mucous tubules (Fig. **A 1**) are weakly stained with hemalum and eosin, whereas the serous acini (Fig. **A 2**) are strongly stained. The mixed elements are distinguished by the presence of dark serous demilunes or crescents of Giannuzzi (Fig. **A 3**) at the end of the mucous tubules. Figure **B** corresponds to the inset in Fig. **A**.

The cytoplasm of the columnar mucous cells (Fig. **B 1**) contains such large numbers of mucous droplets that the dark, flattened nucleus is displaced to the basal lamina. These cells develop by a particular process which is characteristic of mixed salivary glands, whereby cells of the intercalated ducts (Fig. **B 2**) transform into mucous cells. For this reason, intercalated ducts are rare or even absent in seromucous glands. If present, they also produce a mucous secretion. Owing to the high viscosity of the product secreted by the mucous cells, the lumen of the tubule is relatively broad.

At the end of the mucous tubule, the ergastoplasm-rich serous cells (Fig. **B 3**) are arranged in a hemispherical fashion.

These cells are connected to the tubular lumen by means of intercellular canaliculi (Fig. **B 4**). In this manner, seromucous saliva is produced, and its expulsion is accelerated by contraction of the myoepithelial cells (Fig. **B 5**).

Blood capillaries (Fig. **B 6**) and unmyelinated nerve fibers (Fig. **B 7**) are found close to the terminal portions of mixed glands. (See Plate 161 in KRSTIĆ 1979.)

Magnifications: Fig. **A**, × 400; Fig. **B**, × 2,800

REFERENCES

Al Yassin TM, Toner PG (1977) Fine structure of squamous epithelium and submucosal glands of human oesophagus. J Anat 123:705–721

Barka T (1980) Biologically active polypeptides in submandibular glands. J Histochem Cytochem 28:836–859

Gibson MHL (1983) The prenatal human submandibular gland: a histological, histochemical and ultrastructural study. Anat Anz 153:91–105

Murphy RA, Watson AY, Metz J, Forssmann WG (1980) The mouse submandibular gland: an exocrine organ for growth factors. J Histochem Cytochem 28:890–902

Tandler B (1983) Ultrastructure of mink submandibular gland. J Submicrosc Cytol 15:519–530

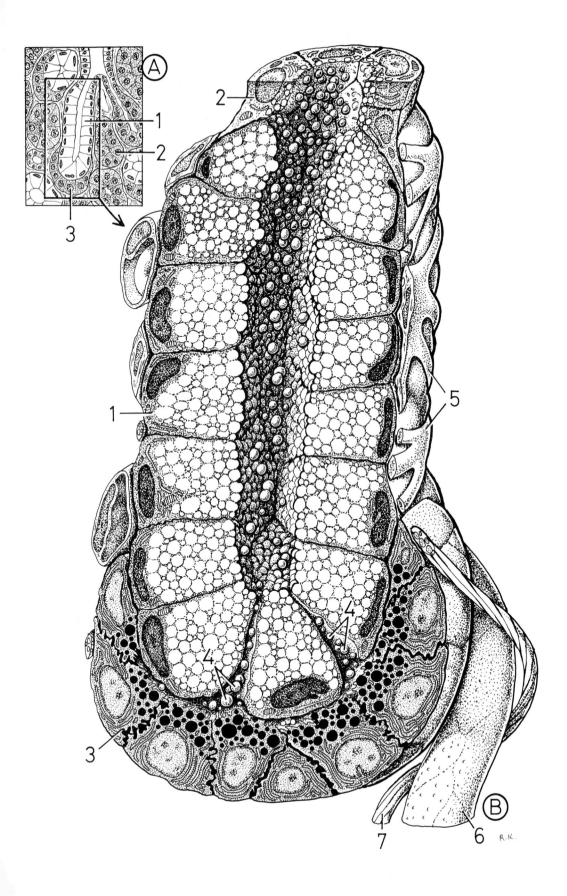

Plate 40. Glandular Epithelia. Terminal Portion of an Apocrine Alveolar Gland. Example: Lactating Rat Mammary Gland

Figure **A** illustrates the light-microscopic image of a lactating mammary gland of the rat. The cells directed toward the lumen are characterized by the presence of pale fat vacuoles (Fig. **A 1**). A single alveolus is drawn in Fig. **B**.

The epithelium of the alveolus is composed of a layer of columnar cells, rich in ergastoplasm. In the apical pole, the cytoplasm of these cells contains two types of vacuole: Small, osmiophilic, proteinaceous (black) casein granules (Fig. **B 1**) are found in the first type; the second, significantly larger, type of vacuole contains fat droplets (Fig. **B 2**). Expulsion of the protein granules occurs by means of the eccrine mechanism. The secretion of the fat droplets, however, is associated with a loss of apical cytoplasm and is termed apocrine secretion.

In this manner, milk (Fig. **B 3**) is formed in the broad alveolar lumen; expulsion of the milk is assisted by contraction of myoepithelial cells (Fig. **B 4**). (See Plate 156 in KRSTIĆ 1979.)

Magnifications: Fig. **A**, × 500; Fig. **B**, × 2,300

REFERENCES

Bargmann W, Knoop A (1959) Über die Morphologie der Milchsekretion. Z Zellforsch 49:344–388

Brooker BE (1983) Pseudopod formation and phagocytosis of milk components by epithelial cells of the bovine mammary gland. Cell Tissue Res 229:639–650

Kurosumi K, Kawabata I (1977) Transmission and scanning electron microscopy of the human ceruminous apocrine gland. Arch Histol Jpn 40:203–244

Schaumburg-Lever G, Lever WF (1975) Secretion from human apocrine glands: an electron microscopic study. J Invest Dermatol 64:38–41

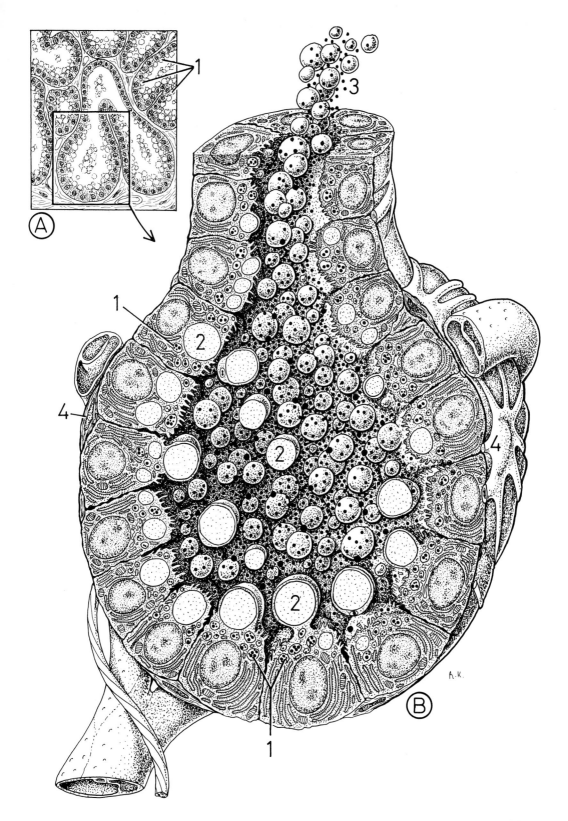

Plate 41. Glandular Epithelia. Sebaceous Sac of a Holocrine Pluristratified Alveolar Gland. Example: Human Sebaceous Gland

The majority of sebaceous glands (Fig. A 1) open into hair follicles (Fig. A 2). They are composed of numerous sacs or alveoli (Fig. A 3). A sebaceous alveolus is enlarged in Fig. B.

As in the epidermis, the deepest or basal cells (Fig. B 1) in contact with the basal lamina (Fig. B 2) are capable of division. They form a germinal layer from which the dead cells that have been transformed into sebum are replaced by mitosis (Fig. B 3). The internal cells (Fig. B 4) produced by the germinal layer gradually become sebaceous cells by the accumulation of lipid droplets in the cytoplasm; however, all cells remain firmly connected to one another by means of numerous desmosomes.

At a later stage of sebum development, the cells gradually die; the nuclei become very dense, i.e., pyknotic, and shrink (Fig. B 5, 6). At the same time, the cell borders disappear, and the sebaceous vacuoles leave the now amorphous cytoplasm, the end product of which is an anuclear fatty mass. The sebum (Fig. B 7) reaches the skin surface via the hair follicle.

Pluristratified alveolar glands are, without exception, holocrine. This term signifies a process in which synthesis of the secretory product is accompanied by death of the cell. Thus, such glands constantly require replacement of cells from the germinal layer. (See Plate 157 in KRSTIĆ 1979.)

Magnifications: Fig. A, × 70;
Fig. B, × 2,600

REFERENCES

Bell MA (1971) A comparative study of sebaceous gland ultrastructure in subhuman primates. Anat Rec 170:331–342

Mesquita-Guimarae J (1976) Holocrine cell lysis in the rat preputial sebaceous gland. Evidence of autophagocytosis during cell involution. Anat Rec 186:49–68

Mesquita-Guimarae J, Pignatelli D, Coimbra A (1979) Autophagy during holocrine cell lysis in skin sebaceous glands. J Submicrosc Cytol 11:435–447

Rupec M (1969) Zur Ultrastruktur der Talgdrüsenzelle. Arch Klin Exp Dermatol 234:273–292

Strauss JS, Pochi PE (1969) Histology, histochemistry and electron microscopy of sebaceous glands in man. In: Gans O, Steigleder GK (eds) Normale und pathologische Anatomie der Haut. Springer Berlin Heidelberg New York

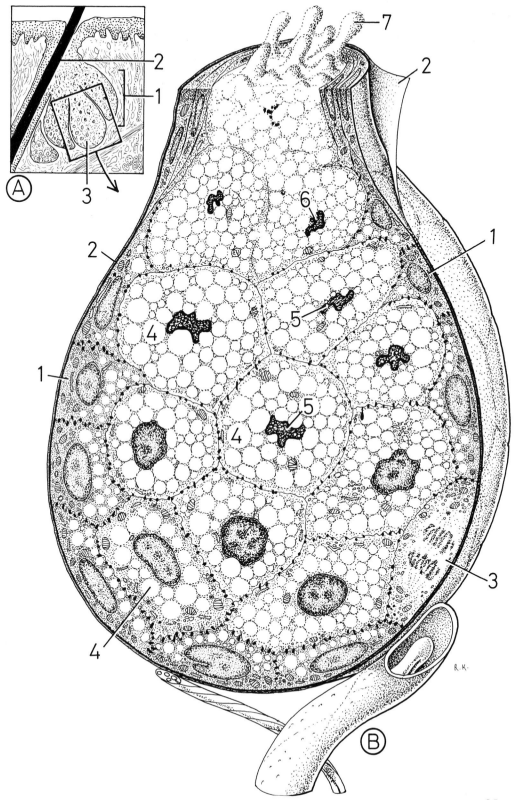

Plate 42. Glandular Epithelia. Scheme of a Compound Tubuloacinar Gland

This plate is solely intended to clarify the general organization of a compound exocrine gland. For this reason, the acinar lumen, which is usually narrow, has been drawn as a wide space.

The acini (**1**) are connected via narrow intercalated ducts (**2**) to the striated ducts (**3**).

Several striated ducts unite outside the lobule (**4**) to form one interlobular or extralobular duct (**5**); several extralobular ducts unite to form the principal excretory duct (not represented), which in the section reaching the surface is lined by pseudostratified epithelium.

Each gland is supplied with arteries (**6**) and veins (**7**). Nervous impulses are transmitted by bundles of unmyelinated nerve fibers (**8**).

Magnification: ×280

REFERENCE

Banerjee SD (1977) Basal lamina of embryonic salivary epithelia. Production by the epithelium and role in maintaining lobular morphology. J Cell Biol 73:445–463

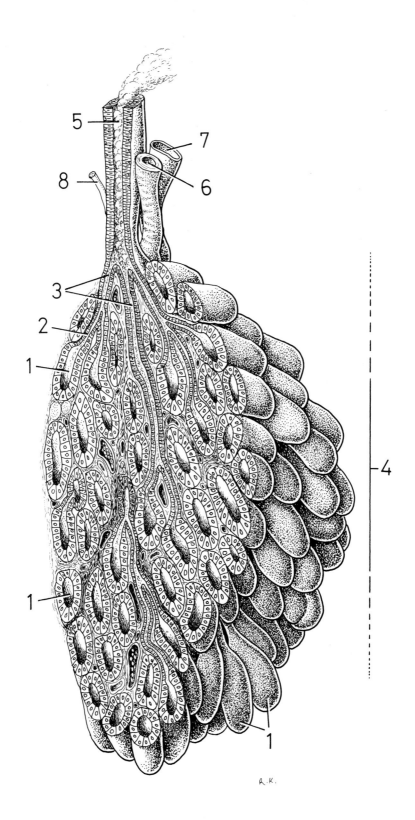

EPITHELIAL TISSUE

Plate 43. Glandular Epithelia. Endocrine Glands.
Examples: Pancreatic Islets and Thyroid Gland

The pancreas (Fig. **A 1**), which fits into a concavity of the duodenum (Fig. **A 2**), is made up of two types of glandular epithelium. Acini (Fig. **B 1**), mentioned in Plate 38, with their excretory channel system (Fig. **B 2**), compose the greater part of the pancreatic parenchyma. In the interior of the exocrine lobules (Fig. **B 3**) are located small, well-defined epithelial cell groups, the islets of Langerhans or pancreatic islets (Fig. **B 4**), which do not possess excretory ducts and therefore release their products (hormones) directly into the blood. The islet in the inset appears three-dimensionally in Plate 44.

The lobulated thyroid gland (Fig. **C 1**) is located on the anterior side of the larynx (Fig. **C 2**). The lobules are fairly well defined and contain numerous vesicular or tubular cavities, follicles (Fig. **D 1**), which are lined with a simple epithelium. Colloid, carrier of the thyroid hormones, is stored in the interior of the follicles. The outer surface of the epithelial cells is in close contact with blood capillaries since, like the pancreatic islets and all other endocrine glands, the thyroid has no excretory ducts. The follicle in the inset is presented in greater detail in Plates 45 and 46.

Magnifications: Fig. **B**, × 100; Fig. **D**, × 150

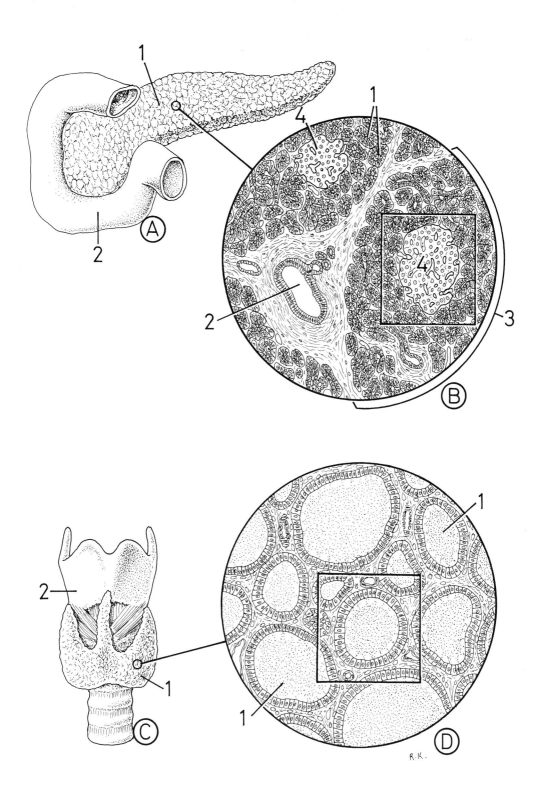

Plate 44. Glandular Epithelia. Endocrine Glands. Pancreatic Islets. Continuation of Plate 43

The pancreatic islets are minute endocrine glands scattered within the exocrine pancreas. Their epithelial or islet cells (**1**) are polygonal elements with lobate, ellipsoidal nuclei and a variable number of secretory granules depending on the functional stage; they form branched interconnected strands. Many blood capillaries (**2**) run between the epithelial strands, often accompanied by unmyelinated nerve fibers (**3**) which frequently contact the epithelial cells.

Several types of islet cell have so far been identified (see histology texts for further information); for example, B cells synthesize and secrete insulin into the blood capillaries. This vital hormone reduces the level of sugar in the blood and deficiency leads to diabetes mellitus. An increase in blood sugar level induces release of the hormone glucagon, which is produced in the A cells. (See Plates 163, 164 in KRSTIĆ 1979.)

Magnification: × 1,000

REFERENCES

Capella C, Solcia E (1972) The endocrine cells of the pig gastrointestinal mucosa and pancreas. Arch Histol Jpn 35:1–29

Cooperstein JS, Watkins D (eds) (1981) The islets of Langerhans. Academic, New York

Orci L (1976) Morphofunctional aspects of the islets of Langerhans. Metabolism 25 (Suppl. 1):1303–1313

Orci L (1977) The islet of Langerhans: a multihormonal micro-organ. In: Bajaj JS (ed) Insulin and metabolism. Elsevier, Amsterdam

Zimny ML, Blackard WG (1975) The surface structure of isolated pancreatic islet cells. Cell Tissue Res 164:467–471

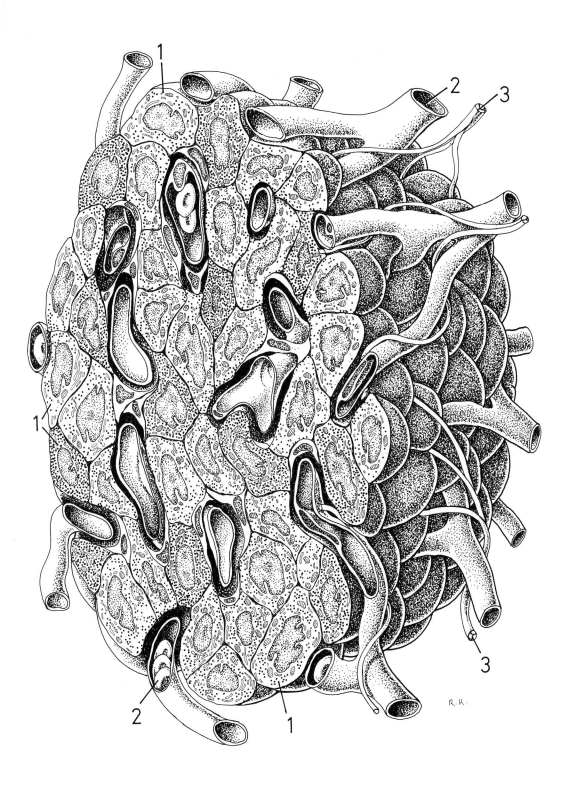

Plate 45. Glandular Epithelium. Endocrine Glands. Thyroid Gland of the Rat. Continuation of Plate 43

The thyroid follicles (1) are spherical, ovoid, or tubular structures, lined by a simple epithelium (2) of varying thickness. Unmyelinated nerve fibers (3) and arterioles (4) occur in the interfollicular connective tissue. The arterioles form a dense capillary network (5) around the follicle, through which every epithelial or thyroid follicular cell is in contact with the blood circulatory system.

The epithelial cells synthesize and discharge homogeneous colloid into the lumen of the follicle. Two iodine-containing thyroid hormones (tri- and tetraiodothyronine) are bound to the globulins of colloid. The epithelial cells thus ensure production and storage of thyroid hormones. If thyroid hormones are required by the organism, the hitherto viscous colloid becomes more fluid, the hormone-protein complexes are split, and the hormones are released into the blood.

The second type of epithelial cells in the thyroid, the C cells (6), stem from the ultimobranchial body and do not form colloid. Their product – the hormone calcitonin – is released directly into the blood.

For the sake of clarity, the basal laminae of the capillaries and epithelial cells have not been drawn.

It should be noted that the majority of endocrine epithelial cells possess a small intracellular hormone reserve.

Magnification: × 1,000

REFERENCES

Chan AS (1983) Ultrastructural observations on the formation of follicles in the human fetal thyroid. Cell Tissue Res 233:693–698

Fujita H, Murakami T (1974) Scanning electron microscopy on the distribution of the minute blood vessels in the thyroid gland of the dog, rat, and rhesus monkey. Arch Histol Jpn 36:181–188

Ketelbant-Balasse P, Rodesch F, Neve P, Pasteels MJ (1973) Scanning electron microscope observations of apical surfaces of dog thyroid cells. Exp Cell Res 79:111–119

Kobayashi S (1973) Rasterelektronenmikroskopische Untersuchungen der Schilddrüse. Arch Histol Jpn 36:107–117

Sobrinho-Simoes M, Johanessen JV (1981) Scanning electron microscopy of the normal human thyroid. J Submicrosc Cytol 13:209–222

Young BY, Baker TG (1982) The ultrastructure of rat thyroid glands under experimental conditions in organ culture. J Anat 135:407–412

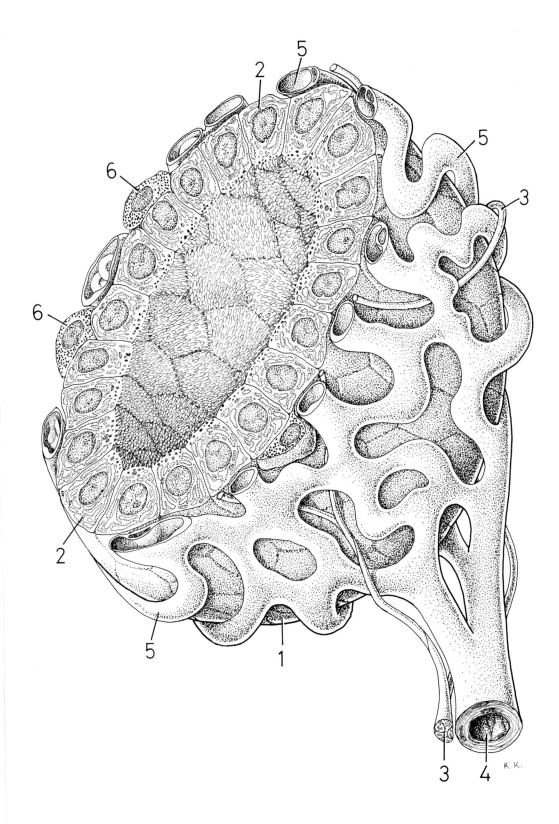

Plate 47. Regeneration and Transplantation of Epithelial Tissue

The functions of surface and many glandular epithelia are associated with a continuous loss of cells; thus, these cells have to be constantly replaced by new ones. Mitoses play a major role in compensating for the lost cells, and they usually occur as far as possible from the sites of consumption, e.g., in stratified squamous epithelia mitoses only occur in the germinal layer (Fig. A, arrows). The regeneration of enterocytes takes place solely in the intestinal crypts (Fig. B, arrows; see Plate 118). The basal cells (Fig. C, arrows) of sebaceous glands are responsible for replacing sebaceous cells that disintegrate during holocrine secretion (see Plate 41).

Epithelial regeneration that proceeds in a cyclic manner, as in the monthly changes that take place in the endometrium of the uterus during the reproductive period of women, is termed physiological or cyclic regeneration. During the first 4–5 days of a menstrual cycle, a large part of the endometrium is rejected (menstrual hemorrhage) if no blastocyst has implanted. An approximately 1-mm-thick basal layer or pars basalis (Fig. D, asterisk) and bases of the uterine glands (Fig. D1) are all that persist in the uterus. Regeneration of the epithelial coating and subendothelial connective tissue of the endometrium takes place in the course of the next 9–10 days, proceeding from the epithelium of the glandular remnants. During this proliferative stage, there is a high degree of mitoses (Fig. D2) in the glandular epithelium and neighboring connective tissue (Fig. D3), which regenerates from the remains of the interglandular tissue. The time from the 1st day of menstrual hemorrhage until complete regeneration of the desquamated endometrium is about 14 days.

Following a surface wound of the skin (Fig. E), the germinal layer of the wound periphery begins to cover the lesion, accompanied by a great deal of mitotic activity, leading to epithelial migration (Fig. F, arrows). The cells of the hair follicles (Fig. F1) and excretory ducts of sweat glands (Fig. F2) also take part in epidermal regeneration.

With larger wounds, the defect is filled with regenerated vascular connective tissue, called granulation tissue (Fig. G1), from subepidermal layers. Since the epidermis moves slowly over the granulation tissue from the wound periphery, a skin transplant is necessary to accelerate this process. In Fig. H, as an example, a patch of epidermis is shown being grafted (with underlying loose connective tissue of the stratum papillare) according to the method of Ollier-Thiersch.

From the examples given, it is clear that surface epithelia and many glandular epithelia regenerate well under normal and pathological conditions. Highly differentiated glandular epithelia, however, regenerate significantly poorer than surface epithelia.

Magnifications: Fig. A, × 550; Figs. B, C, × 700; Fig. D, × 70; Figs. E–H, × 40

REFERENCES
Gabbiani G, Montandon D (1977) Reparative processes in mammalian wound healing: the role of contractile phenomena. Int Rev Cytol 48:187–219
Odland G, Ross R (1968) Human wound repair: I. Epidermal regeneration. J Cell Biol 39:135–151
Potten CS, Allen TD (1975) The fine structure and cell kinetics of mouse epidermis after wounding. J Cell Sci 17:413–447
Reeve DRE (1974) Mitotic activity of epithelial cells in wounded rectal mucous membrane. J Anat 118:231–239

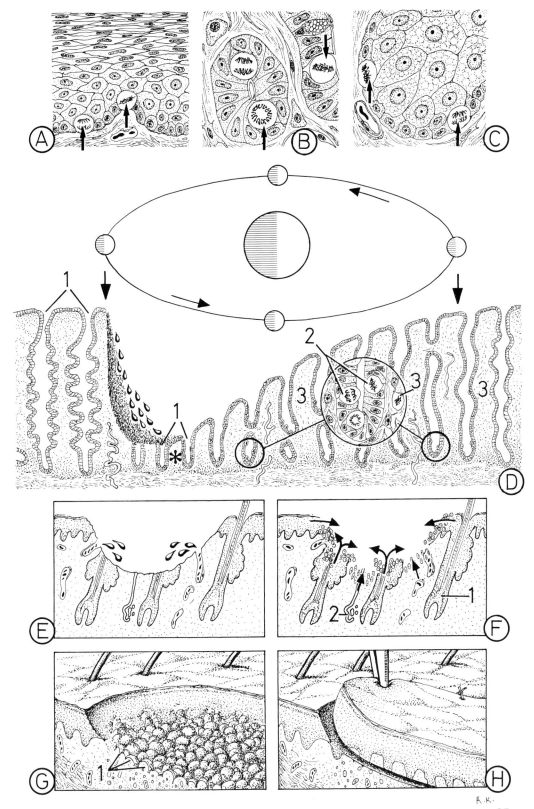

Plate 48. Origin of Connective and Supporting Tissues

The mesenchyme, from which – with the exception of the notochord – all connective and supporting tissues develop, stems largely from the middle germ layer or mesoderm. Only a small part of the mesenchyme derives from the ectoderm of the neural crest (see embryology texts for further information).

In Fig. A, the trilaminar stage of an 18-day-old human embryo is illustrated at the level of the primitive streak (unshaded arrow). All three germ layers are visible in the section: ectoderm (1), endoderm (2), and the interjacent mesoderm (3).

The initially closed epitheliumlike cell layer of the mesoderm becomes less compact due to the uptake of fluid and transforms into the embryonic connective tissue, the mesenchyme. The cells of the mesenchyme differentiate and form, among others:

– Stem cells (4) of all blood elements
– Practically all smooth muscle cells (5)
– Immature connective tissue cells, fibroblasts (6), which differentiate into mature connective tissue cells, fibrocytes (7)
– Immature bone-forming cells, osteoblasts (8), which become osteocytes (9) during the process of ossification. Fibroblasts (6) can also differentiate into osteoblasts. The dentin-forming elements, odontoblasts (10), similarly develop from mesenchymal cells
– All cartilage cells, chondrocytes, of hyaline (11), elastic (12), and fibrous (13) cartilage
– Polynuclear chondroclasts and osteoclasts (14)
– Mast cells (15)
– Lipoblasts (16) – immature – and adipocytes (17) – mature fat cells, which are able to differentiate into one another
– Reticular cells (18) of the reticular tissue [fat cells (16, 17) can develop by metaplastic transformation of reticular cells]
– Connective tissue macrophages or histiocytes (19); reticular cells also have the ability to transform into macrophages
– Endothelial cells (20) of blood and lymphatic vessels.

As is evident from the above examples, mesenchyme is a pluripotential tissue.

Magnifications: Fig. A, × 120; Figs. 4–20, × 800

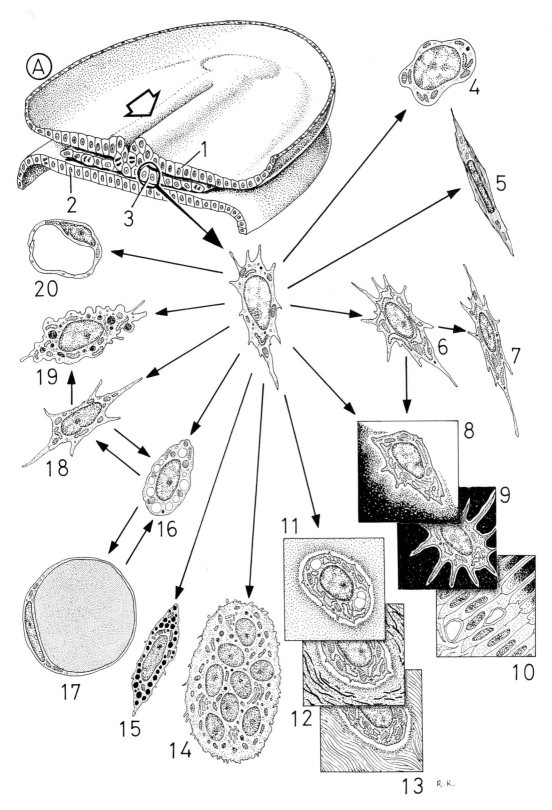

CONNECTIVE AND SUPPORTING TISSUES

Plate 49. Classification of Connective and Supporting Tissues

This plate provides a summary of the very varied forms of connective and supporting tissues. The main tissue-groups are shown in rectangles, and the subdivisions in circles.

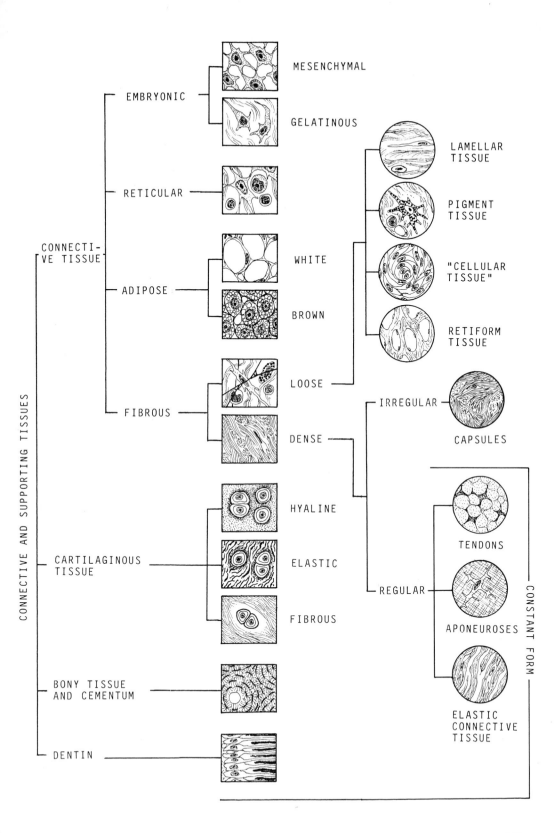

103

Plate 50. Mesenchyme or Embryonic Connective Tissue from the Dorsal Mesogastrium of a 14-Day-Old Mouse Fetus

The low cells (1) at the upper edge of the illustration will in the course of development, after flattening, differentiate into mesothelium, i.e., epithelium of mesodermal origin. Like all surface epithelia, it forms a boundary to the underlying amorphous mesenchymal tissue.

The mesenchymal cells (2) are connected to one another by fine cell processes (3), invisible under the light microscope. Uptake of fluid into the intercellular spaces causes the structure to become less compact and a three-dimensional lattice develops, the interstices of which contain capillaries (4).

Mesenchymal tissue is characterized by large numbers of mitoses (5). Cells that are about to divide round themselves off, though they still remain connected to neighboring cells by fine processes, which can be observed in the electron microscope. After mitosis, the daughter cells adopt the form of other cells in the lattice.

The interstices in mesenchymal tissue are filled with an intercellular substance which does not contain specifically differentiated structures.

In addition to their ability to proliferate, mesenchymal cells tend to form localized clusters of cells, blastemas. All connective and supporting tissues, musculature, and other organs of mesodermal origin (e.g., part of the kidney, adrenal cortex) develop from these areas of tightly packed cells. Though morphologically very similar to gelatinous connective tissue, mesenchymal tissue is characterized by this special histoplastic ability to form other tissues. (See Plate 112 in KRSTIĆ 1979.)

Magnification: ×2,500

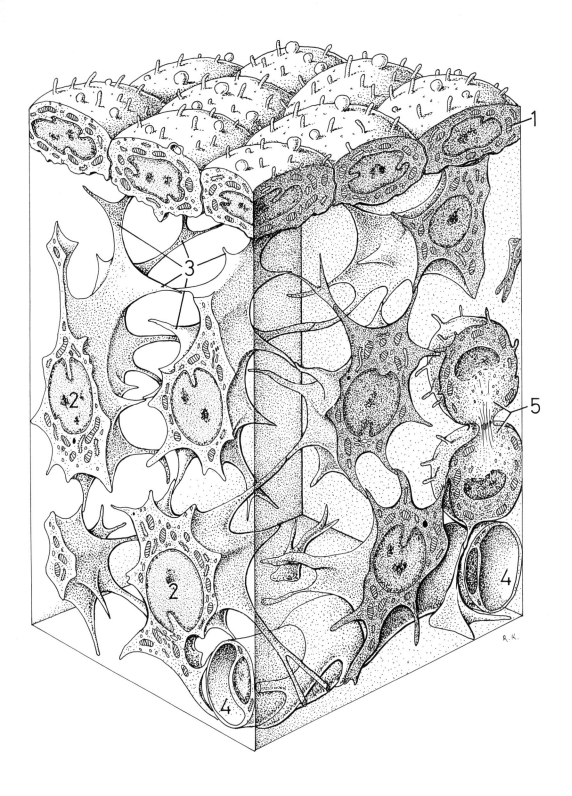

CONNECTIVE AND SUPPORTING TISSUES

Plate 51. Mesenchymal Cell. Same Preparation as in Plate 50

Mesenchymal cells are stellate elements. Their numerous, fairly regular cell processes (**1**) connect with those of other cells in the connective tissue meshwork. The great diversity in external appearance is in contrast to the primitive internal organization.

The nuclei of mesenchymal cells are large, predominantly ellipsoidal, and with deep indentations. The appearance of these nuclei is characteristic of very active cells, where the area of contact between nucleus and cytoplasm has to be as great as possible. The nuclei contain dispersed chromatin, though they do contain a large nucleolus, which is displaced toward the nuclear membrane.

Mesenchymal cells are very poor in organelles: Apart from a few mitochondria (**2**), some cisternae of rough endoplasmic reticulum (**3**), and a medium-sized Golgi apparatus (**4**), the cytoplasm only contains a small amount of free ribosomes. Sporadic lysosomes and lipid droplets occasionally appear.

The motility of mesenchymal cells is limited. They can leave the meshwork in order to accumulate at particular sites, where they serve as an anlage for other tissues or organs.

The intercellular substance of mesenchyme is fluid. It contains occasional microfibrils (**5**) and an amorphous, moderately osmiophilic material. (See Plate 166 in Krstić 1979.)

Magnification: ×7,000

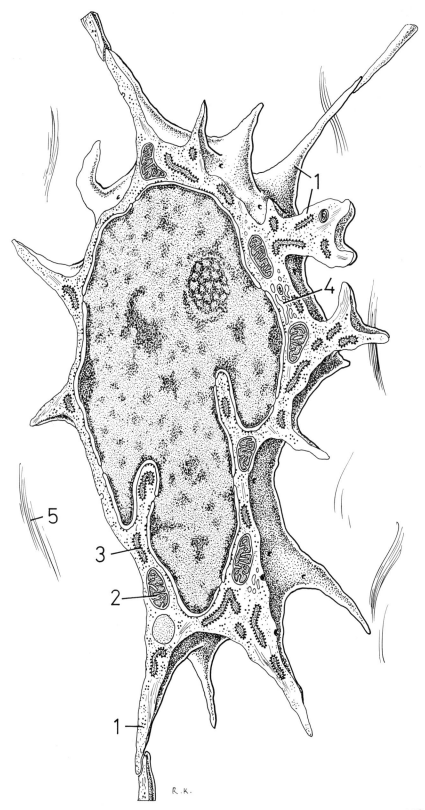

Plate 52. Gelatinous or Mucous Connective Tissue. Example: Umbilical Cord of a Human Neonate

Gelatinous tissue is morphologically very similar to mesenchyme. There is, however, a fundamental difference between the two: Whereas mesenchyme represents a pluripotential tissue, gelatinous tissue is already mature and incapable of further differentiation.

Gelatinous tissue occurs as Wharton's jelly in the umbilical cord (Fig. A1) and chorionic plate (Fig. A2), where it surrounds the fetal blood vessels (Fig. A3). At the bottom of the illustration, part of the placenta with the basal plate (Fig. A4) can be seen. Varicosities (Fig. A5), which are common in the umbilical vessels, can also be distinguished. The light-microscopic appearance of a section through the umbilical cord is shown in Fig. B.

The cells of gelatinous tissue (Fig. B1) are stellate like mesenchymal cells, though less numerous. It is not possible using the light microscope to discern whether the thin cell processes connect with one another.

In the spaces between the cells, there is a feltwork composed of delicate collagen fibrils (Fig. B2) and a proteoglycan ground substance. A section from Fig. B (inset), including the amniotic epithelium, is presented three-dimensionally in Fig. C.

Electron-microscopic studies have demonstrated that the stellate cells (Fig. C1) of gelatinous tissue form a three-dimensional lattice. The bundles of collagen microfibrils (Fig. C2), which cross in all directions, provide the tissue, and thus the umbilical cord, with a certain tensile strength. The ability of the very polymerized hyaluronic acid of the ground substance to absorb water imparts turgidity to the gelatinous tissue and thus increases the mechanical resistance of the cord. This helps to prevent kinking of the umbilical cord, which could lead to interruption of blood circulation.

The outer surface of the umbilical cord is lined with amniotic epithelium (Fig. C3). The epithelium is nonsecretory in this section since Wharton's jelly contains neither capillaries nor nerve fibers. The structure of adult dental pulp is reminiscent of gelatinous tissue, though the former is well vascularized and supplied with numerous nerve fibers.

Mesenchyme and gelatinous tissue occur only temporarily during the course of prenatal development.

Magnifications: Fig. A, × 1; Fig. B, × 450; Fig. C, × 1,000

REFERENCE
Parry EW (1970) Some electron microscope observations on the mesenchymal structures of full-term umbilical cord. J Anat 107:505–518

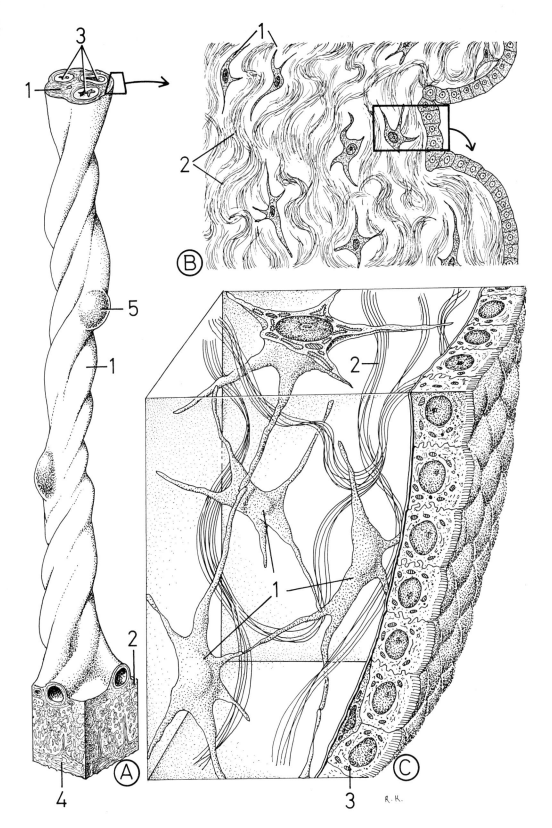

CONNECTIVE AND SUPPORTING TISSUES

Plate 53. Reticular Connective Tissue. Human Lymph Node

Reticular connective tissue forms the basic framework of and fulfills specific functions in the bone marrow, spleen, and lymph nodes. It also occurs in the tonsils, solitary lymphatic nodules, and Peyer's patches of the ileum. In addition, it separates the lobuli of glands and accompanies blood and lymph capillaries.

A lymph node (Fig. **A**) has been chosen to elucidate the structure of reticular tissue. The capsule (Figs. **A1**, **B1**) is composed of dense connective tissue, through which numerous afferent lymphatic capillaries (Figs. **A2**, **B2**) pass; they open into the subcapsular sinus (Figs. **A3**, **B3**).

The lymphatic or lymphoreticular tissue (lymphocytes + reticular tissue) of the organ is divided into two zones: a peripheral zone, the cortex characterized by large numbers of lymphatic follicles (Fig. **A4**); an inner zone, the medulla, distinguished by medullary cords (Fig. **A5**).

An artery (Fig. **A6**) enters the node at the hilus (Fig. **A7**), as veins (Fig. **A8**) and an efferent lymphatic vessel (Fig. **A9**) leave the organ at the same site. Thus, lymph streams constantly through the network of reticular cells in the direction of the arrows toward the hilus.

A detailed light-microscopic view of the capsule, subcapsular sinus, and cortex is illustrated in Fig. **B**. This zone of the lymph node is particularly suitable for a study of reticular tissue, since the reticular cells form here a loose cell union. An afferent lymphatic capillary (Fig. **B2**) with valves (Fig. **B4**) enters the capsule (Fig. **B1**) and brings lymph into the subcapsular sinus (Fig. **B3**), which is lined with a monolayer of littoral cells (Fig. **B5**). Stellate reticular cells (Fig. **B6**) are stretched transversely across the sinus, and a network of branching reticular fibers (Fig. **B7**) is in close contact with these cells. Large numbers of small lymphocytes (Fig. **B8**) and macrophages (Fig. **B9**) are present in the spaces between the reticular cells.

Part of Fig. **B** is three-dimensionally presented in Plate 54.

Magnifications: Fig. **A**, ×15; Fig. **B**, ×800

REFERENCES

Belisle C, Sainte-Marie G (1981) Tridimensional study of the deep cortex of the rat lymph node: I. Topography of the deep cortex. Anat Rec 199:45–59

Belisle C, Sainte-Marie G (1981) Tridimensional study of the deep cortex of the rat lymph node: II. Relation of deep cortex units of afferent lymphatic vessels. Anat Rec 199:61–72

Belisle C, Sainte-Marie G (1981) Tridimensional study of the deep cortex of the rat lymph node: III. Morphology of the deep cortex units. Anat Rec 199:213–226

Clark SL Jr (1962) The reticulum of lymph nodes in mice studied with the electron microscope. Am J Anat 110:217–224

Forkert PG, Thliveris JA, Bertalanffy FD (1977) Structure of sinuses in the human lymph node. Cell Tissue Res 183:115–130

Fossum S, Vaaland JL (1983) The architecture of rat lymph nodes: I. Combined light and electron microscopy of lymph node cell types. Anat Embryol 167:229–246

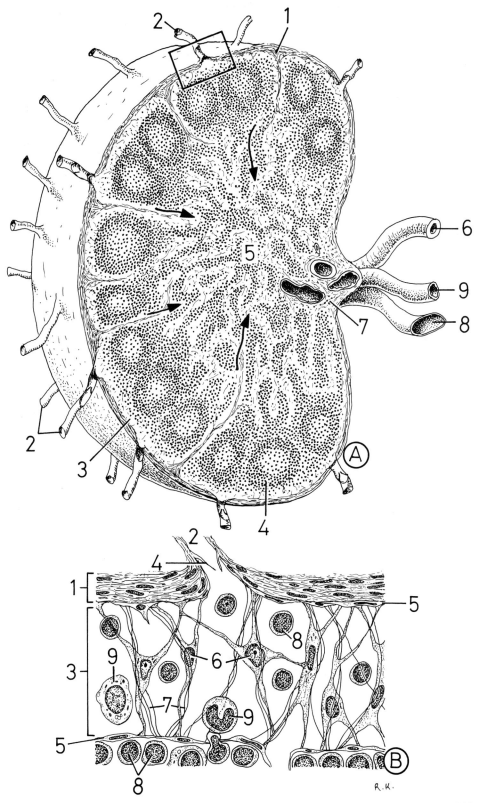

Plate 54. Reticular Connective Tissue. Continuation of Plate 53

The capsule (1) of the lymph node has been drawn thinner and with only a few collagen bundles so that the course of an afferent lymphatic capillary (2) and the point where it opens into the subcapsular sinus (3) can be seen more clearly. The capillary endothelial cells (4) are continuous with the very flattened reticular cells, littoral cells (5), which line all sinuses of the lymph node.

The ramified, stellate, reticular cells (6) are usually stretched transversely through the sinus. They are connected with neighboring cells of the same type by means of thin cell processes. Many bundles of reticular microfibrils (7), which can be visualized by silver staining, are in close contact with the reticular cells; such bundles are frequently surrounded by processes of the reticular cells.

Several wandering cells move freely in the lymph of the sinus: a monocyte (8; see Plate 71), a histiocyte (9; see Plate 66), a lymphocyte (10; see Plate 70), and a plasma cell (11; see Plate 72). A few lymphocytes (12) by virtue of their plasticity and a certain degree of ameboid movement are able to enter the sinus by passing between the littoral cells (13) that line the cortex. Some of these littoral cells cover the inner surface of a small drainage sinus (14), which allows the lymph to drain slowly in the direction of the arrow through the network of reticular cells to the efferent capillary. Lymph thus comes into contact with every reticular cell.

Magnification: × 3,500

REFERENCES

Carr I (1970) The fine structure of the mammalian lymphoreticular system. Int Rev Cytol 27:283–348

Fujita T, Kashimura M (1981) The "Reticulo-endothelial system" reviewed by scanning electron microscopy. Biomed Res (Suppl) 2:159–171

Fujita T, Miyoshi M, Murakami T (1972) Scanning electron microscope observation of the dog mesenteric lymph node. Z Zellforsch 133:147–162

Vizioli MR, Valdrighi L (1974) The appearing and evolution of reticulin: observations on the repair of post extraction marmoset sockets. Ann Histochim 19:65–72

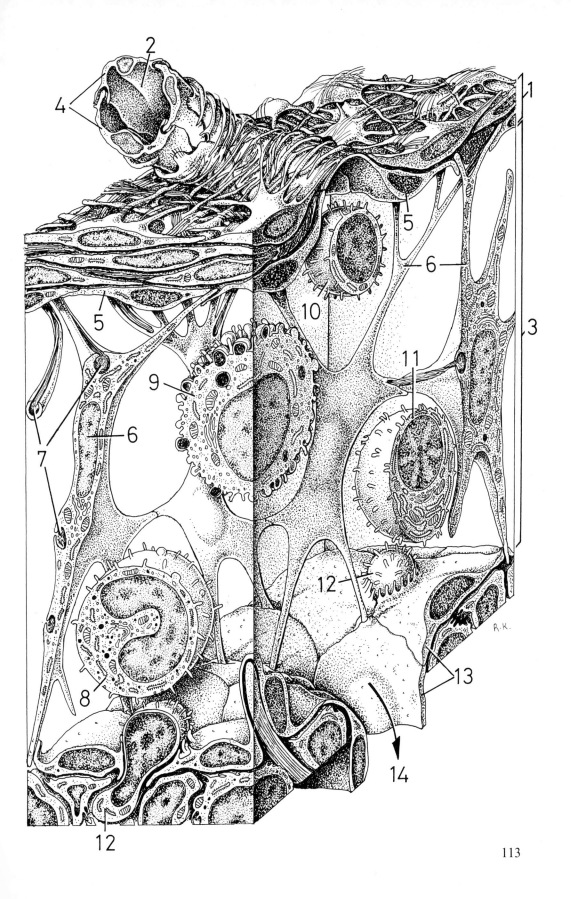

Plate 55. Reticular Connective Tissue. Continuation of Plate 54

The left drawing shows a section through a reticular cell, revealing the ultrastructure.

A large part of the cell body is occupied by a voluminous, spherical or ellipsoidal, indented nucleus (**1**) with a conspicuous nucleolus (**2**). Several mitochondria (**3**) and a simple or multiple Golgi apparatus (**4**) are found in the cytoplasm. The cisternae of rough endoplasmic reticulum (**5**) have moderately osmiophilic contents and are broad or narrow according to cytophysiological stage. Lysosomes (**6**), free ribosomes, fine bundles of microfilaments, and glycogen particles occur in variable amounts in reticular cells. The cell presented in this drawing corresponds to a poorly differentiated form because of its relative paucity of organelles.

If a reticular cell synthesizes fibers, the rough endoplasmic reticulum becomes broader owing to the intensive protein synthesis. In this plate, the normally very fine reticular microfibrils (**7**) are depicted somewhat thicker. Reticular microfibrils combine to form bundles, i.e., reticular fibers, which are in close contact with the reticular cells. The bundles are frequently surrounded by very thin processes (arrows) of the reticular cells and a glycoprotein coating (not shown), probably responsible for special staining properties of the reticular fibers. Reticular cells do not possess a basal lamina.

Under certain circumstances, a reticular cell can transform into a phagocyte (cell on the right). Particles of dye (**8**) are phagocytized and stored by reticular cells when subject to vital staining (injection of stains into the connective tissue which penetrate or are taken up by the cells without causing damage), e.g., with trypan blue. During this process, the reticular cells detach themselves from the microfibrillar lattice (**7**) and, by means of an undulating membrane (**9**), migrate through the tissue. The cell forms several pseudopodia (**10**) with which it draws the particles of dye into the cytoplasm; it also develops a large number of microvilli (**11**) and filopodia (**12**). Phagocytosis of bacteria, viruses, cell debris, etc. proceeds according to the same mechanism. The majority of littoral cells have similar abilities.

Reticular cells of lymph nodes are also able to bind antigens to their cell membranes. The lymphocytes that are in contact with such cells differentiate into antibody-producing plasma cells (see Plates 72, 115).

Reticular cells of the spleen as macrophages, phagocytize old red blood cells. Bilirubin and iron result from disintegration of the erythrocytes. Bilirubin leaves the cell by diffusion; iron is transferred to the erythroblasts (see Plate 68), from which red blood cells subsequently develop (see histology texts for further information).

Reticular tissue is thus primarily involved in defense of the organism.

Magnification: × 5,000

REFERENCES

Fukuta K, Mochizuki K (1982) Formation of reticular fibers in the developing spleen of the chick embryo. Arch Histol Jpn 45:181–189

Klika E, Antalikova L (1983) Licht- und elektronenmikroskopische Untersuchungen über sich differenzierende Retikulin-Fasern. Verh Anat Ges 77:611–613

Montes GS, Krisztan RM, Shigihara KM, Tokoro R, Mourao PAS, Junqueira LCU (1980) Histochemical and morphological characterization of reticular fibers. Histochemistry 65:131–141

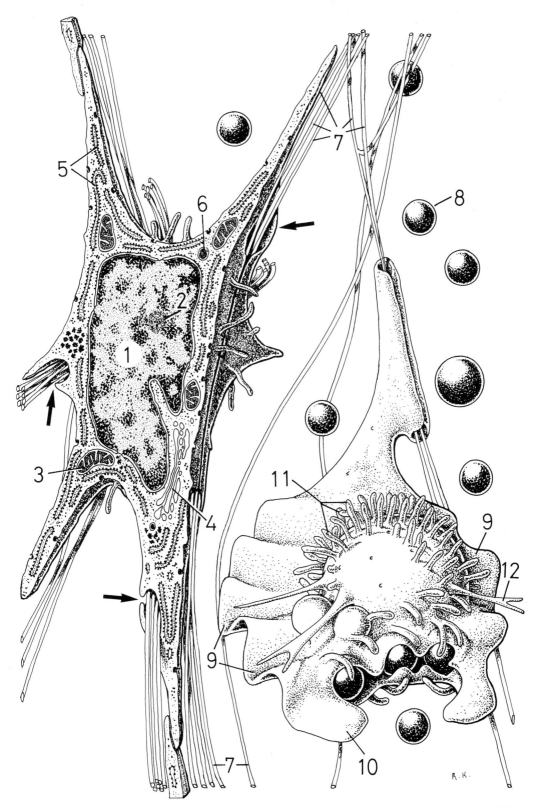

Plate 56. White Adipose Tissue. Human Epiploic Appendices of the Large Intestine

Lobular reticular tissue of the so-called primitive organs forms the basis for adipose tissue: By the deposition of fat it becomes transformed into lobular adipose tissue; it can thus be regarded as a storage form of reticular tissue. The close similarity between reticular and adipose tissue explains the frequent metaplastic transformations that occur between the two (see Plate 5).

Adipose tissue consists of white and brown types (see Plates 59–61). Two kinds of white adipose tissue are found in the organism.

1. Structural adipose tissue fulfills several functions: It forms envelopes providing mechanical support for organs (e.g., kidneys, lymph nodes, eyes) and elastic pads in areas subject to pressure (e.g., palms, soles, buttocks, joints). Structural adipose tissue also occupies the space where unformed organs will develop (e.g., mammary glands). A large reduction in weight has little influence on structural adipose tissue. The major sites in the human body where structural adipose tissue is found are marked black in Fig. **A**.

2. Storage adipose tissue functions as a caloric reserve material and provides thermal insulation. Through its ability to bind water it plays an important part in water balance. Storage adipose tissue is found particularly in the subcutis and abdominal cavity (subserosa, omenta, mesenteries, epiploic appendices, etc.).

In Fig. **B**, grapelike complexes of adipose tissue, epiploic appendices (Fig. **B1**), of the large intestine (Fig. **B2**) are illustrated. The inset is enlarged in Fig. **C**.

All epiploic appendices are covered with mesothelium (Fig. **C1**), visceral peritoneum. Lobules (Fig. **C2**) composed of groups of fat cells (adipocytes or lipocytes, Fig. **C3**) can be observed on the surface of the section. Several blood and lymphatic vessels run through the connective tissue septa. The light-microscopic image of the inset in Fig. **C** is shown in Fig. **D**. Adipocytes (Fig. **D1**) are 40- to 120- μm-large cells containing a voluminous fat droplet or vacuole, (Fig. **D2**). In unilocular (containing one droplet) adipocytes, the droplet displaces the cytoplasm and nucleus toward the plasmalemma. In routine histological preparations, the contents of the fat cells are dissolved due to the treatment with alcohol, benzene, xylene, etc. The fat cells then have the appearance of signet rings, with thin cytoplasm, flattened nuclei, and voluminous, empty vacuoles. To prevent dissolution of the fat, the adipose tissue is fixed in formalin, frozen, and cut with a freezing microtome. The sections are then treated with liposoluble dyes (scarlet red, Sudan III), which penetrate and stain the fat droplets.

Magnification: Fig. **D**, × 400

REFERENCES

Cahill GF, Renold AE (1983) Adipose tissue – a brief history. In: Angel A, Hollenberg CH, Roncari DAK (eds) Adipocyte and obesity. Raven, New York

Cushman SW (1970) Structure-function relationship in the adipose cell: I. Ultrastructure of the isolated adipose cell. J Cell Biol 46:326–341

Hausberger FX (1964) Influence of nutritional state on size and number of fat cells. Z Zellforsch 64:13–18

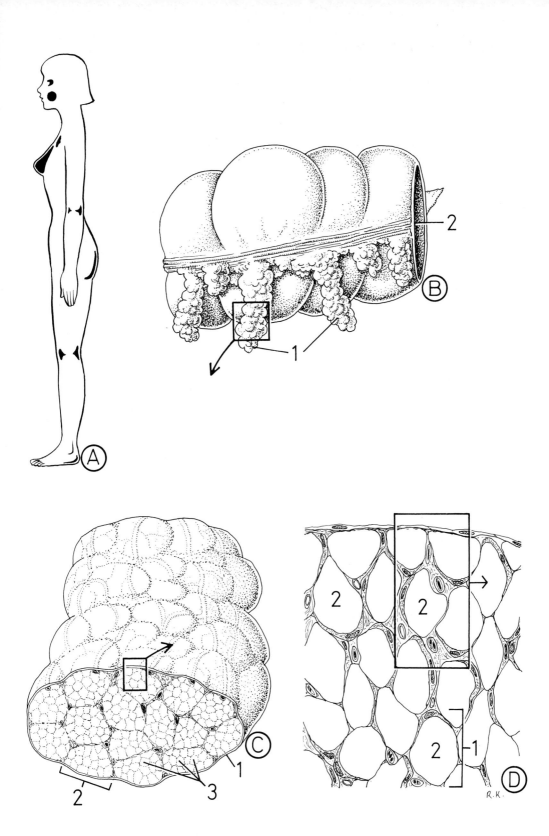

Plate 57. White Adipose Tissue. Continuation of Plate 56

As indicated in the previous plate, the epiploic appendices are covered with simple squamous epithelium, the peritoneal mesothelium (**1**). Directly beneath this epithelium and connected only by a few reticular microfibrils (arrow) are the globular, densely packed adipocytes (**2**). They contain large, homogeneous, moderately osmiophilic lipid droplets (**3**), which develop during the histogenesis of adipose tissue by the fusion of several small intracellular lipid droplets. The cytoplasm is confined to a narrow, sickle-shaped border and contains the nucleus (**4**).

Unlike routine light-microscopic preparations, where the vacuoles of fat are dissolved, fixation with O_sO_4 (osmium tetroxide) for electron microscopy preserves the vacuolar contents. In this process, the fat is stained dark as a result of the formation of osmium esters.

Adipocytes are densely packed cells surrounded by a basal lamina (here omitted) and a well-developed basket of reticular microfibrils (**5**). These microfibrils and the plasticity of the fatty material together give adipose tissue its cushioning properties. When adipose tissue is compressed, the fat cells become ellipsoidal. The original globular form is reinstated by the feltwork of entwined reticular fibers upon removal of the mechanical stress. Tensile stress is largely absorbed by the interlobular collagen fibers.

Adipose tissue is very well vascularized, which makes substantial demands upon the circulation. Numerous capillaries (**6**), accompanied by nerve fibers (**7**), run between the fat cells. The thick accumulation of fat cells and rich supply of blood give adipose tissue a certain similarity to epithelial tissue (see Plate 44).

Adipose tissue is subject to hormonal influences like all tissues. Histogenesis, for example, is largely controlled by pituitary and sex hormones. Two hormones responsible for the metabolism of adipose tissue, lipotropins, have been isolated in the sheep. Epinephrine and norepinephrine exert similar mobilizing effects. Exophthalmos-producing factor (EPF), of pituitary origin, induces an increase in the volume of the orbital adipose tissue and this gives rise to abnormal protrusion of the eyeballs (exophthalmos). (See Plate 79 in KRSTIĆ 1979.)

Magnification: × 2,000

REFERENCES
Napolitano L (1963) The differentiation of white adipose cells. J Cell Biol 18:663–679
Schemmel K, Weisbecker L, Kahl H, Uthenannt H, Kreysing G, Zepf S (1972) Exophthalmogener Effekt durch endogenes und exogenes thyreotropes Hormon im Tierexperiment. Schweiz Med Wochenschr 102:667–669
Slavin BG (1972) The cytophysiology of mammalian adipose cells. Int Rev Cytol 33:297–334
Slavin BG (1979) Fine structural studies on white adipocyte differentiation. Anat Rec 195:63–72
Van RLR, Roncari DAK (1978) Complete differentiation of adipocyte precursors. A culture system for studying the cellular nature of adipose tissue. Cell Tissue Res 195:317–329

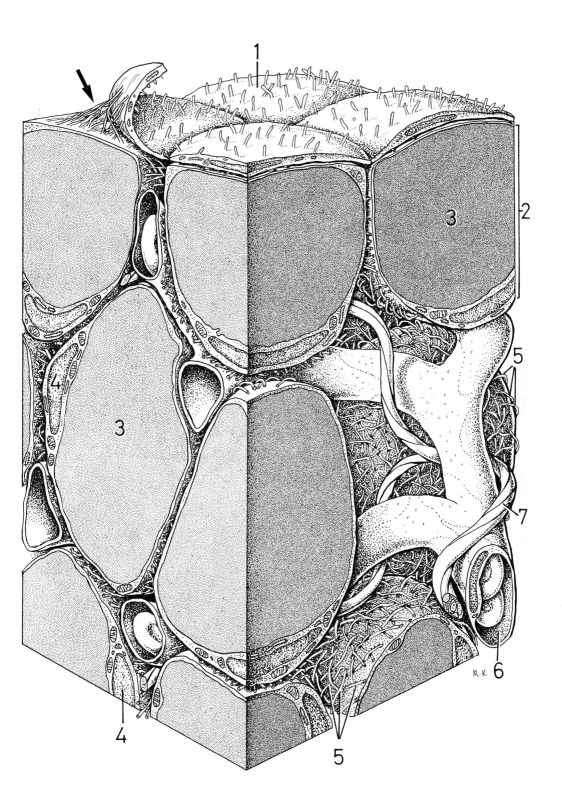

Plate 58. White Adipose Tissue Cell. Continuation of Plate 57

The narrow cytoplasmic rim of fat cells contains the flattened nucleus (**1**), some mitochondria (**2**), flattened cisternae of rough endoplasmic reticulum (**3**), a Golgi apparatus, and a few tubules of smooth endoplasmic reticulum. A giant fat droplet (**4**) dominates the cell body. The droplet is separated from the cytoplasm by numerous microfilaments (**5**), which also stabilize the fatty mass in the cell body; large fat droplets do not possess a limiting membrane of their own.

Every adipocyte is surrounded by a basal lamina (**6**). In the upper part of the plate, the basal lamina has been folded back so that the micropinocytotic vesicles (**7**) can be discerned. A feltwork of ramified reticular microfibrils (**8**; drawn somewhat thicker than normal) surrounds the fat cell.

Adipose tissue is a very active, dynamic tissue, which is constantly being built up and broken down.

Fatty acids, which are formed by hydrolysis of triglycerides from chylomicrons (see Plate 31) or serum lipoproteins, are taken up by the fat cells and resynthesized to triglycerides. Adipocytes can also synthesize triglycerides from carbohydrates. Lipids are stored in the cells as neutral triglycerides. Animal and human fat comprises a mixture of oleic, palmitic, and stearic acids. Exogenous liposoluble pigments, e.g., lipochromes and carotenoids, which are ingested with the food, give adipose tissue a yellowish color. Reduction in body weight leads to fat depletion in the cells; the fat droplets become smaller or disappear completely from the cytoplasm, and several cell processes form of various length. Adipocytes that have lost their fat content thus come to resemble reticular cells. In extreme cases, the body of the emptied cell becomes filled with a viscous fluid (so-called serous fat cells).

The hormones epinephrine and norepinephrine influence lipolysis by means of cAMP (see Plate 163). Stimulation by these enzymes leads to activation of hormone-sensitive lipase, which effects hydrolysis of the stored triglycerides. Fatty acids are released, they enter the circulation, become attached to albumins, and are transported to other cells of the organism, where they are utilized as a high-calorie material. (See Plate 80 in KRSTIĆ 1979.)

Magnification: × 5,000

REFERENCES

Frowein J (1972) Wirkungsmechanismen von Hormonen aus molekularbiologischer Sicht. Dtsch Med Wochenschr 97:1918–1922

Imaizumi M (1969) On the fine structure of the surface of lipid droplets in adipose cells. Arch Histol Jpn 30:353–365

Motta P (1975) Scanning electron microscopic observations of mammalian adipose cells. J Microsc 22:15–20

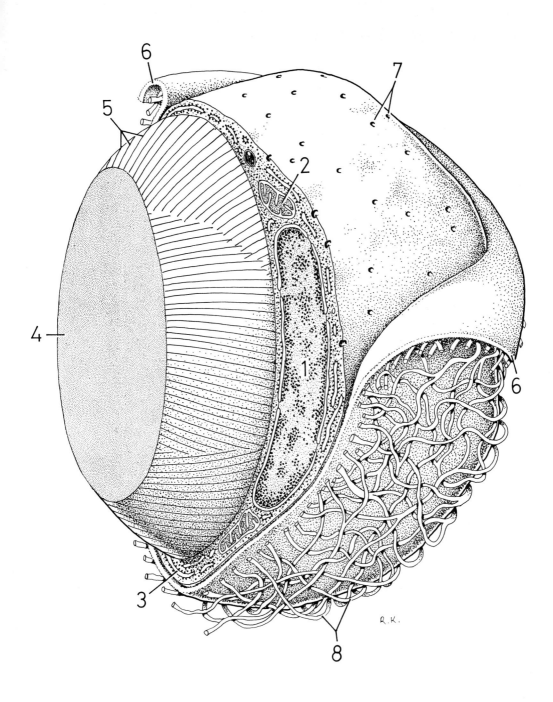

Plate 59. Brown Adipose Tissue.
Interscapular Brown Adipose Tissue of the Rat

Brown or multilocular adipose tissue is relatively infrequent in man and is found predominantly in newborns. It occurs in the neck, armpit (Fig. A 1), in the vicinity of the subclavian artery (Fig. A 2) and kidney (Fig. A 3), dorsal skin, mediastinum (Fig. A 4), and mesenteries. Brown adipose tissue is more widespread in many animals, particularly hibernators ("hibernation gland"). In addition to the retroperitoneal space it is regularly found as the interscapular fat organ (Fig. B 1). An area from this organ is depicted light-microscopically in Fig. C.

Brown adipose tissue is arranged in lobules, like white adipose tissue. Blood and lymphatic vessels and nerve fibers enter the tissue through interlobular septa (Fig. C 1). Brown fat cells (Fig. C 2) are polygonal elements with a central nucleus; the cytoplasm contains several small fat droplets, hence the name "multilocular" adipose tissue. White adipocytes (Fig. C 3) are occasionally found in brown fat and are distinctly larger than the brown adipocytes. Numerous blood capillaries (Fig. C 4) can be seen between the brown adipocytes.

Magnification: Fig. C, ×800

REFERENCES

Barnard T (1969) The ultrastructural differentiation of brown adipose tissue in the rat. J Ultrastruct Res 29:311–332

Hahn P, Novak M (1975) Development of brown and white adipose tissue. J Lipid Res 16:79–91

Schmidt FG, Donat K, Budras KD (1982) Über Entspeicherung normalen und entnervten braunen Fettgewebes beim Gerbil (Meriones unguiculatus M.-E. 1867). Z Mikrosk Anat Forsch 96:885–897

Suter E (1969) The fine structure of brown adipose tissue. Lab Invest 21:246–258

Tanuma Y, Ohata M, Ito T, Yokochi C (1976) Possible function of human brown adipose tissue as suggested by observation on perirenal brown fats from necropsy cases of variable age groups. Arch Histol Jpn 39:117–145

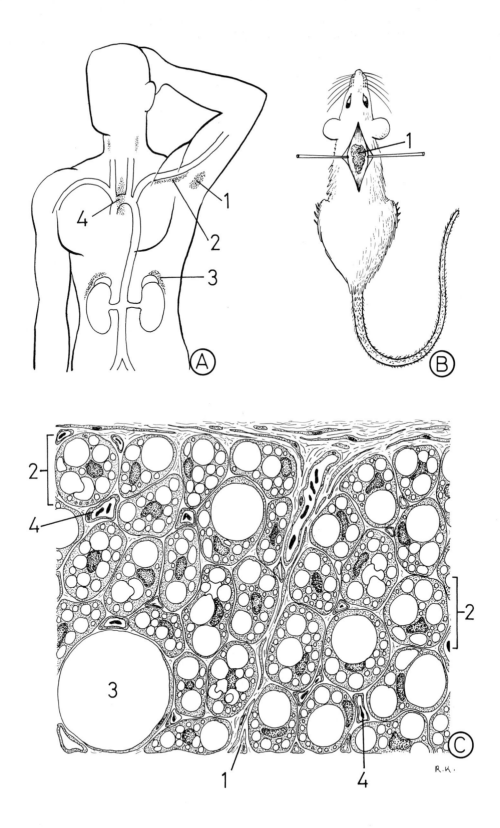

CONNECTIVE AND SUPPORTING TISSUES

Plate 60. Brown Adipose Tissue of the Rat. Continuation of Plate 59

This plate presents a three-dimensional electron-microscopic image of a lobule composed of a small number of brown fat cells. The brown fat cells (1) are closely packed and form an epitheliumlike structure (see Plate 59). In the cytoplasm are located a spherical nucleus (2) and a conspicuous number of mitochondria (3). The previously mentioned multiple fat droplets (4) occur in the cell body as inclusion products. Rich ramifications of the blood vessels ensure that each cell comes into contact with capillaries (5). The large numbers of unmyelinated nerve fibers (6) supply practically every cell with a club-like nerve ending (7). Brown adipocytes are surrounded by a basket of reticular and collagen microfibrils (8). For the sake of clarity, the basal laminae have not been drawn on this plate.

The major function of brown adipose tissue is thermogenesis, hence the wide distribution in the bodies of hibernators, where the tissue assists in maintaining body temperature during the winter. Oxidation of fatty acids in brown adipocytes leads to a local increase in temperature, such that brown adipose tissue acts as a heating element and warms the perfusing blood. In human neonates, brown adipose tissue is important in adaptation to environmental temperature.

Magnification: $\times 3,200$

REFERENCES

Derry DM, Schönbaum E, Steiner G (1969) Two sympathetic nerve supplies to brown adipose tissue of the rat. Can J Physiol Pharmacol 47:57–63

Dyer RF (1968) Morphological features of brown adipose cell maturation in vivo and in vitro. Am J Anat 123:255–282

Robison GA, Sutherland EW (1971) Cyclic AMP and the function of eukaryotic cells: an introduction. Am NY Acad Sci 185:5–9

Thompson JF, Habeck DA, Nance SL, Beetham KL (1969) Ultrastructural and biochemical changes in brown fat in cold-exposed rats. J Cell Biol 41:312–334

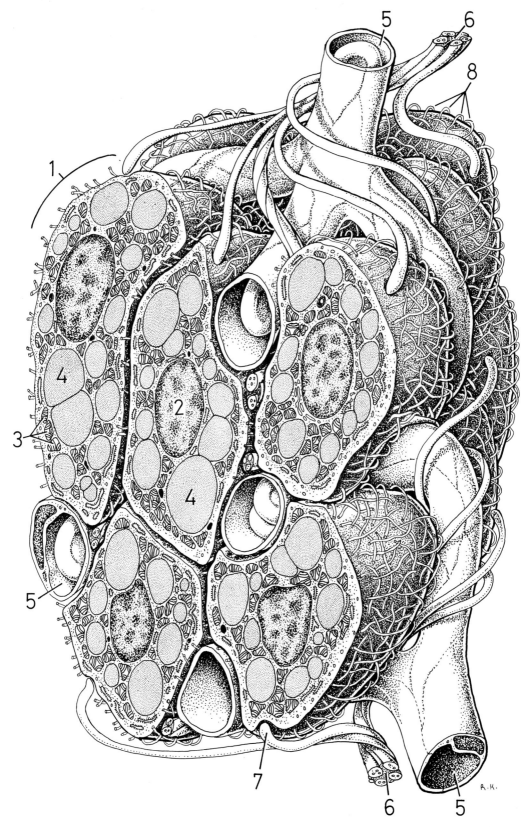

Plate 61. Brown Adipose Tissue Cell. Continuation of Plate 60

A section through a brown adipocyte reveals a protruding nucleus (1) with several nuclear pores (2). The cytoplasm contains a large number of mitochondria (3), responsible for the intensive oxidation, and a few cisternae of rough (4) and tubules of smooth (5) endoplasmic reticulum. The distinctive feature of a brown adipocyte is the presence of several lipid droplets (6). These are – similar to the droplets of white fat cells – confined within a small basket of microfibrils (7) which provide support for the lipid mass and prevent the fusion of several fat droplets into one large droplet. The contents of the droplets are moderately osmiophilic and homogeneous.

Every multilocular cell possesses its own basal lamina (8), which is externally reinforced by a network of reticular and collagen microfibrils (9). Many micropinocytotic vesicles (10) form on the cell membrane.

High concentrations of cytochrome in mitochondria as well as the lipochrome pigment of the lipid droplets give the brown color to brown adipose tissue.

Brown adipose tissue is richly innervated. The plate shows the terminal portion of an unmyelinated nerve fiber (11) with two axons (12). They are accompanied by a process of a Schwann's cell (13). The basal lamina is intercalated between the axons and the brown adipocyte. The axons contain numerous synaptic vesicles, some of which have highly osmiophilic contents. Norepinephrine is released at the axonal endings and, via cAMP, activates the previously mentioned hormone-sensitive lipase. This brings about hydrolysis of the triglycerides to glycerol and fatty acids, oxidation of which leads to thermogenesis. (See Plate 78 in KRSTIĆ 1979.)

Magnification: × 7,000

REFERENCES

Bargmann W, Hehn GV, Lindner E (1968) Über die Zellen des braunen Fettgewebes und ihre Innervation. Z Zellforsch 85:601–613

Kühnau J (1971) Physiologie und Biochemie der endokrinen Regulationen und Korrelationen. In: Bargmann W, Kühnau J, Siebenmann RE, Steiner H, Uehlinger E (eds) Handbuch der allgemeinen Pathologie. Endokrine Regulations- und Korrelationsstörungen, vol VIII/1. Springer, Berlin Heidelberg New York

Linck G, Stoeckel M-E, Porte A, Petrovic A (1973) An electron microscope study of the specialized cell contacts and innervation of adipocytes in the brown fat of the european hamster (Cricetus cricetus). Cytobios 7:431–436

Nedergaard J, Lindberg O (1982) The brown fat cell. Int Rev Cytol 74:187–286

Wood EM (1967) An ordered complex of filaments surrounding the lipid droplets in developing adipose cells. Anat Rec 437–448

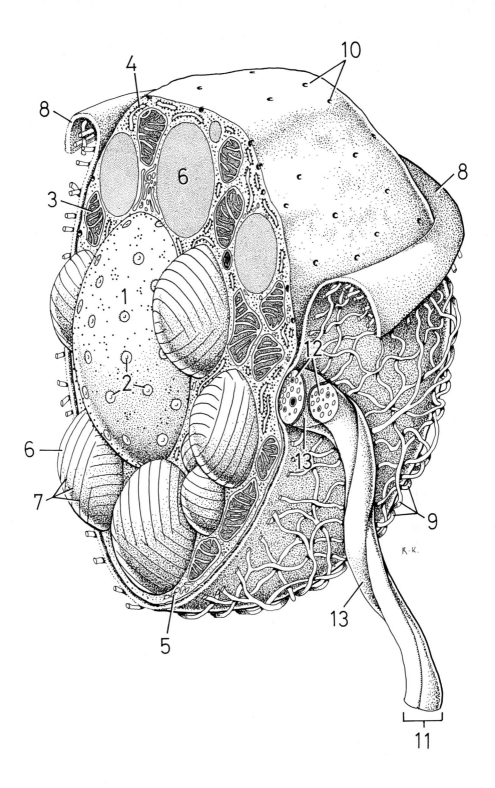

CONNECTIVE AND SUPPORTING TISSUES

Plate 62. Loose Connective Tissue

Loose connective tissue, or areolar tissue, is widely distributed in the organism. It fills the spaces between the skin and musculature, between the muscle fibers, and between the muscles; it surrounds vessels, nerves, and various organs, forms the stroma of the kidneys, liver, glands, testes, ovaries, etc., forms the leptomeninges, choroid of the eye, stratum papillare of the skin, and occurs in the omenta, pleura, and wherever gaps between organs have to be filled. Loose connective tissue also connects various organs or parts of organs. It is found in hollow organs (e.g., esophagus) between the epithelium (Fig. A 1) and lamina muscularis mucosae (Fig. A 2) as the lamina propria (Fig. A 3), as the tela submucosa (Fig. A 4), and as an organ envelope, the tunica adventitia (Fig. A 5).

The components of all connective tissues can be seen particularly well in loose connective tissue (Fig. B).

The following is a broad classification (after BUCHER 1980):

The insets in Fig. A correspond to Fig. B, which has been simplified for the sake of clarity.

Fixed cells in connective tissues consist of fibroblasts (Fig. B 1), fibrocytes (Fig. B 2), fat cells (Fig. B 3), and pericytes (Fig. B 4). Wandering cells comprise the histiocytes (Fig. B 5), mast cells (Fig. B 6), lymphocytes (Fig. B 7), plasma cells (Fig. B 8), eosinophilic granulocytes (Fig. B 9), and monocytes (Fig. B 10). Fat cells have already been described with adipose tissue (see Plates 56–58).

Reticular (Fig. B 11), collagen (Fig. B 12), and branched elastic (Fig. B 13) fibers form a broad network, the interstices of which are filled with a ground substance. Collagen fibers often occur in bundles.

Lymphatic (Fig. B 14) and blood vessels (Fig. B 15) and nerve fibers (Fig. B 16) run through loose connective tissue and most other tissues of this group.

Magnifications: Fig. A, × 5; Fig. B, × 850

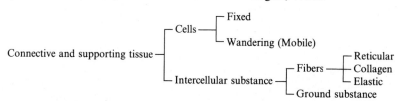

REFERENCES

Bucher O (1980) Cytologie, Histologie und mikroskopische Anatomie des Menschen, 10th edn. Huber, Bern

Deane HS (1964) Some electron microscopic observations on the lamina propria of the gut, with comments on the close association of macrophages, plasma cells and eosinophils. Anat Rec 149:453–474

Snodgrass MJ (1977) Ultrastructural distinction between reticular and collagenous fibers with an ammoniacal silver stain. Anat Rec 187:191–206

Weinstock M (1972) Collagen formation. Observations on its intracellular packaging and transport. Z Zellforsch 129:455–470

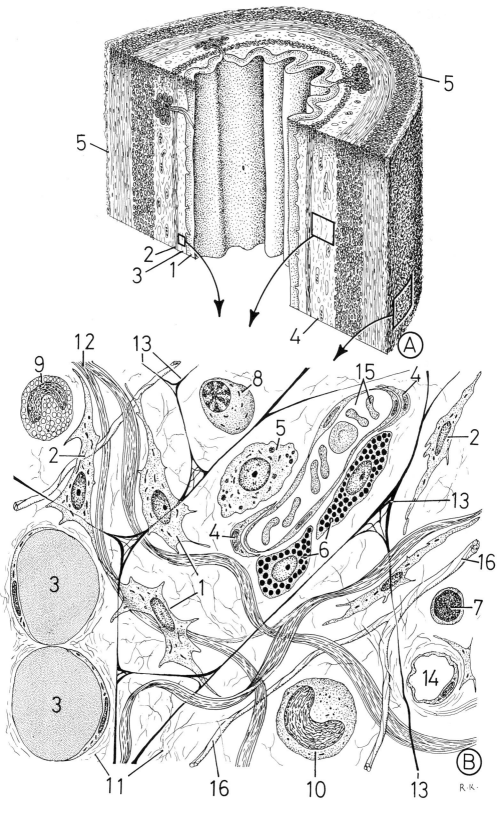

Plate 65. Loose Connective Tissue. Formed and Amorphous Components of the Intercellular Substance

As previously mentioned, fibroblasts are above all protein-synthesizing cells. They form four kinds of protein-containing macromolecule:
A. Collagen
B. Microfibrillar proteins of elastic fibers and elastin
C. Proteoglycans
D. Structural glycoproteins (fibronectin and laminin)

A. Through the uptake of hydroxyproline (arrows) and other amino acids, the first peptides are synthesized on the ribosomes of rough endoplasmic reticulum (1). These are incorporated into the pre- or procollagen polypeptide chains. The first 280-nm-long and 1.5-nm-wide tropocollagen molecules (2) develop from three helically entwined chains. After incorporation of sugar components, they pass from the Golgi apparatus into collagen secretory granules (3). The tropocollagen (2a) is expelled from the cell into the intercellular space, where it is polymerized into collagen (4) or reticular (5) microfibrils. During this process, the heads of the molecules arrange themselves a constant distance of 70 nm apart, which gives rise to the regular transverse striation of both fibrous elements.

Collagen microfibrils (4) have a diameter of about 10–200 nm; reticular microfibrils (5) are considerably thinner with a diameter of about 5–15 nm. The collagen structures unite to form fibrils, fibers, and fiber bundles, and the latter can attain a diameter of 12 µm. Recticular microfibrils combine to form reticular fibers, which are, unlike collagen fibers, thin, disposed in a network in the tissues, and argyrophilic, i.e., they can be impregnated and stained intensely black by silver nitrate. Reticular microfibrils occur between the basal laminae and the collagen microfibrils (see Plates 24, 132). Biochemically and morphologically, there is a gradual transition between the two types of fiber.

B. After a process which has not been adequately investigated, fibroblasts produce elastic fibers. This occurs, according to the current understanding of events, by the expulsion of short microfibrillar proteins (6), which unite in the vicinity of the cell to form a feltlike structure (oxytalan fibers). The proelastin molecules (7), also formed in the fibroblasts, accumulate within the feltwork, fuse, and become the electron-microscopically homogeneous elastic fibers (8). Unlike the practically inextensible collagen structures, elastic fibers can be stretched to 150% of their original length as a result of their coiled molecular structure.

C. Connective tissue cells synthesize the proteoglycans (9) of the ground substance. The hypothetical picture of a highly enlarged detail from the circular inset is also shown in Fig. C.

The proteoglycans are very large molecules, consisting of a core protein (10) to which glycosaminoglycans (11) are laterally connected. The latter are made up of repeating disaccharide units, chondroitin 4- and 6-sulfates, dermatan, heparan, and keratan sulfates. By a link glycoprotein (12), the proteoglycans are bound to a filiform hyaluronic acid molecule (13) to form a dense molecular feltwork, which structures large volumes of extracellular water.

As a result of the water-binding properties of proteoglycans, the ground substance acts as a reserve of extracellular water. It also contains numerous ions, enzymes, hormones, vitamins, and antibodies. Increased water content in the ground substance caused by diseases of the heart and kidney as well as by damaged capillary permeability is termed edema. The ground substance acts as a filter in the diffusion of various substances from the capillaries into the cell and vice-versa, retaining larger molecules and allowing smaller ones to pass through.

D. Fibronectin and laminin are involved in cell interactions as well as in the adhesion of cells to their collagen support via basal laminae. (See Plates 118, 132–137 in KRSTIĆ 1979.)

REFERENCES

Hay ED (1981) Extracellular matrix. J Cell Biol 91:205s–223s

Hynes RO, Yamada KM (1982) Fibronectins: multifunctional modular glycoproteins. J Cell Biol 95:369–377

Lennarz WJ (ed) (1980) The biochemistry of glycoproteins and proteoglycans. Plenum, New York

Varma RS, Varma R (eds) (1982) Glycosaminoglycans and proteoglycans. Karger, Basal

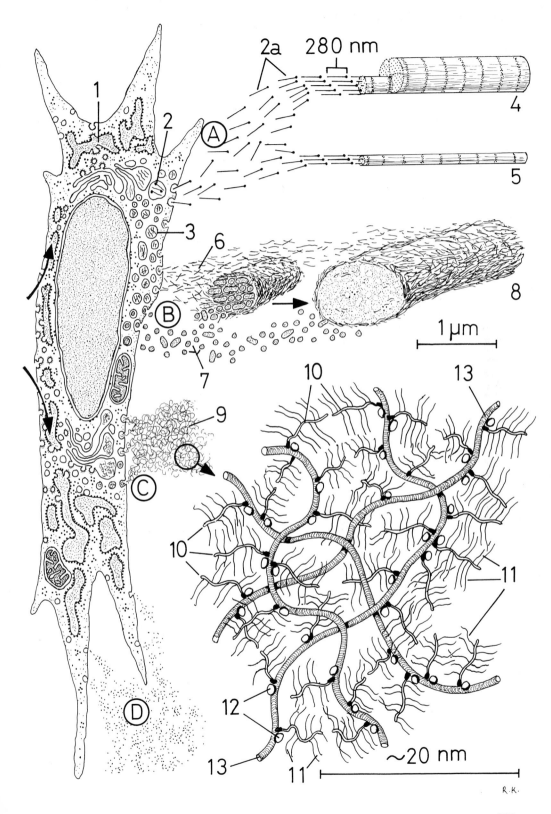

Plate 66. Loose Connective Tissue. Wandering Cells: Histiocyte

The relatively large (10–20 µm) histiocyte has an irregular outline and is a permanent component of loose connective tissue. The nucleus is mainly spherical or ellipsoidal and sometimes displays deep indentations. It contains finely dispersed chromatin and a distinct nucleolus.

The cytoplasm contains a few mitochondria, sparse flattened cisternae of rough endoplasmic reticulum, a well-developed Golgi apparatus, and fairly abundant free ribosomes. Histiocytes are characterized by a variable number of lysosomes at various stages of development, depending on the functional state of the cell. In addition to small primary lysosomes (**1**), which are mainly found around the Golgi apparatus, secondary lysosomes (**2**), phagolysosomes (**3**), and residual bodies (**4**) are found in the interior of active histiocytes. The external morphology of these cells is extremely variable. The surface often bears a profusion of irregular microvilli (**5**) or globular protruberances (**6**), which correspond to the more extensive phagolysosomes in the cell interior. Invaginations in the plasmalemma are formed by thin folds (**7**) of cytoplasm. When these folds close, small amounts of intercellular fluid are engulfed by the cell, a process termed macropinocytosis. Histiocytes also contain of course large numbers of micropinocytotic vesicles (**8**).

Histiocytes move vigorously by ameboidism. They usually follow the course of vessels and occur in large numbers where the penetration of foreign particles is likely, e.g., in the loose connective tissue of the subepithelial layer (lamina propria) of the intestinal tract.

Histiocytes are distinguished by their marked capacity to phagocytize and store. They are capable of engulfing large particles and/or dead cells and digesting them intracellularly, for which reason they are also termed macrophages. In vital staining with trypan blue (see Plate 55), dye particles are stored in the cytoplasm. (See Plates 32, 33, 47, 56–58, 168, 169 in KRSTIĆ 1979.)

Magnification: × 10,000

REFERENCES

Anderson H, Mattheissen ME (1966) The histiocyte in human foetal tissues. Its morphology, cytochemistry, origin, functions and fate. Z Zellforsch 72:193–211

Han SS, Avery JK, Hale LE (1965) The fine structure of differentiating fibroblasts in the incisor pulp of the guinea pig. Anat Rec 153:187–210

Jackson DS, Bentley JP (1975) Connective tissue cell. In: Beck F, Lloyd B (eds) The cell in medical science, vol. 3. Academic, London New York San Francisco

Kisher CW (1974) Fibroblasts of the hypertrophic scar, mature scar and normal skin: study by scanning and transmission electron microscopy. Tex Rep Biol Med 32:699–709

Ten Cate AR, Deporter DA (1975) The degradative role of the fibroblast in the remodelling and turnover of collagen in soft connective tissue. Anat Rec 182:1–14

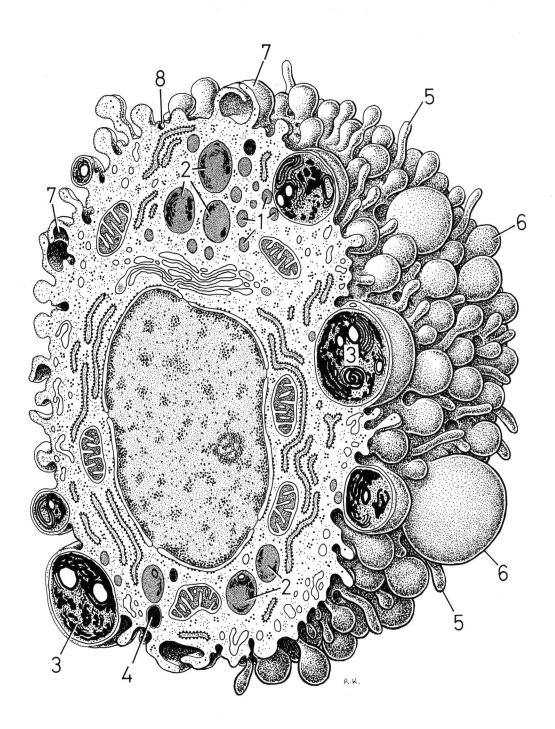

Plate 67. Loose Connective Tissue. Wandering Cells: Macrophage in Tissue Culture

Macrophages from loose connective tissue can be easily cultivated and experimentally manipulated in vitro.

The phagocytic process can be observed by adding a few red blood corpuscles to a macrophage culture. The macrophages soon develop a broad, thin, undulating membrane (**1**) and approach the erythrocytes (**2**). This veil of cytoplasm then glides over the red blood cell, which gradually becomes engulfed by the macrophage.

Numerous microvilli (**3**) and filopodia (**4**) are evident on the prominent nuclear region of the macrophage. It is possible that the filopodia are important in the ameboid movement of macrophages.

Large phagolysosomes (**5**) or previously phagocytized erythrocytes can be easily observed in the very flattened body of the macrophage.

Magnification: $\times 7,500$

REFERENCES

Carr I (1968) Some aspects of the fine structure of the reticuloendothelial system: the cells which clear colloids from the blood system. Z Zellforsch 89:355–370

Pearsall NN, Weiser RS (1970) The macrophage. Lea and Febiger, Philadelphia

Takayama H (1976) Electron microscopic studies on macrophages: I. Effect of colchicine and cytochalasin B on the cell cortex as revealed by scanning electron microscopy. J Electron Microsc (Tokyo) 25:75–82

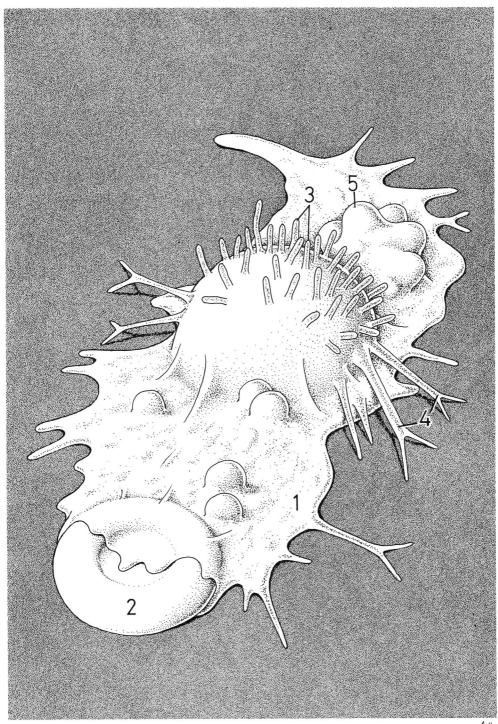

Plate 68. Reticulohistiocytic or Reticuloendothelial System

The reticuloendothelial system (RES), according to the classic interpretation of Aschoff (1924), combines those cells of the body whose major functions are phagocytosis and storage. It thus represents a powerful defense system of the organism. The RES comprises the following cells.

– Reticular cells (1) and, deriving from the reticular cells, endothelial macrophages or littoral cells (2) lining various sinuses of the lymph nodes (3; see Plate 54), spleen (4), and bone marrow (5)

– Histiocytes (6) and monocytes (7) of loose connective tissue (8); because of the presence of histiocytes the RES is also termed the reticulohistiocytic (RHS) system

– Some endothelial cells (9) of the capillaries of the adrenal (10) and pituitary (11) glands

– Kupffer's cells (12) of the liver (13)

– In a broader sense, the microglia (14) of the central nervous system.

The main functions of the RES may be summarized as follows:

A. Phagocytosis and immunological activity (see Plate 115)

B. Storage of exogenous foreign bodies and/or vital dyes

C. Metabolism of hemoglobin and iron [old red blood cells are phagocytized; their iron is transferred (arrow) to the erythroblasts]

D. Extrahepatic synthesis of bile pigments. (Bile pigments develop as a result of digestion of erythrocytes and diffuse out of the phagocytes)

E. Lipid metabolism (reticular cells can store lipids and transform into adipocytes).

The above details represent a classic description of the RES. Recent electron-microscopic studies have revealed that phagocytosis is not performed by the sinus endothelia of the pituitary and adrenal capillaries but by perivascular macrophages. The phagocytotic activity of spleen endothelia has also been questioned following new research. With the elimination of the endothelial components, the term "reticulohistiocytic system" becomes preferable to "reticuloendothelial system."

Modern investigations have shown that macrophages originate from stem cells of the bone marrow (except microglia); for this reason and because of their morphological and functional similarities, a new concept of the mononuclear phagocyte system has been elaborated to replace the RES/RHS concept.

REFERENCES

Aschoff L (1924) Das Retikulo-endotheliale System. Ergeb Inn Med Kinderheilkd 26:1–118

Carr I, Deames WT (eds) (1980) The reticuloendothelial system, vol 1. Morphology. Plenum, New York

Friedman H, Escobar M, Sherwood R (eds) (1984) The reticuloendothelial system. A comprehensive treatise, vol 6. Immunology; vol 7 A Physiology. Plenum, New York

Rose NR, Siegel BV (eds) (1982) The reticuloendothelial system: A comprehensive treatise, vol 4. Plenum, New York

Sbarra AJ, Strauss RR (eds) (1980) The reticuloendothelial system, vol 2. Biochemistry and metabolism. Plenum, New York

Van Furth R (ed) (1980) Mononuclear phagocytes: Functional aspects. Martinus Nijhoff, The Hague

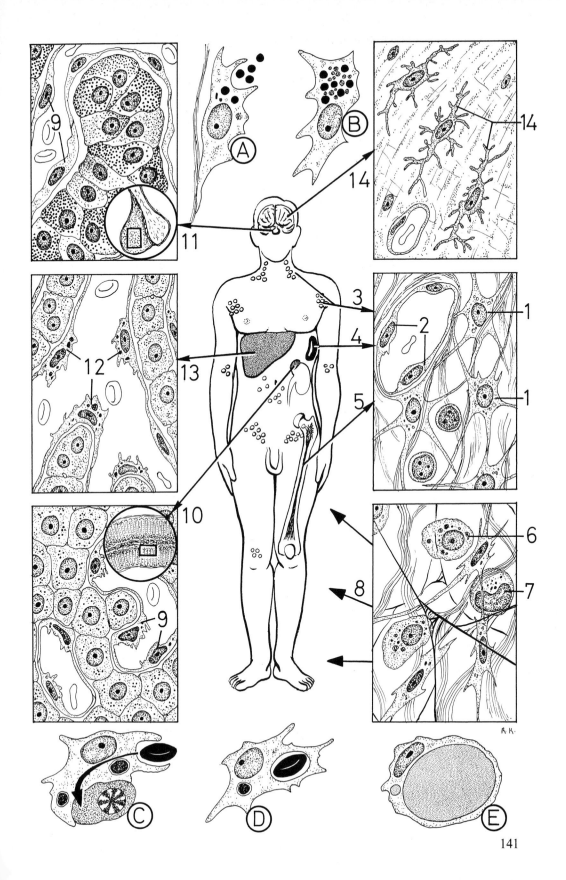

Plate 69. Loose Connective Tissue. Wandering Cells: Mast Cell or Mastocyte

Mast cells are wandering cells found largely along the small vessels of loose connective tissue (see Plates 62, 63). Occasionally, they occur in small groups in the stroma of various organs. They are relatively large cells (up to 20 µm) which can adopt a globular or elongated form.

Mast cells are not evident in normally stained histological sections. Special techniques, involving basic dyes, such as methylene blue, toluidine blue, azure II, are required, whereby the granules appear metachromatic. Metachromasia signifies a substrate staining a different color to that of the dye applied. Thus, e.g., the blue solutions of the above dyes stain mast cell granules reddish-violet.

Mast cells usually contain an ellipsoidal nucleus. The cells are characterized by numerous large (up to 2 µm), electron-dense granules (Figs. **A1**, **B**), surrounded by a unit membrane. In many animals, these granules are filled with crystalline inclusions; in man, they contain odd, rolled cylindrical structures termed "scrolls" (Fig. **B1**). The intergranular cytoplasm of mast cells comprises a few cisternae of rough endoplasmic reticulum, a well-developed Golgi apparatus, and some mitochondria and ribosomes.

The external morphology of mast cells is very variable. At rest, the surface is smooth and bears only a few cell processes. In an active mast cell, the plasmalemma forms numerous irregular microvilli (Fig. **A2**), folds (Fig. **A3**), invaginations (Fig. **A4**), and large numbers of protruding globular structures. The openings (Fig. **A5**) between the processes are probably residual traces of expelled granules.

Mast cell granules contain:

– Heparin, a proteoglycan inhibiting blood coagulation, which is also responsible for the metachromasia of the connective tissue ground substance
– Histamine, a low-molecular-weight tissue hormone that is involved in local inflammatory reactions and edema by increasing capillary permeability
– Serotonin (5-hydroxytryptamine, 5-HT; only in the rat and mouse) effects a constriction of the small blood vessels. Mast cells also produce leukotrienes, prostaglandins, and other important factors in immunity and regulation of the composition of the ground substance. (See Plate 165 in KRSTIĆ 1979 and physiology texts for further details.)

Magnifications: Fig. **A**, ×10,000; Fig. **B**, ×40,000

REFERENCES

Behrendt H (1980) Mastzelle und basophiler Granulozyt (Referat). Verh Anat Ges 74:261–268
Combs JW (1966) Maturation of rat mast cells. An electron microscope study. J Cell Biol 31:563–575
Goth A, Johnson AR (1975) Current concepts on the secretory function of mast cells. Life Sci 16:1201–1214
Kiernan JA (1979) Production and life span of cutaneous mast cells in young rats. J Anat 128:225–238
Kobayasi T, Midtgard K, Asboe-Hansen G (1968) Ultrastructure of human mast cell granules. J Ultrastruct Res 23:153–165
Lawson D, Raff MC, Gomperts B, Fewtrell C, Gilula NB (1977) Molecular events during membrane fusion. A study of exocytosis in rat peritoneal mast cells. J Cell Biol 72:242–259
Nemeth A, Roehlich P (1982) Early membrane retrieval following exocytosis in rat mast cells. Eur J Cell Biol 28:39–46

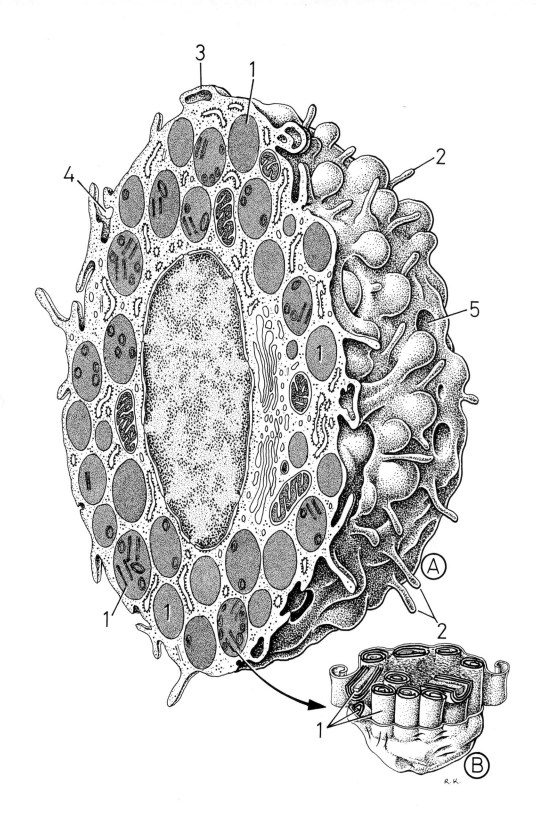

Plate 70. Loose Connective Tissue. Wandering Cells: Lymphocytes

Lymphocytes, with a diameter of 7–10 µm, are the smallest mobile cells of the loose connective tissue. Histochemical evidence has long suggested that there are two types of lymphocyte. Unfortunately, studies employing transmission and scanning electron microscopy were unable to provide definitive confirmation of these suspicions. Immunological studies were the first to prove that there are in fact two functionally different types of lymphocyte – B and T lymphocytes.

B lymphocytes in mammals stem from the bone marrow, and in birds from the bursa of Fabricius, whence the term "B" lymphocyte derives. During development, T lymphocytes spend a short amount of time in the thymus (hence "T" lymphocyte), where they receive programmed information that enables them to recognize substances foreign to the body.

Both types of lymphocyte contain a large spherical nucleus with very dense chromatin. In the thin belt of cytoplasm are found only a few mitochondria, cisternae of rough endoplasmic reticulum, and a poorly developed Golgi apparatus. The diplosome is indicative of the high mitotic activity of these cells. Large numbers of free ribosomes are also scattered in the cytoplasm.

According to the functional condition, the surface of the lymphocyte is smooth or covered with numerous microvilli.

Small lymphocytes account for about 20%–35% of white blood cells. They are able to leave the blood capillaries and enter the connective tissue by virtue of their plasticity. From the connective tissue, they penetrate the lymph capillaries and finally reenter the blood circulation.

The number of lymphocytes in the interstices of connective tissue is small; upon the appearance of foreign elements or substances, however, they increase considerably. T lymphocytes are the carriers of so-called cellular or cell-bound immunity (see Plate 115). Under certain circumstances, B cells can transform into gamma-globulin-producing plasma cells and thus are important in humoral immune reactions.

Magnification: × 10,000

REFERENCES

Alexander EL, Wetzel B (1975) Human lymphocytes: similarity of B and T cell surface morphology. Science 188:732–734

Bach F, Bonavida B, Vitetta E, Fox FC (1979) T and B lymphocytes: Recognition and function. Academic, New York

Bhalla DK, Braun J, Karnovsky MJ (1979) Lymphocyte surface and cytoplasmic changes associated with translational motility and spontaneous capping of Ig. J Cell Sci 39:137–147

Hilschmann N (1969) Die molekularen Grundlagen der Antikörperbildung. Naturwissenschaften 56:195–205

Kataoka K, Minowada J (1979) Human T and B lymphoid cells: An electron microscopic study on thymocytes, peripheral blood lymphocytes, mitogen-stimulated cells and lymphoid cell lines. Arch Histol Jpn 42:355–374

Lennert K, Müller-Hermelink HK (1975) Lymphocyten und ihre Funktionsformen – Morphologie, Organisation und immunologische Bedeutung. Verh Anat Ges 69:19–62

Liebich H-G (1972) Elektronenmikroskopische Untersuchungen an kleinen Lymphocyten aus dem Ductus thoracicus des Menschen. Res Exp Med (Berl) 159:87–98

Rydgren L, Norberg B, Håkansson CH, Mecklenburg v C, Söderström N (1976) Lymphocyte locomotion: I. The initiation, velocity, pattern and path of locomotion in vitro. Lymphology 9:89–96

Vos JG, Roholl PJM, Leene W (1980) Ultrastructural studies of peripheral blood lymphocytes in T cell-depleted rabbits. A scanning- and transmission electron microscopic analysis. Cell Tissue Res 213:221–235

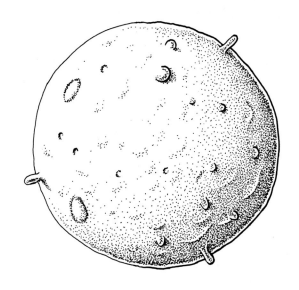
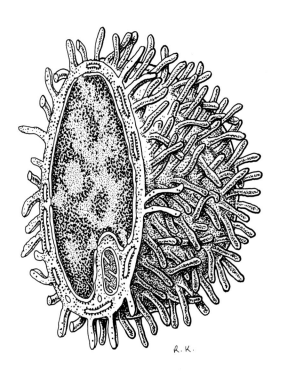

CONNECTIVE AND SUPPORTING TISSUES

Plate 71. Loose Connective Tissue. Wandering Cells: Monocyte

Monocytes are 12- to 20-µm-large, basophilic, spherical cells with a voluminous, kidney-shaped nucleus, which is rich in chromatin and has one or two conspicuous nucleoli. The Golgi apparatus is frequently located in the nuclear concavity. Several small, dense granules (**1**), just visible under the light microscope, are found in the vicinity of the Golgi apparatus. Following combined May-Grünwald and Giemsa staining, they appear purple-red (azurophilic). Electron-microscopic investigations have shown these granules to be primary lysosomes. Centrioles (**2**) are also seen near the Golgi apparatus. Several mitochondria and flattened cisternae of rough endoplasmic reticulum are scattered throughout the cell body.

The outer surface of inactive monocytes is smooth apart from the presence of a few microvilli. With increased activity, the plasmalemma forms numerous globular, digitiform, and microvillous protuberances, such that the monocyte can no longer be differentiated from a histiocyte (see Plate 66).

Monocytes are definite macrophages. They wander through the interstices of connective tissue by means of vigorous ameboid movements and phagocytize bacteria as well as larger particles and dead cells. Like histiocytes, monocytes store vital dyes.

Monocytes comprise about 2%–8% of all white blood cells. As a result of their mobility, they are able to leave the blood circulation and enter the connective tissue. They subsequently alter their morphology, for which reason they are also termed polyblasts. Since monocytes cannot ultimately be differentiated from histiocytes, they can be regarded as a reserve of macrophagic elements in the connective tissue.

Magnification: × 10,000

REFERENCES

Liu HM (1972) Monozytopoese und Kinetik der Blutmonozyten beim Menschen. Blut 24:337–345

Nichols BA, Bainton DF, Farquhar MG (1971) Differentiation of monocytes. Origin, nature, and fate of their azurophil granules. J Cell Biol 50:498–515

Ogawa T, Koerten HK, Brederoo P, Daems WT (1983) A comparative study of primary and secondary granules in monocytopoiesis and myelopoiesis of mouse bone marrow. Cell Tissue Res 228:107–115

Sutton JS, Weiss LV (1966) Transformation of monocytes in tissue culture into macrophages, epithelioid cells and multinucleated giant cells. An electron microscope study. J Cell Biol 28:303–332

Vernon-Roberts B (1976) The macrophage. In: Beck F, Lloyd JB (eds) The cell in medical sciences, vol 4. Academic, London

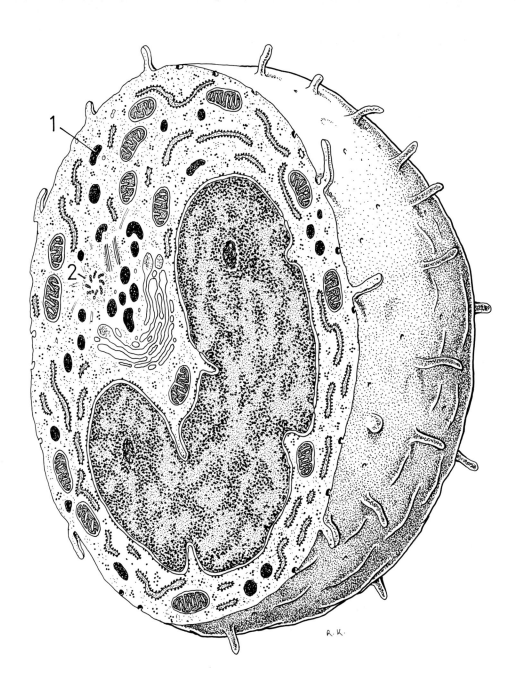

Plate 72. Loose Connective Tissue. Wandering Cells: Plasma Cell

Plasma cells are 10- to 20-µm-large, ovoid, basophilic elements that predominantly occur in the vicinity of small vessels, stroma of several glands (e.g., salivary and lacrimal glands), bone marrow, lymphatic system, omenta, lamina propria of the intestinal tract, and remnant of the endometrium of the uterus during menstruation. Exceptionally, they appear in the blood (rubella).

The nucleus of plasma cells is spherical and eccentric. The clumps of chromatin adopt a characteristic wheel-like pattern ("cartwheel nucleus"). This morphological property is useful in the light-microscopic identification of these cells.

The ultrastructure of plasma cells is distinguished by highly developed parallel cisternae of rough endoplasmic reticulum, ergastoplasm (**1**), which contains a moderately osmiophilic, fine-granular material (reticuloplasm).

A well-developed Golgi apparatus, flanked by several smaller vacuoles and the centriole (**2**), is found close to the nucleus. Under the light microscope, the Golgi complex appears as a pale zone. Several mitochondria and vast numbers of free ribosomes are found between the cisternae of the ergastoplasm. The ribosomes, together with the ergastoplasm, are responsible for the basophilia of plasma cells, visible with the light microscope.

Plasma cells are capable of weak ameboid movement. It is probably for this reason that the surface bears only a few microvilli and globular processes.

Plasma cells are important producers of protein. They synthesize and secrete serum albumins and gamma globulins (immunoglobulins) by a kind of reverse micropinocytosis, i.e., in small vesicles (**3**). Plasma cells are thus intensively involved in the defense of the organism.

Overproduction of proteins as a result of an inflammation leads to distension of the ergastoplasmic cisternae. This is caused by the accumulation of a material termed Russell bodies, which can be visualized light-microscopically after eosin or fuchsin staining.

Plasma cells can develop from B lymphocytes in the lymph nodes (see Plate 115) and from lymphocytelike elements in the region of blood vessels. (See Plates 35, 45, 54, 55 in KRSTIĆ 1979.)

Magnification: × 10,000

REFERENCES

Leduc EH, Avrameas S, Bouteille M (1968) Ultrastructural localization of antibody in differentiating plasma cells. J Exp Med 127:109–118

Pimenta PFP, De Souza W (1982) Ultrastructure and cytochemistry of the cell surface of eosinophils. J Submicrosc Cytol 14:227–237

Sainte Marie G (1964) Study on plasmocytopoiesis: I. Description of plasmocytes and of their mitoses in the mediastinal lymph nodes of ten-week-old rats. Am J Anat 114:207–233

Weiss L (1972) The cells and tissues of the immune system. Structure, functions, interactions. Prentice-Hall, Englewood Cliffs

Wulfhekel U (1980) Die Morphogenese des eosinophilen Granulozyten beim Menschen. Verh Anat Ges 74:223–227

Zagury D, Uhr JW, Jamieson JD, Palade GE (1970) Immunoglobulin synthesis and secretion: II. Radioautographic studies of sites of addition of carbohydrate moieties and intracellular transport. J Cell Biol 46:52–63

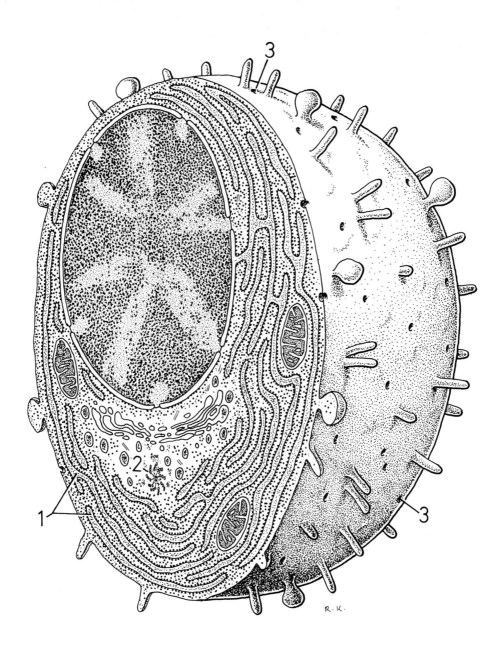

Plate 73. Loose Connective Tissue. Wandering Cells: Eosinophilic Granulocyte

Eosinophilic granulocytes, which are about 11–14 µm in diameter, comprise approximately 1%–4% of all white blood cells in the circulation. They are easily identified since the cytoplasm contains granules (Fig. **A 1**) that stain red with eosin and are visible under the light microscope.

Eosinophilic granulocytes are particularly numerous in the lamina propria of the intestinal tract, especially when parasites (worms) are present. With the exception of the thymus, these cells are less common in the connective tissue of other organs.

Under the light or electron microscope, the nucleus is seen to be in the shape of a dumbbell or a pair of glasses since it is made up of two interconnected lobes. The nucleus contains condensed chromatin and one or two nucleoli. A well-developed Golgi apparatus is located in the space between the nuclear lobes; here are also found smaller vesicles with a low osmiophilic content. Mitochondria and rough endoplasmic reticulum are moderately developed.

The granules, which are up to 1.5 µm in diameter and surrounded by a unit membrane (Fig. **B 1**), contain one or two protein crystals (Figs. **A 2**, **B 2**) embedded in a dense, osmiophilic matrix. These granules are now considered lysosomes.

Eosinophilic granulocytes can leave the capillaries by virtue of their ameboid movement and enter the connective tissue. They are apparently also able to differentiate in situ directly from immature, pluripotential cells. Recent studies have shown that eosinophilic granulocytes are capable of weak phagocytosis.

Eosinophilic granulocytes are sensitized to the tissue hormone histamine. Adrenocorticotropic hormone (ACTH) and cortisol reduce the number of these cells in the blood and tissues. (See Plate 59 in KRSTIĆ 1979.)

Magnifications: Fig. **A**, × 10,000; Fig. **B**, × 40,000

REFERENCES

Bainton DF, Farquhar MG (1970) Segregation and packaging of granule enzyme in eosinophilic leucocytes. J Cell Biol 45:54–73

Beeson PB, Bass DA (1977) The eosinophil. Saunders, Philadelphia

Duhamel G (1976) Physiologie des ′éosinophiles. Nouv Presse Med 5:1054–1056

Miller F, De Harven E, Palade GE (1966) The structure of eosinophil leucocyte granules in rodents and in man. J Cell Biol 31:349–362

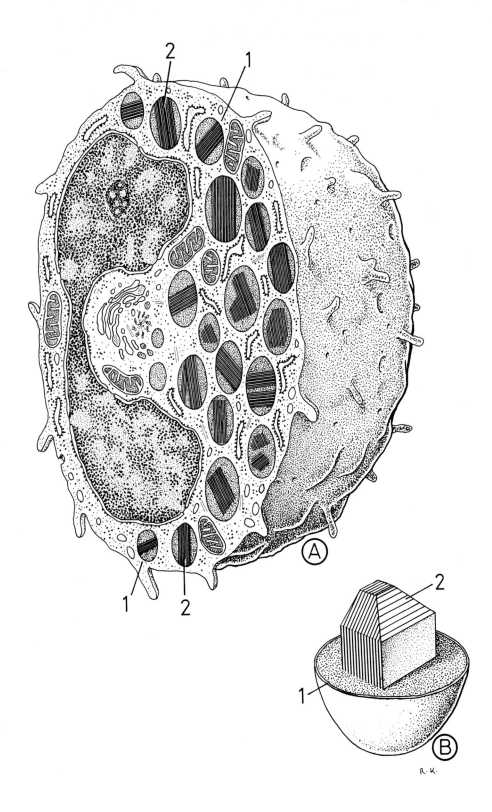

Plate 74. Loose Connective Tissue.
Special Form: Pigment Connective Tissue

Pigment connective tissue is a type of loose connective tissue. In man, it is found in the iris (Fig. A1), choroid (Fig. A2), leptomeninges, and genital skin. A section from the iris is presented in Fig. B.

Pigment connective tissue is partly built up of stellate, serrated, or lobular pigment cells of ectodermal origin, melanocytes (Fig. B1, see Plate 24). These cells contain large numbers of granules of the endogenous pigment melanin and frequently contact one another by means of their processes. The framework thus created also accomodates numerous fibrocytes (Fig. B2).

Collagen (Fig. B3) and elastic (Fig. B4) fibers, lymphatic and blood capillaries (Fig. B5), and nerve fibers (Fig. B6) run through the interstices of the pigment connective tissue. Thus, it has all the features of a loose fibrous connective tissue, whose intercellular fluid also contains wandering cells.

The genetically determined number of melanocytes in the iris is responsible for the color of the eyes. Many densely packed pigment cells give the iris a dark-brown to black color. The appearance of green or brown shades is due to fewer numbers of melanocytes. Blue and gray eyes do not contain pigment cells; the iris stroma in front of the black pigment epithelium (see Plate 13) of the retina appears bluish. In albinos, this retinal layer is also free of melanin, and the iris appears pink owing to reflection of the many choroidal blood vessels of the fundus of the eye.

Magnification: Fig. B, $\times 2{,}500$

REFERENCES

Dietrich CE (1972) Zur Feinstruktur der Melanocyten in der menschlichen Iris. Albrecht Von Graefes Arch Ophthalmol 183:317–333

Poirier J, Nunez-Dispot C (1968) Le mélanocyte: II. Histogenèse. Presse Med 76:1525–1526

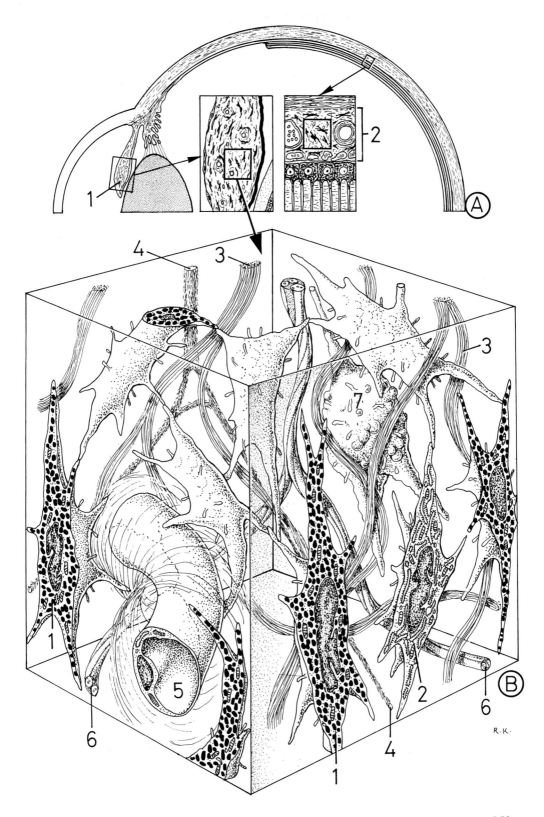

CONNECTIVE AND SUPPORTING TISSUES

Plate 75. Loose Connective Tissue.
Special Form: Pigment Connective Tissue.
Melanocyte of the Human Iris

Melanocytes (1) are, as seen in the previous plate, stellate or spindle-shaped cells similar to fibrocytes. Numerous collagen microfibrils (2) are in contact with the cell body and its processes (3). Melanocytes are basically fixed cells.

The nucleus of a melanocyte is indented, ellipsoidal, and appears under the light microscope as a pale zone in the pigment-rich cell body. The unit membrane-bound, 1- to 2-µm-wide granules (4) are mostly made up of mature melanin. Melanosomes (5), precursors of the melanin granules, also occur in melanocytes (see Plate 24).

Magnification: ×7,000

REFERENCE

Low FN (1962) Microfibrils, fine filamentous components of the tissue space. Anat Rec 142:131–137

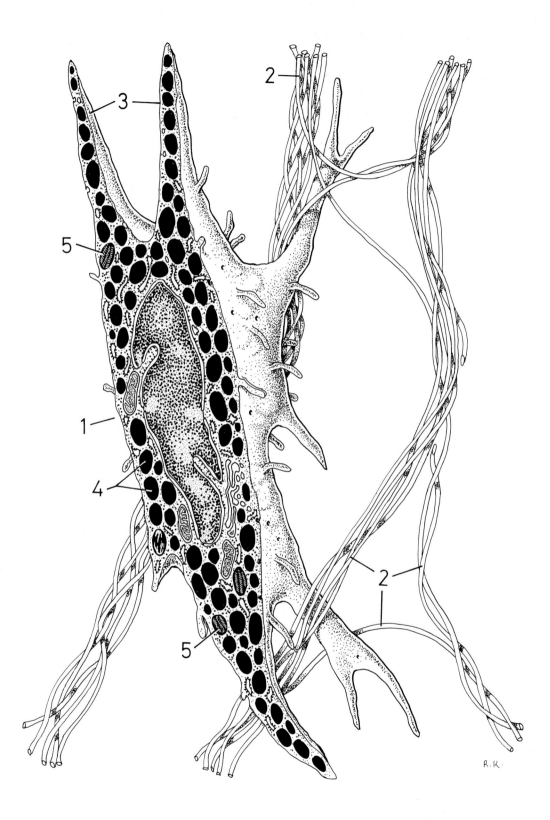

Plate 76. Loose Connective Tissue.
Special Form: "Cellular" Connective Tissue of the Ovary

A special form of loose connective tissue is found in the cortical zone of the ovary (Fig. **A 1**). Beneath the germinal epithelium (Fig. **A 2**), a fibrous layer or tunica albuginea (Fig. **A 3**) is found, followed by an exceptionally cell-rich loose connective tissue, which is represented three-dimensionally in Fig. **B**.

The spindle-shaped fibrocytes of this tissue, which have a plump appearance owing to the small number of processes, and the intervening reticular and collagen fibers are arrayed in irregular whorls (Figs. **A 4, B**).

The narrow intercellular spaces contain fibrous structures and permit only a small degree of movement for any free elements of connective tissue (Fig. **B**). It is unclear whether the cells of this tissue are able to transform into steroid hormone-producing cells of the theca interna (Fig. **A 5**).

Another, similarly cell-rich, loose connective tissue is found in the endometrium (see Plate 47). Its cells differentiate during pregnancy into large, epithelioid, pale, glycogen-rich decidual cells.

The term "cellular" in the context of this kind of loose connective tissue is only intended to signify its richness in fixed cells.

Magnifications: Fig. **A**, ×250;
Fig. **B**, ×2,500

REFERENCES

Glauert AM, Mayo CR (1973) The study of the three-dimensional structural relationships in connective tissues by high voltage electron microscopy. J Microsc 97:83–94

Kaiser E, Stockinger L (1971) Morphologie und Biochemie des Bindegewebes. Münch Med Wochenschr 113(10):321–333

Kivirikko KI; Risteli L (1976) Biosynthesis of collagen and its alteration in pathological states. Med Biol 54:159–186

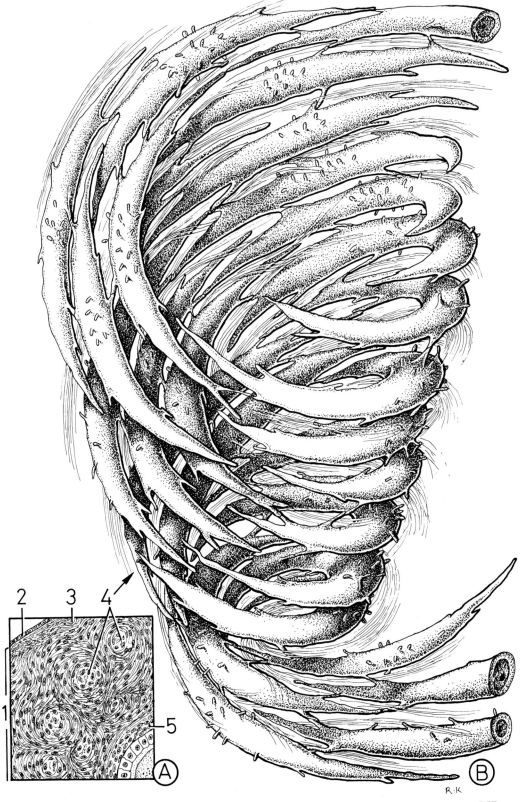

Plate 77. Loose Connective Tissue. Special Form: Retiform Connective Tissue of the Human Omentum Majus

The loose connective tissue of the omenta (Fig. **A 1**) of the abdominal cavity displays a special type of organization, shown at low magnification in Fig. **B**.

It is a loose connective tissue comprised of strands (Fig. **B 1**) of various thicknesses, connected to one another in a kind of network. Blood and lymphatic vessels (Fig. **B 2**) run through the broad strands. A section from Fig. **B** is enlarged in Fig. **C** and shows the three-dimensional structure of this tissue.

Collagen (Fig. **C 1**), elastic (Fig. **C 2**), and reticular (Fig. **C 3**) fibers form the skeleton of the strands, which are covered with polygonal, flattened mesothelial cells (Fig. **C 4**), bearing large numbers of microvilli. The interior of the strands contains fibrocytes (Fig. **C 5**), histiocytes (Fig. **C 6**), adipocytes (Fig. **C 7**), mast cells (Fig. **C 8**), and other cellular elements of loose connective tissue. In addition, there are capillaries (Fig. **C 9**) with pericytes (Fig. **C 10**), lymphatic vessels (not shown), and nerve fibers (Fig. **C 11**).

The lymphocytes and histiocytes of this special form of loose connective tissue can be concentrated along vessels in groups which have a milky appearance in fresh specimens (so called milky spots).

The histiocytes and lymphocytes of this tissue become active upon the penetration of microorganisms into the abdominal cavity. They then wander over the surface of the strands, where they phagocytize the bacteria or produce antibodies against the pathogen. Histiocytes that die in the process are replaced by differentiation of fibroblasts or pericytes.

Mesothelial cells react to toxic effects by mitosis and by swelling and loosening of the cell union. Some of the elements that develop as a consequence of the action of toxins differentiate into fibrocytes, others into polynuclear giant cells. Gaps in the strands brought about by death of mesothelial cells are filled by histiocytes. Whether mesothelial cells are capable of phagocytosis is open to debate.

The leptomeninges (pia mater and arachnoid; see Plate 153) are also classed as loose connective tissue.

Magnifications: Fig. **B**, × 150; Fig. **C**, × 2,000

REFERENCE

Mironov VA, Gusev SA, Baradi AF (1979) Mesothelial stomata overlying omental milky spots: scanning electron microscopic study. Cell Tissue Res 201:327–330

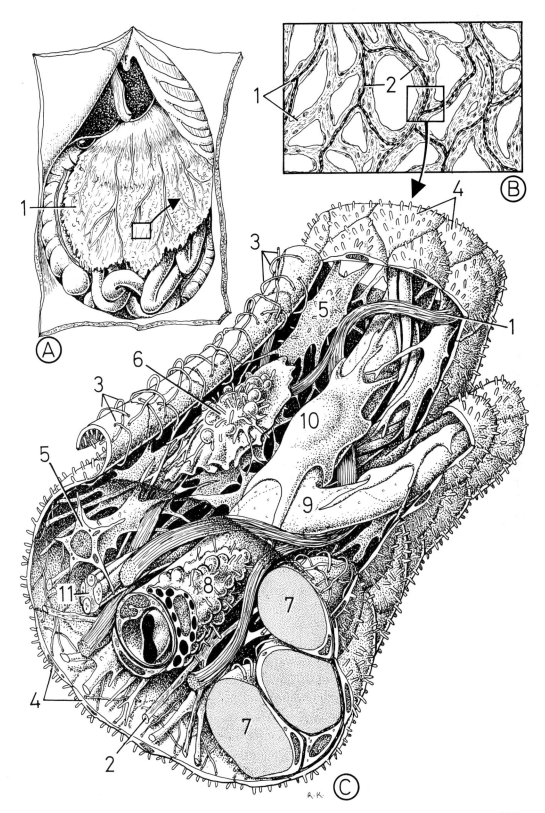

Plate 78. Irregular Dense Connective Tissue. Section from an Articular Capsule

As a result of an increase in fibrillar structures and a reduction in cells and ground substance, a dense fiber-rich connective tissue forms over areas subject to a high degree of mechanical stress. Dense connective tissue forms various organ capsules (testis, kidney, spleen, liver, etc.), the dura mater, the sclera, the stratum reticulare of the corium, the periosteum, and perichondrium, the pericardium, the skeleton of the cardiac valves, and the articular capsules. This tissue is characterized by very thick interwoven collagen fibers and fiber bundles (**1**). These fibers are accompanied by occasional elastic fibers (**2**), which are responsible for dense connective tissue reverting to its original form following deformation.

A few, very flattened fibrocytes (**3**) are found in the spaces between the fibrous structures. Both cells and ground substance are quantitatively reduced in this tissue.

Dense fibrous connective tissue only contains a small number of blood and lymphatic vessels and nerve fibers. These histological properties are typical of a bradytrophic tissue, i.e., one with slow metabolism.

Magnification: × 3,500

REFERENCES

Brown IA (1972) Scanning electron microscopy of human dermal fibrous tissue. J Anat 113:159–168

Goldberg B, Green H (1964) An analysis of collagen secretion by established mouse fibroblast lines. J Cell Biol 22:227–257

Gotte L, Giro MG, Volpin D, Horne RW (1974) The ultrastructural organisation of elastin. J Ultrastruct Res 46:23–33

Robert L (1975) Les glycoprotéines dans l'organisation de la matrice intercellulaire. Aspects ontogénétiques, phylogénétiques et pathologiques. Biol Med (Paris) 4:1–30

Plate 79. Regular Dense Connective Tissue. Tendon

The structure of tendons results from the particular organization of fibrillar material in response to tension exerted in one defined direction.

Macroscopically, the large number of collagen bundles give tendons the appearance of whitish cables stretched between the muscles and bones (see Plate 125). Tendons can lie on a bony base, and this gives rise to synovial vaginae. Such a case is presented in Fig. **A**.

The majority of tendons are surrounded by a loose fibrous connective tissue, the paratendineum (Figs. **A1**, **B1**). This material serves to connect the tendon with its synovial vagina (Fig. **A2**), which facilitates sliding of the tendon over the hard base.

On the outer surface, the synovial vagina is covered by a fibrous layer, called the fibrous vagina (Fig. **A3**). The inner surface of the synovial vagina is lined by simple squamous epithelium, similar to mesothelium and rich in nerves and vessels. Each synovial vagina consists of two sheets – an inner sheet (Fig. **A4**), directly in contact with the tendon, and an outer sheet (Fig. **A5**), lining the fibrous vagina. Both layers are continuous with one another at the ends of the synovial vagina.

The serous cavity delimited by the two sheets is a narrow cleft, containing only a small amount of synovial fluid. In Fig. **A**, the width of this cavity has been exaggerated for the sake of clarity.

Lymphatic and blood vessels (Fig. **A6**) and nerve fibers (Fig. **A7**) supplying the synovial vagina and the tendon run through a connective tissue plate enveloped by mesothelium, termed the mesotendineum (Fig. **A8**).

A tendon is made up of many primary fascicles (Fig. **A9**), each consisting of numerous tendon fibers (fibrae tendinae). Figure **B** shows a detail of the transverse section from the circular inset at higher magnification. The paratendineum (Fig. **B1**) envelops the outer surface of the tendon and is continuous with a connective tissue layer, the epitendineum (Fig. **B2**), on its inner surface. The primary fascicles (Fig. **B3**) are separated from one another by the endotendineum (Fig. **B4**), which contains blood vessels and nerve fibers. Several primary fascicles constitute a secondary fascicle. The tendon cells (Fig. **B5**) appear as small stellate structures compressed by the tendon fibers (Fig. **B6**).

Part of the paratendineum and epitendineum in the rectangular inset in Fig. **A** is enlarged in Fig. **C** to show the light-microscopic structure of a longitudinal section of the tendon.

The primary fascicles (Fig. **C1**) of tendon fibers that are not stretched run in a wavelike manner. This is of great importance in movements which need to be initiated smoothly, since the tendon fibers have to be taut before muscular contraction can exert its effect on the bones. The endotendineum (Fig. **C2**) and the tendon cells (Fig. **C3**) arranged in rows, are also evident in longitudinal section.

Magnifications: Fig. **A**, × 10;
Fig. **B**, × 800;
Fig. **C**, × 550

REFERENCE

Jozsa L, Balint JB, Reffy A, Demel ZS (1979) Histochemical and ultrastructural study of adult human tendon. Acta Histochem 65:250–257

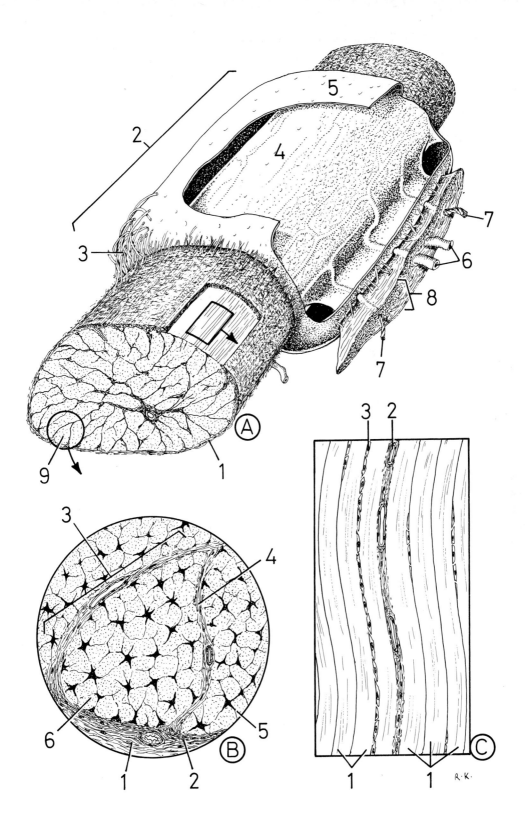

Plate 80. Regular Dense Connective Tissue. Tendon Cell

As a result of the highly developed tendon fibers (Fig. A 1), tendon cells are laterally compressed, and this leads to the formation of thin, winglike cytoplasmic processes (Fig. A 2). Tendon cells are fibrocytes which have predominantly adapted to synthesize collagen. The cell body contains an ovoid nucleus with condensed chromatin. The cytoplasm displays the features of a cell highly involved in protein synthesis, i.e., well-developed rough endoplasmic reticulum and great numbers of free ribosomes.

In addition to collagen tendon microfibrils (Figs. A 4, B 1), which run in a helicoidal fashion, unite and form tendon fibers, tendon cells also synthesize elastic fibers (Fig. A 3), which, as described in Plate 65, are enveloped by a feltwork of 10-nm-thick microfibrils.

In man, tendon microfibrils with diameters of 65–175 nm are not uncommon. The microfibrils are connected to one another by means of interfibrillar bridges (Fig. B 2), which occasionally extend into the interior of the microfibrils (Fig. B 3). Collagenous material is practically inextensible (it can only be stretched up to 4% –5% of its original length), and this gives tendons very high tensile strength (500–1,000 kg/cm^2). Consequently, the danger of rupture in tendons is considerably lower than in muscles.

Tendons are mainly connected to the fibrous layer of the periosteum (see Plate 106). Many tendon fibers also penetrate the bony substance in a brushlike manner (Sharpey's fibers).

Bone formation is occasionally found to a limited extent in tendons (so-called sesamoid bone). The best-known example is the knee cap (patella).

Tendons are basically good at regeneration and being transplanted (see Plate 114). The parts of the tendon surrounded by the synovial vaginae are technically more difficult to graft and do not functionally adapt so well because the mesothelial layers tend to fuse together. Fibroblasts, which are necessary for regeneration and synthesis of tendon fibers, originate from the connective tissue of the epitendineum and endotendineum.

Magnifications: Fig. A, × 19,000; Fig. B, × 26,500

REFERENCES

Dyer FR, Enna CD (1976) Ultrastructural features of adult human tendon. Cell Tissue Res 168:247–259

Fahrenbach WH, Sandberg LD, Cleary EG (1966) Ultrastructural studies on early elastogenesis. Anat Rec 155:563–576

Greenlee TK Jr, Ross R (1967) The development of the rat flex or digital tendon, a fine structure study. J Ultrastruct Res 18:354–376

Torri Tarelli L, Petruccioli MG (1971) Studies on the ultrastructure of collagen fibrils: II. Filamentous structure with negative staining. J Submicrosc Cytol 3:153–170

Trelstad RL, Hayashi K (1979) Tendon collagen fibrillogenesis: intracellular subassemblies and cell surface changes associated with fibril growth. Dev Biol 71:228–242

Yamada M (1976) Ultrastructural and cytochemical studies on the calcification of the tendon-bone joint. Arch Histol Jpn 39:347–378

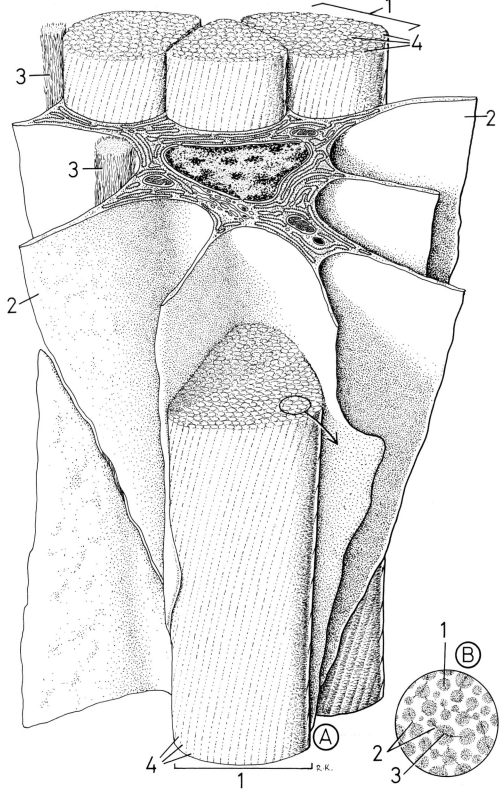

Plate 81. Regular Dense Connective Tissue. Aponeurosis. Example: Stroma of the Cornea

In the "tensile tendons" just described, the traction lies only in one direction of muscular contraction, unlike the flattened tendons or aponeuroses which are exposed to traction from several directions. The structure of these flattened tendons is demonstrated here in the stroma of the cornea, which, though without muscular connections, is an example of a very specialized aponeurosis.

Figure A is a transverse section through the cornea. Between the corneal epithelium (Fig. A1) and corneal endothelium (Fig. A2) is located the connective tissue of the corneal stroma (Fig. A3). The structures in the rectangular inset correspond to those in Fig. B.

Nonkeratinized stratified squamous epithelium (Fig. B1), as described in Plate 22, lies on the thick Bowman's membrane (Fig. B2). The densely packed, collagen microfibrils (Fig. B3), arranged in layers, cross practically at right angles in the corneal stroma. Extremely flattened cells, keratocytes (Fig. B4), are flattened between the layers. These cells are fibrocytes differentiated to synthesize collagen and ground substance. Seen in plan view, keratocytes are characterized by long processes, giving rise to smaller branches, which follow the intersection of the collagen layers.

Keratocytes also form elastic fibers (Fig. B5) in addition to collagen microfibrils. The cytophysiological differences between keratocytes and fibrocytes are due to the different properties of the secreted proteoglycans of the ground substance.

Whereas normal fibrocytes mainly produce hyaluronic acid, keratocytes are specialized in the secretion of large amounts of keratan sulfate and other glycosaminoglycans. This gives three important morphofunctional properties to the cornea:

– Since the ground substance contains five times more sulfur-containing glycoproteins than the sclera, the cornea traps water and becomes more resistant to compression

– The high concentration of keratan sulfate confers transparency to the corneal stroma

– The corneal stroma can receive nutrients without the presence of blood vessels as a result of the high diffusive capacity of its proteoglycans. Neighboring scleral cells and cells of other aponeuroses in the body synthesize only very little keratan sulfate, for which reason they have a whitish color.

The avascular corneal stroma is unable to regenerate because of its high degree of specialization, however, it can within certain limits be removed from corpses, preserved, and homografted (keratoplasty). Other aponeuroses of the organism which are less differentiated than the cornea can regenerate quite well; they can be successfully auto- and homografted (see Plate 114).

Magnifications: Fig. A, × 150; Fig. B, × 1,500

REFERENCES

Craig AS, Parry DAD (1981) Collagen fibrils of the vertebrate corneal stroma. J Ultrastruct Res 74:232–239

Hamada R (1975) Aspect ultrastructural des cellules et du tissu conjonctif cornéen normal. Arch Ophthalmol 35:23–26

Robert L, Robert B (1975) The macromolecular structure of normal cornea. Arch Ophthalmol (Paris) 35:11–22

Sandler SS (1974) Direct three-dimensional reconstruction of a corneal stromal lamella from electron micrographs. J Theor Biol 48:207–213

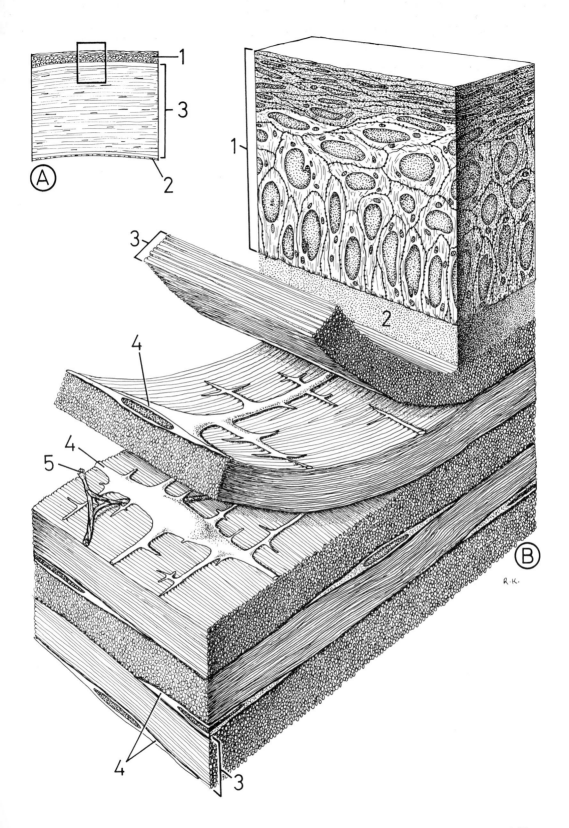

Plate 82. Elastic Connective Tissue or Elastic Ligaments

The major elements of the elastic connective tissue (yellow ligaments) are the densely packed elastic fibers. The strongest elastic ligament is the ligamentum nuchae (Fig. A 1) of ungulates, from which the heavy head of these animals "hangs." In humans, elastic connective tissue is less widely distributed; it occurs largely as the ligamenta flava (Fig. B 1) between the vertebrae and in the vocal cords (Fig. C 1).

In transverse section (Fig. D 1), the elastic connective tissue is seen to be made up of large numbers of homogeneous polygonal structures between which are scattered a few fibrocytes (Fig. D 2), some collagen fibrils, and occasional blood vessels. With resorcin-fuchsin and orcein, elastic fibers can be stained blue-black and dark brown, respectively (see Plate 86).

On the basis of electron-microscopic observations, part of an elastic ligament is reconstructed in Fig. E. The strong elastic fibers (Fig. E 1) are, as previously described (see Plates 65, 80), surrounded by a microfibrillar feltwork and run almost parallel as a result of constant tension. At many sites, however, the fibers branch or run into one another at oblique angles. Between the fibers, there is a network of reticular and collagen fibrils and microfibrils (Fig. E 2) and a small number of fibrocytes (Fig. E 3).

The function of elastic ligaments is to reduce the work of the whole organism, or of certain muscles when parts of the body are maintained in a particular position. Elastic connective tissue regenerates poorly since its synthesis demands a high degree of cellular differentiation.

Magnifications: Fig. D, × 300;
Fig. E, × 2,000

REFERENCES

Bradamante Z, Švajger A (1977) Pre-elastic (oxytalan) fibres in the developing elastic cartilage of the external ear of the rat. J Anat 123:735–743

Ichimura T, Hashimoto PH (1982) Three-dimensional fine structure of elastic fibers in the perivascular space of some circumventricular organs as revealed by high-voltage electron microscopy. J Ultrastruct Res 81:172–183

Kewley MA, Steven FS, Williams G (1977) The presence of fine elastin fibrils within the elastin fibre observed by scanning electron microscopy. J Anat 123:129–134

Robert AM, Robert D (1980) Biology and pathology of elastic tissue. Frontiers of matrix biology, vol 8. Karger, Basle

Ross R (1973) The elastic fiber. A review. J Histochem Cytochem 21:199–208

Serafini-Fracassini A, Field MJ, Smith JW, Stephens WGS (1977) The ultrastructure and mechanics of elastic ligaments. J Ultrastruct Res 58:244–251

Serafini-Fracassini A, Field JM, Hinnie J (1978) The primary filament of bovine elastin. J Ultrastruct Res 65:190–193

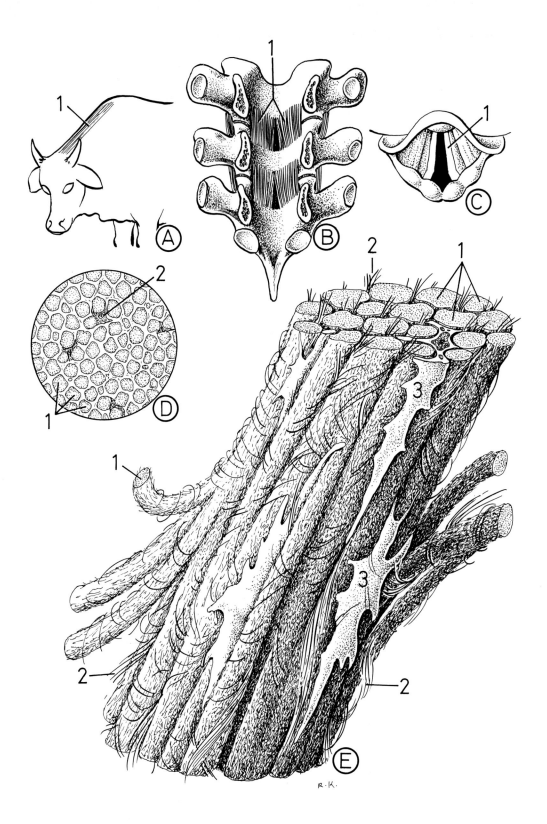

Plate 83. Cartilaginous Tissue.
Histogenesis of Hyaline Cartilage

Cartilaginous tissue differentiates from the mesenchyme (see Plate 48). This histogenesis of hyaline cartilage is illustrated here by a section at the level of the heart (Fig. A 1) and lungs (Fig. A 2) of an 18-day-old mouse fetus.

Accumulations of mesenchymal cells corresponding to hyaline cartilage anlagen (arrows) are evident at several sites in the section. The rectangular inset is magnified in Fig. B.

Mesenchyme (Fig. B 1) forms a dense aggregation of cells called precartilage or protochondral tissue, in which mitoses (Fig. B 2) are frequent. Prior to morphological differentiation, these blastema cells undergo functional differentiation: They synthesize intercellular substance or cartilage matrix, which consists of collagen microfibrils and such a quantity of ground substance that the fibrillar structures become concealed for ordinary light-microscopic observation. As a result of the increase in the cartilage matrix, the cells become devoid of processes and flattened. In this phase they are termed chrondroblasts (Fig. B 3). At the center of these precartilaginous areas, the cells continue to grow and gradually differentiate into mature cartilage cells or chondrocytes (Fig. B 4) in small cavities of the intercellular substance.

Cartilage growth is the result of two parallel processes:

1. The volume of the anlage increases by the apposition of new mesenchymal cells onto the periphery of the piece of cartilage.
2. Mitotic divisions (Fig. B 5) of the cartilage cells in the intercellular substance (interstitial growth) also lead to an enlargement of the primordial cartilage, since secretion of the intercellular substance causes the daughter cells to move apart.

At a later stage, the mesenchyme lying on the surface of the skeletal area forms the perichondrium, the chondroblasts of which continue forming cartilage.

The development and growth of elastic and fibrous cartilage takes place basically according to the same mechanism.

Magnifications: Fig. A, ×20;
Fig. B, ×250

REFERENCES

Anderson DR (1964) Ultrastructure of hyaline and elastic cartilage of the rat. Am J Anat 114:403–434

Hall BK (ed) (1983) Cartilage, vol 2. Development, differentiation, and growth. Academic, New York

Knese KH (1979) Die Initialstadien der Bildung des Knochenkerns und die Gliederung des Skeletorgans. Der metamorphosierende Knorpel: I. Mitteilung. Gegenbaurs Morphol Jahrb 125:758–778

Thyberg J (1977) Electron microscopy of cartilage proteoglycans. Histochem J 9:259–266

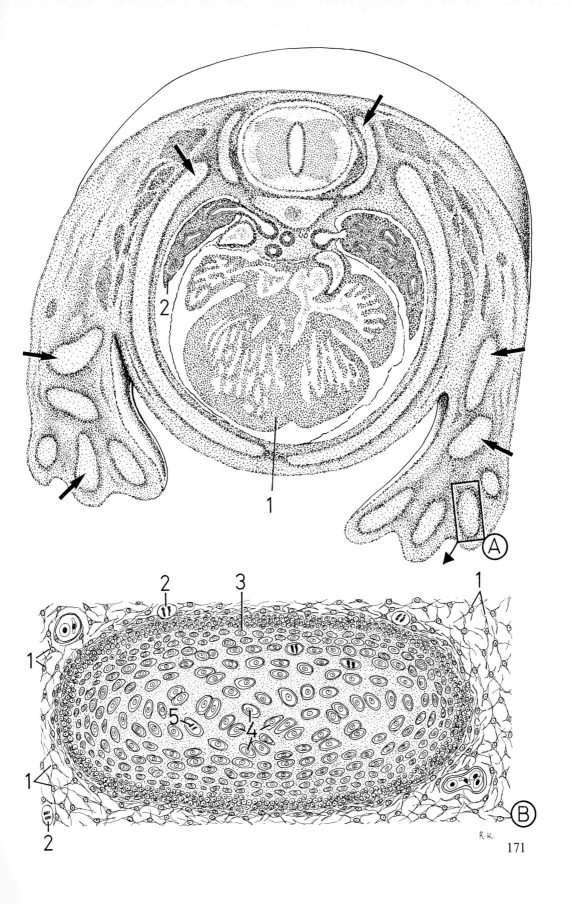

Plate 84. Cartilaginous Tissue.
Hyaline Cartilage of the Trachea

Fresh hyaline cartilage has a distinctive bluish-white opaque appearance and in thinner sections appears transparent. In the human body, it is the most widespread type of cartilaginous tissue. It is found in the embryonic skeleton, the cartilage of the ribs, joints, and nose, the large laryngeal cartilages, and the cartilages of the trachea (Fig. **A**) and large to medium-sized bronchi. A cartilaginous crescentic ring (Fig. **A1**) from the trachea has been chosen to elucidate the structure of hyaline cartilage. The inset as viewed under ordinary light and polarizing microscopes is presented in Figs. **B–D**.

Every piece of cartilage in the body (apart from articular cartilage) is surrounded by a relatively dense network of connective tissue with elastic fibers, the perichondrium (Figs. **B1, D1**). This tissue also contains vessels and nerve fibers, which do not enter the cartilage mass. There is a smooth transition between perichondrium and cartilage. Many single cartilage cells (Fig. **B2**) and groups of cartilage cells, chondrones (Fig. **B3**), appear in the homogeneous cartilage matrix, which is completely devoid of structures and has a glassy (hyaline) appearance. Since chondrones stem from one mother cell they are also termed isogenic groups. The axis of these cell groups generally runs perpendicular to the perichondrium. Four cells of an isogenic group are enlarged in Fig. **C**.

Cartilage cells (Fig. **C1**) are located in small cavities or lacunae (Fig. **C2**), delimited by a thin, highly stained zone, the capsular or territorial matrix, also called the cartilage capsule (Fig. **C3**). The high chondroitin sulfate content causes this zone to stain much more metachromatically. The capsular matrix also surrounds the whole chondrone and gradually passes into the poorly staining zone which is termed the interterritorial matrix (Fig. **C4**).

The collagen fibrils are masked by a considerable amount of proteoglycans rich in chondroitin sulfates concentrated in the cartilage ground substance, but can be discerned in polarized light (Fig. **D**). It is then possible to see, especially in the tracheal cartilages, two S-shaped, crisscrossing systems of fibrils (Fig. **D5**), stretching from one perichondrium to another. As in every type of cartilage, the course of the collagen fibrils is dependent on functional circumstances.

In the interstices are located the chondrones (not visible in polarized light), which function as elastic pads.

Magnifications: Fig. **A**, × 5;
Figs. **B, D**, × 150;
Fig. **C**, × 800

REFERENCES

Bucher O (1942) Beitrag zum funktionellen Bau des hyalinen Knorpels (auf Grund von Untersuchungen im polarisierten Lichte). Z Zellforsch 32:281–300

Cooper GW, Prockop DJ (1968) Intracellular accumulation of protocollagen and extrusion of collagen by embryonic cartilage cells. J Cell Biol 38:523–537

Hall BK (ed) (1983) Cartilage, vol 1. Academic, New York

Rodan GA, Bourret LA, Cutler LS (1977) Membrane changes during cartilage maturation. Increase in 5′-nucleotidase and decrease in adenosine inhibition of adenylate cyclase. J Cell Biol 72:493–501

Sheldon H, Kimball FB (1962) Studies on cartilage: III. The occurrence of collagen within vacuoles of the Golgi apparatus. J Cell Biol 12:599–613

Stofft E, Graf J (1983) Rasterelektronenmikroskopische Untersuchung des hyalinen Gelenkknorpels. Acta Anat 116:114–125

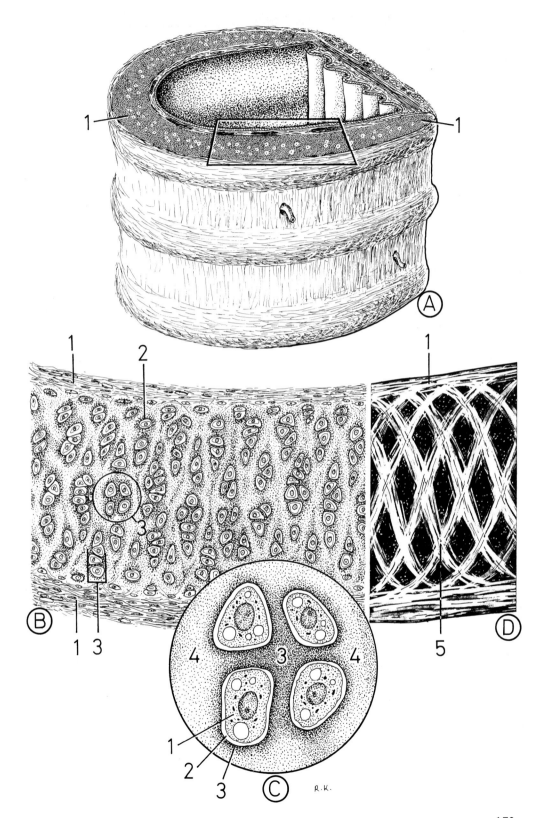

Plate 85. Cartilaginous Tissue. Hyaline Cartilage. Continuation of Plate 84

The ultrastructural organization of a peripheral piece of young hyaline cartilage is demonstrated here in a three-dimensional figure.

Surrounded by several flattened fibrocytes (1) and collagen fibrils (2), a capillary (3) can be seen in the upper part of the picture. This zone corresponds to the perichondrium (4), which for the sake of clarity has been drawn thinner and with fewer fibers. Perichondrial cells (5) resemble fibroblasts; they differentiate into cartilage cells (6) by rounding off and increasing in size. This process, as previously described, is appositional growth.

Below the upper zone, two isogenic groups (7) can be seen, and in this example they each comprise two chondrocytes. A spindle apparatus (8) indicates the mitotic activity of young cartilage cells, which leads to interstitial cartilage growth.

Cartilage cells are located in lacunae (9) in the glassy ground substance (10) which conform to their spherical shape. Collagen fibrils (11) run from the perichondrium through the ground substance. Like the fibrocytes of the perichondrium, chondrocytes synthesize collagen microfibrils (12). The ability of cartilage to produce considerable amounts of metachromatic sulfated proteoglycans into the ground substance is of great importance in the metabolism of this tissue and its ability to withstand compression.

Since hyaline and other types of cartilage are avascular, they obtain nutrients from the perichondrium by diffusion by means of proteoglycans of the ground substance. Oxygen pressure in the vicinity of chondrocytes is thus extremely low, which naturally gives rise to very slow, partly anaerobic processes. Hence, cartilage is classed as a bradytrophic tissue. The diffusion of metabolites into central areas of the hyaline cartilage is sometimes inadequate, particularly in the aged, and degenerative processes can frequently be observed here (reduction in ground substance and appearance of collagen fibers under the ordinary light microscope = asbestos cartilage). Hyaline and fibrous cartilage can also calcify and/or become replaced by bone with advanced age.

Damaged mature cartilaginous tissue does not regenerate; a connective tissue scar develops in its place. Conversely, cartilage is occasionally able to form new tissue and be successfully transplanted if the perichondrium remains intact (see Plate 114).

Magnification: × 4,000

REFERENCES

Caplan AI (1984) Cartilage. Sci Am 251/4:82–90
Hough AJ, Mottram FC, Sokoloff L (1973) The collagenous nature of amiantoid degeneration of human costal cartilage. Am J Pathol 73:201–216
Kuettner KE, Pauli BU, Gall G, Memoli VA, Schenk RK (1982) Synthesis of cartilage matrix by mammalian chondrocytes in vitro: I. Isolation, culture characteristics, and morphology. J Cell Biol 93:743–750
Takagi M, Parmley RT, Toda Y, Austin RL (1982) Ultrastructural cytochemistry and immunocytochemistry of sulfated glycosaminoglycans in epiphyseal cartilage. J Histochem Cytochem 30:1179–1185

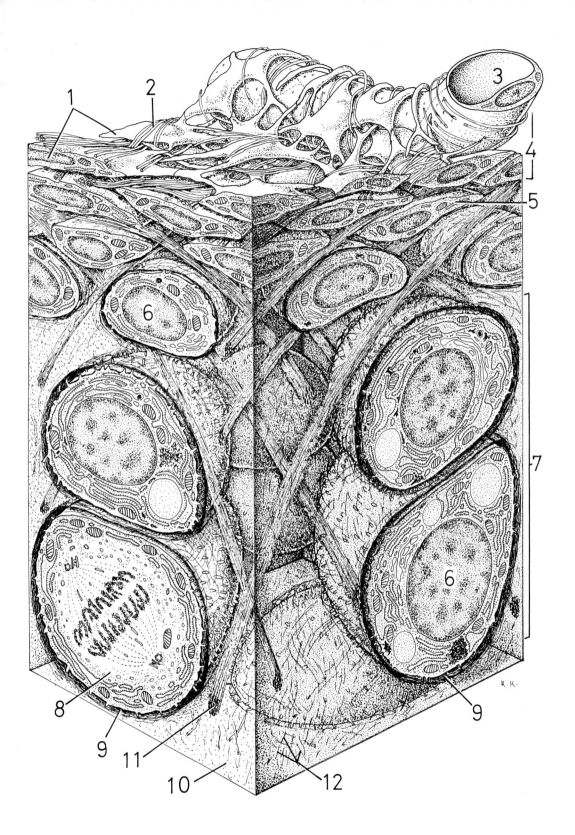

Plate 86. Cartilaginous Tissue. Elastic Cartilage

The slightly yellowish elastic cartilage is much less widely distributed in the body than hyaline cartilage. It is found in the auricle (Fig. **A**), external auditory meatus, Eustachian tube, some laryngeal cartilages, e.g., the epiglottis (Fig. **B**), and the smallest bronchi.

Light-microscopic images of the insets in Figs. **A** and **B** are reproduced in Figs. **C** and **D**. As in hyaline cartilage, the cartilage surface of elastic cartilage is covered by a perichondrium (Figs. **C1**, **D1**). In the interior of the cartilage, chondrocytes (Figs. **C2**, **D2**) and isogenic groups (Figs. **C3**, **D3**) occur, surrounded by cartilage matrix. Isogenic groups are less common in elastic than in hyaline cartilage. Collagen fibrils masked by proteoglycans rich in chondroitin sulfates also exist in elastic cartilage, though strong, coarse, intertwined elastic fibers distinguish this cartilage. They are found in the intercellular substance and form interterritorial filamentous networks (Figs. **C4**, **D4**), which can be discerned following staining with resorcin-fuchsin or orcein (Fig. **D**).

Elastic cartilage contains less ground substance than hyaline cartilage, which renders it softer, and because of the many elastic fibers it is very supple. Elastic cartilage is incapable of regeneration; but it neither calcifies nor ossifies, even with advanced age.

Magnifications: Figs. **C**, **D**, $\times 150$

REFERENCES

Cox RW (1977) The fine structure of developing elastic cartilage. J Anat 123:283–296

Horwitz AL, Dorfman A (1968) Subcellular sites for synthesis of chondromucoprotein of cartilage. J Cell Biol 38:358–368

Kostović-Knežević L, Bradamante Ž, Švajger A (1981) Ultrastructure of elastic cartilage in the rat external ear. Cell Tissue Res 218:149–160

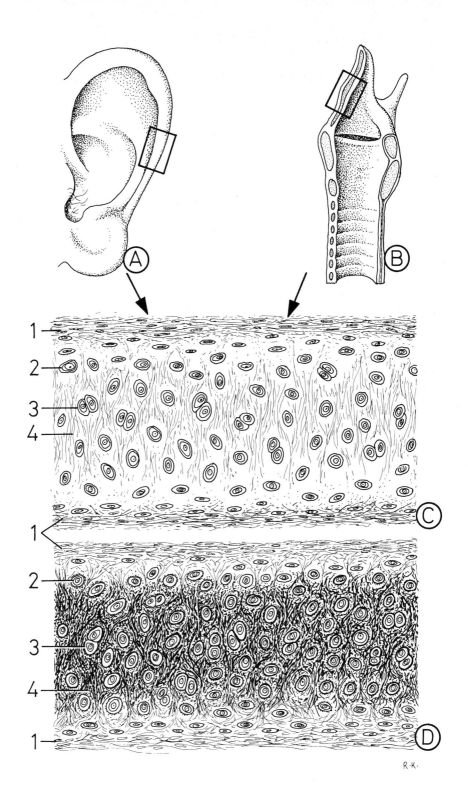

Plate 87. Cartilaginous Tissue. Elastic Cartilage

In elastic cartilage, as in hyaline cartilage, lymphatic and blood vessels (1) are found exclusively in the perichondrium (2). Collagen microfibrils (3) can be seen in the glassy cartilage matrix between the peripheral, flattened cells (4).

The chondrocytes (5) of elastic cartilage (sometimes also termed elastocytes) are spherical elements contained within lacunae. Large numbers of elastic fibers (6), which become thicker from the periphery toward the interior, are evident in the interterritorial matrix. Ramifications and cross connections of these fibers give rise to the elastic fibrous network characteristic of this type of cartilage. The presence of fenestrated elastic membranes (7) does not appear to be uncommon.

The high refractive index of elastic fibers entails that elastic cartilage can be distinguished under the ordinary light microscope without elastica staining.

Magnification: ×3,000

REFERENCES

Nielsen EH (1976) The elastic cartilage in the normal rat epiglottis: I. Fine structure. Cell Tissue Res 173:179–191

Nielsen EH, Bytzer P (1979) High resolution scanning electron microscopy of elastic cartilage. J Anat 129:823–831

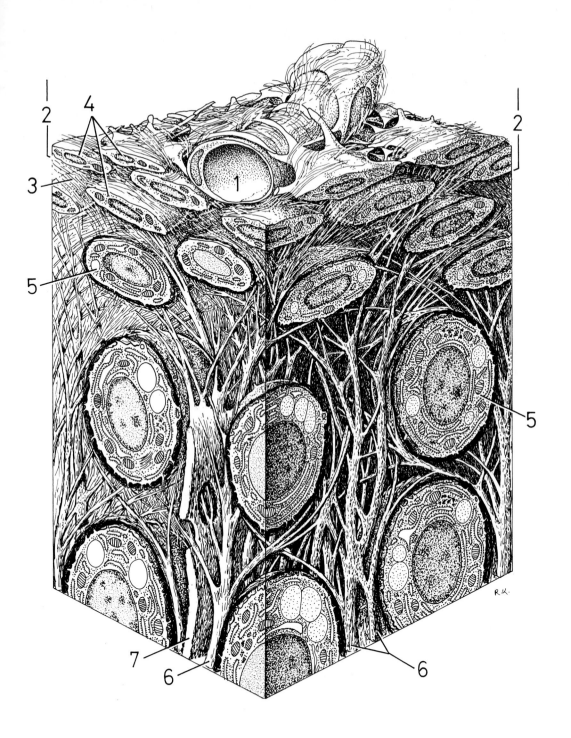

Plate 88. Cartilaginous Tissue.
Chondrocyte of Hyaline or Elastic Cartilage

Chondrocytes are spherical or ellipsoidal cells with a correspondingly shaped central or eccentric nucleus with condensed chromatin. In order to increase maximally the area of contact with the ground substance, cartilage cells have a serrated outline formed by numerous irregular microvilli (**1**).

Younger chondrocytes have a strongly developed rough endoplasmic reticulum (**2**), a voluminous Golgi apparatus (**3**), and several mitochondria (**4**). Inclusions in cartilage cells consist largely of glycogen (**5**) and lipid droplets (**6**). In older cartilage cells, the lipid droplets can fuse to form a single vacuole. These inclusions constitute an extremely important energy-rich reserve material for the bradytrophic cartilaginous tissue.

Cartilage cells are completely surrounded by cartilage matrix (**7**). The collagen fibrils (**8**) and elastic fibers (**9**) are not in direct contact with the plasmalemma but are found at a distance of 1–2 μm from the cell body. This, together with the shrinkage of cartilage cells during fixation for light-microscopic observation, makes the space around the chondrocyte very distinct.

Chondroytes, like fibroblasts (see Plate 64), synthesize collagen microfibrils, which explains the highly developed rough endoplasmic reticulum. Only the chondrocytes of elastic cartilage are differentiated to produce elastic fibers. It is assumed that this synthesis proceeds according to the scheme described in Plate 65.

Radioactive sulfur (^{35}S) labeling has demonstrated that synthesis of the ground substance glycosaminoglycans and their binding with sulfur to form chondroitin sulfate takes place in the Golgi apparatus. The radioactive material was initially evident in the cytoplasm, and subsequently in the cartilage matrix. The protein components for its proteoglycans are of course provided by the rough endoplasmic reticulum.

Irregularly dispersed dense matrix vesicles or spherulites (**10**), about 20–200 nm across and enclosed by a unit membrane, also occur in the cartilage matrix. They presumably become detached from the microvilli (**1**) and can act as crystallization foci, inducing calcification of the cartilage matrix in certain cartilages.

Chondrocytes are very sensitive to hormonal effects. Production of the cartilage matrix is stimulated by growth hormone, thyroxine, and testosterone. Cortisol, cortisone, and the female sex hormone estradiol induce the opposite effects. (See Plate 175 in KRSTIĆ 1979.)

Magnification: × 14,000

REFERENCES

Bonucci E (1970) Fine structure and histochemistry of calcifying globules in epiphyseal cartilage. Z Zellforsch 103:192–217

Nelboeck M (1972) Cyclisches Adenosinphosphat in der Kausalkette hormoneller Wirkungen. Naturwissenschaften 59:209–213

Palfrey AJ, Davies DV (1966) The fine structure of chrondocytes. J Anat 100:213–226

Stockwell RA (1979) Biology of cartilage cells. Cambridge University Press, Cambridge

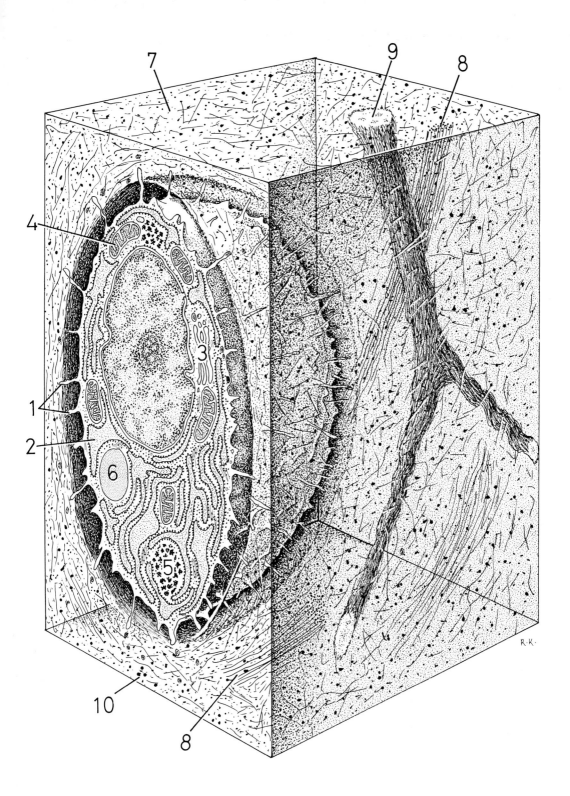

Plate 89. Cartilaginous Tissue.
Fibrous Cartilage or Fibrocartilage

Fibrous cartilage occurs only in a few sites in the organism; certain parts of the intervertebral discs (Fig. A), symphysis pubica, and some articular cartilages are composed of this tissue. Hyaline cartilage is usually found at its periphery, and there is a gradual transition between the two.

In Fig. B, a detail from an intervertebral disc is presented as seen under light-microscopic magnification. The chondrocytes in fibrous cartilage are basically much more sparse than in the other two types of cartilage described. As in all cartilage, the chondrocytes are located within a zone of narrow territorial matrix (Fig. B1). It is only in this area that the ground substance masks the collagen fibers.

The highly entwined collagen fibers (Fig. B2) of the interterritorial matrix are not masked by the sparse ground substance and are thus distinct even in routinely stained sections.

Figure C is a three-dimensional reconstruction of the two cells in the inset in Fig. B. As with other types of cartilage, the chondrocytes are located in small lacunae, in the walls of which varying numbers of thin collagen microfibrils (Fig. C1) can be observed. The strong collagen fibers (Fig. C2) leave little space for ground substance.

Fibrous cartilage cells (Fig. C3) are mostly ellipsoidal and have a spherical nucleus. They display an extensive rough endoplasmic reticulum, like other cartilage cells. The articular disc and articular meniscus are very similar to fibrous cartilage, though the cells do not have the properties of chondrocytes (production of sulfated proteoglycans) and thus are without a territorial matrix.

Fibrous cartilage is the most resistant of all types of cartilage hitherto discussed. Although only sparsely present, the cartilage matrix bears the compressive load of the body. The collagen fibers, which run in a spiral fashion, are subjected to strain during rotary movements of the spinal column.

Magnifications: Fig. B, × 500;
Fig. C, × 3,500

REFERENCE

Matukas VJ, Panner BJ, Orbison JL (1964) Studies on ultrastructural identification and distribution of protein polysaccharide in cartilage matrix. J Cell Biol 32:365–378

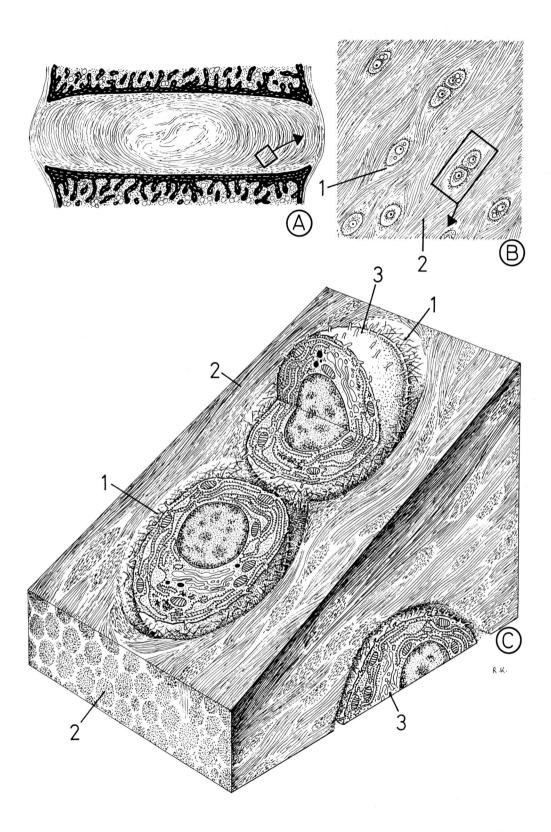

Plate 90. Notochordal Tissue. Fetus of the Mouse

In mammals, the notochord only forms during embryonic development. Here it functions primarily as an organizer for the surrounding, more highly developed types of tissue and not, as in lower animals, as a kind of supporting organ. In addition, it induces the development of the neural tube.

The notochord is remarkable in that it does not derive from the mesenchyme, but from the chorda-mesodermal plate (see embryology texts for further information), and also because it can perform its functions without developing special intercellular substances.

In young mammals, chordal tissue (Fig. A, inset) is located within the hyaline cartilage anlage (Fig. A 1) for the intravertebral discs. The spinal cord (Fig. A 2) and vertebral arches (Fig. A 3) are evident in this figure. Figure B is a three-dimensional reconstruction of the area within the inset in Fig. A.

The notochord is basically a rod-shaped body tapered at both ends which becomes enveloped by hyaline cartilage (Fig. B 1) at a very early stage. For this reason, the chordal cells (Fig. B 2) are located in a canal (Fig. B 3) formed of hyaline cartilage that becomes segmented at a later stage. In mammals, the chordal cells are stellate and connected to one another by means of their processes. The cell within the inset in Fig. B is drawn three-dimensionally in Fig. C.

In the flattened cell body, the chordal cells contain an ellipsoidal, moderately osmiophilic nucleus with a distinct nucleolus. Within the cytoplasm are found some mitochondria, a few cisternae of rough endoplasmic reticulum, and a moderately developed Golgi apparatus. Bundles of microfilaments (Fig. C 1), connected to desmosomes (Fig. C 2), run through the cell body. The desmosomes form the sites of attachment between the chordal cells. Very fine collagen microfibrils (Fig. C 3) occur in the intercellular substance.

With further development of the vertebral column, the notochordal tissue becomes progressively confined to the intervertebral discs (Fig. D 1), where it eventually becomes the nucleus pulposus (Fig. D 2). The cells then become vesicular and secrete a mucous intercellular substance. In humans, the chordal cells remain beyond the 7th year of life in the form of a chordal reticulum within the gelatinous nucleus pulposus.

Magnifications: Fig. A, $\times 70$;
Fig. B, $\times 400$;
Fig. C, $\times 8,500$

REFERENCES

Bancroft M, Bellairs R (1976) The development of the notochord in the chick embryo, studied by scanning and transmission electron microscopy. J Embryol Exp Morphol 35:383–401

Jurand A (1974) Some aspects of the development of the notochord in mouse embryos. J Embryol Exp Morphol 32:1–33

Kenney MC, Carlson E (1978) Ultrastructural identification of collagen and glycosaminoglycans in notochordal extracellular matrix in vivo and in vitro. Anat Rec 190:827–850

Smith JW, Serafini-Fracassini A (1968) The distribution of the protein-polysaccharide complex in the nucleus pulposus matrix in young rabbits. J Cell Sci 3:33–40

Trout JJ, Buckwalter JA, Moore KC (1982) Ultrastructure of the human intervertebral disc: II. Cells of the nucleus pulposus. Anat Rec 204:307–314

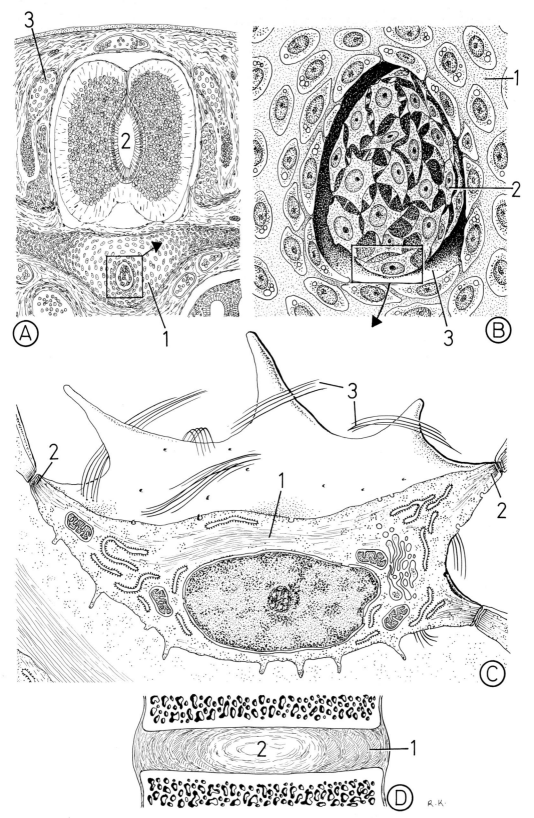

Plate 91. Bony Tissue. Direct or Intramembranous Bone Formation

Initially, woven or immature bone develops directly or indirectly from the mesenchyme. This woven bone gives rise to definitive, mature, lamellar bone.

1. In direct or intramembranous ossification, the mesenchymal cells transform directly into bone cells. The bones of the dome of the cranium, face, and clavicula develop according to this pattern. Bone that forms in this way is also termed membranous bone, since it differentiates from the embryonic connective tissue.

2. The process of indirect ossification is more complicated. The cartilage skeleton initially develops from the mesenchyme (see Plate 83); this becomes destroyed and bone is then secondarily built up onto the debris (endochondral or cartilage bone).

Direct ossification proceeds as follows:

A. In presumptive genetically defined areas, mesenchymal cells (Fig. **A1**) differentiate into osteoblasts (Fig. **A2**), which secrete an organic proteoglycan-containing material, osteoid (Fig. **A3**). Polymerization of tropocollagen into collagen microfibrils occurs in this material.

B. Osteoid becomes enriched with calcium ions and organic phosphate due to the presence of chondroitin sulfates in proteoglycans. The osteoblasts then induce calcification of the osteoid (Fig. **B1**), which becomes the bone matrix of immature bone. Meanwhile, mesenchymal cells (Fig. **B2**) give rise to more osteoblasts (Fig. **B3**), which secrete osteoid in their immediate vicinity and, after mineralization has been completed, become embedded in the bone matrix.

C. In this manner, the first spicules and, soon after, the trabeculae (Fig. **C1**) of membranous bone develop, in which the cells – now termed osteocytes (Fig. **C2**) – remain in contact with one another. Apposition of the newly differentiated osteoblasts (Fig. **C3**) leads to the formation of an epitheliumlike layer on the surface of the bony trabeculae. The osteoblasts secrete osteoid in the direction of the already mineralized bone matrix, thereby thickening the trabeculae. Capillaries (arrows) can be observed around centers of ossification.

Magnifications: Figs. **A–C**, × 350

REFERENCES

Cabrini RL (1961) Histochemistry of ossification. Int Rev Cytol 11:283–306

Carneiro J, Leblond CP (1959) Role of osteoblasts and odontoblasts in secreting the collagen of bone and dentine as shown by radioautography in mice given tritium-labelled glycin. Exp Cell Res 18:291–300

Owen M (1970) The origin of bone cells. In Rev Cytol 28:213–238

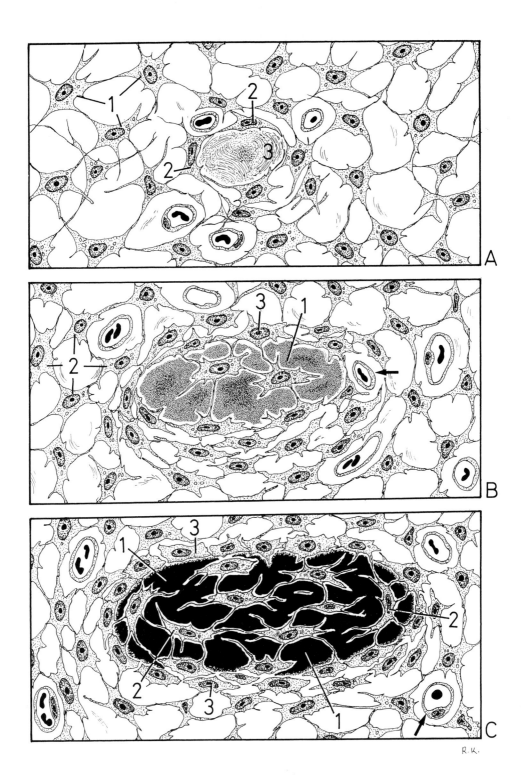

Plate 93. Bony Tissue. Direct or Intramembranous Bone Formation. Continuation of Plate 92

On the left of the figure, the side oriented toward the scalp, several osteoblasts (1) can be seen. They are stellate, plump-looking cells with several short processes and microvilli. Osteoblasts lie on the trabecula (2) and there secrete an osteoid layer (3). Some osteoblasts have in this way become enclosed within the bony mass, for example, in the bottom left corner, fine, winglike bony processes (4) can be seen covering the osteoblast. No cells have been drawn in the lower half of the trabecula so that an almost completely embedded osteoblast (5) can be clearly seen. Osteoblasts are connected to osteocytes (7) by means of cell processes (6) and this gives rise to a wide-meshed cell union.

On the right of the drawing, the side directed toward the dura mater, only a few osteoblasts, a mesenchymal cell (8), and an osteoclast (9) are present. The latter has formed with its osteolytic enzymes a shallow concavity in the bony lamella, the Howship's lacuna. A capillary (10) supplies the cells with the necessary nutrients. (See Plate 138, 139 in Krstić 1979.)

Magnification: ×2,500

REFERENCES

Bernard GW, Pease DC (1969) An electron microscopic study of intial intramembranous osteogenesis. Am J Anat 125:271–290

Jones SJ, Boyde A (1977) The migration of osteoblasts. Cell Tissue Res 184:179–193

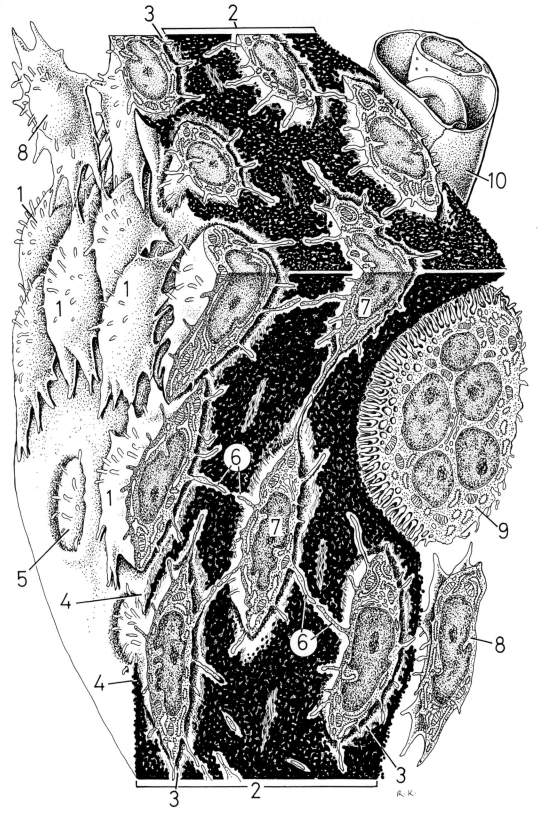

Plate 94. Bony Tissue. Direct or Intramembranous Bone Formation. Osteoblasts. Continuation of Plate 93

An epithelioid osteoblast layer adjacent to a bone trabecula (1) has here been vertically sectioned. The bone-forming cells i.e. osteoblasts (2) are flattened and joined to one another by their processes. The cells contain a spherical or ellipsoidal nucleus with condensed chromatin and a large peripheral nucleolus. Broad cisternae of rough endoplasmic reticulum (3) with moderately osmiophilic contents are conspicuous in the cytoplasm. Around the voluminous Golgi apparatus (4), numerous vesicles (5) are present. Several mitochondria and large numbers of free ribosomes occur between the cisternae of rough endoplasmic reticulum.

Every osteoblast is enclosed within a feltwork of collagen microfibrils (6) of its own manufacture. (For the sake of clarity, the collagen microfibrils have been drawn thicker than normal.) The proteoglycan-containing ground substance occurs in this calcium-free osteoid border. Organic substances of the osteoid synthesized by the osteoblasts, such as tropocollagen molecules and structural proteins, are secreted from the cells in vesicles (7). In addition, the bone-forming cells secrete (probably by detachment of their processes) matrix vesicles (8), which are important in the calcification of the bone matrix as centers of crystallization. Osteoblasts also produce alkaline phosphatase, which is responsible for the concentration of phosphate ions in the ground substance. About 1–2 µm from the cells, osteoid becomes penetrated by hydroxyapatite crystals and appears black in nondecalcified electron-microscopic preparations. Calcification is not carried out homogeneously. There are regions in which some collagen microfibrils and the intervening ground substance remain unmineralized (9).

It has already been mentioned that osteoblasts secrete their organic products in all directions, but particularly toward the trabeculae. They thus become embedded in the bone matrix while still active.

The osteoblast in the center of the drawing is already covered by a round platelet of bone (10), which has been partly formed by the osteoblasts on the surface of the trabeculae.

Normal bone formation is an extremely complex process in which several hormones (growth hormone, parathormone, thyroxine, calcitonin, sex hormones, etc.) and vitamins (A, C, and D 3) are involved. (See biochemistry and physiology texts for further information.)

Magnification: × 10,000

REFERENCES

Bab IA, Muhlrad A, Sela J (1979) Ultrastructural and biochemical study of extracellular matrix vesicles in normal alveolar bone of rats. Cell Tissue Res 202:1–7
Felix R, Fleisch H (1976) Role of matrix vesicles in calcification. Fed Proc 35:169–171
Ornoy A, Atkin I, Levy J (1980) Ultrastructural studies on the origin and structure of matrix vesicles in bone of young rats. Acta Anat 106:450–461
Rohr H-P (1965) Die Kollagensynthese in ihrer Beziehung zur submikroskopischen Struktur des Osteoblasten. Virchows Arch [Pathol Anat] 338:342–354
Thyberg J, Friberg U (1970) Ultrastructure and acid phosphatase activity of matrix vesicles and cytoplasmic dense bodies in the epiphyseal plate. J Ultrastruct Res 33:554–573

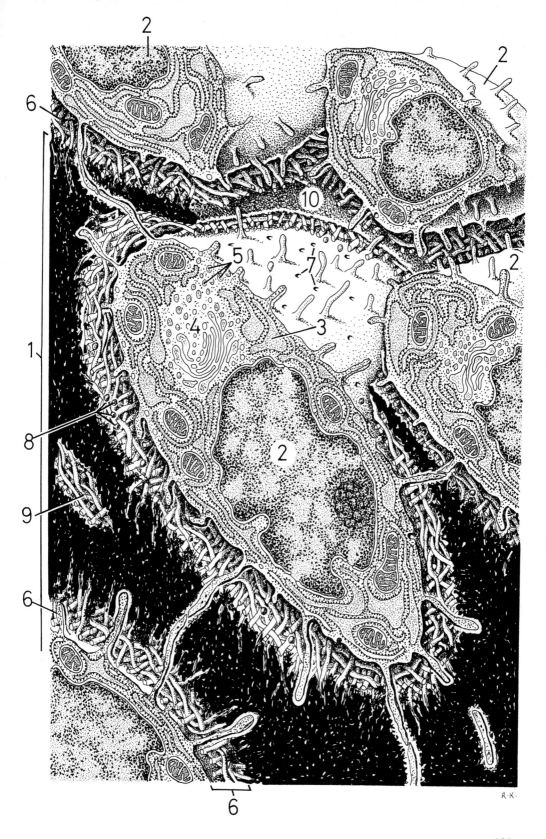

Plate 95. Bony Tissue. Indirect or Chondral Bone Formation

In chondral bone formation, direct perichondrial ossification precedes destruction of the cartilaginous tissue. Replacement bone develops by indirect (or endochondral) ossification where the hyaline cartilage becomes absorbed.

A summary of the processes of ossification and their nomenclature is given in the following scheme:

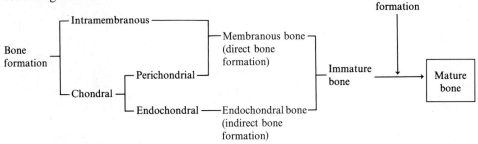

As stated in Plate 83, the skeleton is initially composed of hyaline cartilage. Indirect ossification is described below with the example of a long bone formation of an extremity.

The two ends of a long piece of cartilage, the epiphyses (Figs. A1, B1), can be distinguished from the middle section, the diaphysis or shaft (Figs. A2, B2). By appositional and interstitial growth (see Plate 83) the piece of cartilage enlarges (Fig. B). At a particular moment, which is genetically determined for every mammalian species (in humans it is about the 7th week of embryonic life), perichondrial ossification begins in the form of a collar at the level of the diaphysis. It proceeds according to the same mechanism of direct bone formation as described in Plates 91–94.

The directly ossified periosteal collar, also termed the periosteal bony band (Fig. B3), surrounds the diaphysis and provides mechanical stability for the cartilage until its destruction. However, the periosteal bony band also hinders the diffusive processes which are important in supplying the cartilage with nutrients.

A section of the periosteal bony band is enlarged in Fig. C. On the outer surface, toward the perichondrium, the osteoblasts (Fig. C1) form an epithelioid layer. Several of these cells will soon become enclosed within the calcifying bony matrix and differentiate into osteocytes (Fig. C2). The perichondrium is then termed the periosteum. The surface (Fig. C3) on the cartilage side is smooth.

Magnifications: Figs. A, B, × 65; Fig. C, × 600

REFERENCES

Bagnall KM, Harris PF, Jones PRM (1982) A radiographic study of the longitudinal growth of primary ossification centers in limb long bones of the human fetus. Anat Rec 203:293–299

Vaughan JM (1981) The physiology of bone. Clarendon, Oxford

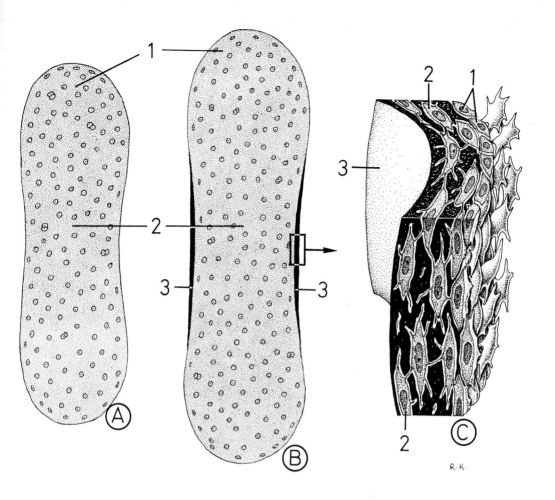

Plate 96. Bony Tissue. Indirect or Chondral Bone Formation. Continuation of Plate 95.

As the periosteal bony band (Fig. **A 1**) extends toward the epiphyses (Fig. **A 2**), the cartilage cells within the diaphysis (Fig. **A 3**) become larger and richer in glycogen and lipids. Although they still produce alkaline phosphatase, sulfated proteoglycans, and matrix vesicles, the cartilage matrix is reduced and soon becomes calcified. This calcification and the periosteal bony band cause the nutritive processes to become completely interrupted in the diaphysis. Thus, the hypertrophied cartilage cells die. Such an area in the hyaline cartilage visible to the naked eye is termed a primary ossification center.

Enlargement of chondrocytes and their lacunae leads to lateral and longitudinal tensions. Since the diameter of the diaphysis initially remains constant due to the periosteal bony band, interstitial proliferation of chondrocytes causes the two epiphyses to extend longitudinally. As a result, the cartilage cells are forced into a columnar arrangement.

A detail from the cartilage and primary ossification center is reproduced in Fig. **B**.

In the upper half of the drawing, enclosed within the cartilage matrix (Fig. **B 1**), the flattened, stacked chondrocytes (Fig. **B 2**) can be seen. The cartilage cells arranged in this manner comprise the proliferation zone (Fig. **B 3**). In the lower half of the illustration, the cartilage cells and their cavities become more hypertrophic and voluminous, and this is therefore termed the hypertrophic zone (Fig. **B 4**). Hydroxyapatite crystals (Fig. **B 5**) become concentrated in the intercellular substance and appear as black points in the transmission electron microscope.

Magnifications: Fig. **A**, ×65; Fig. **B**, ×2,000

REFERENCE

Salvkin HC, Croissant RD, Bringas P, Matosian P, Wilson P, Mino W, Guenther H (1976) Matrix vesicle heterogeneity: possible morphogenetic functions for matrix vesicles. Fed Proc 35:127–134

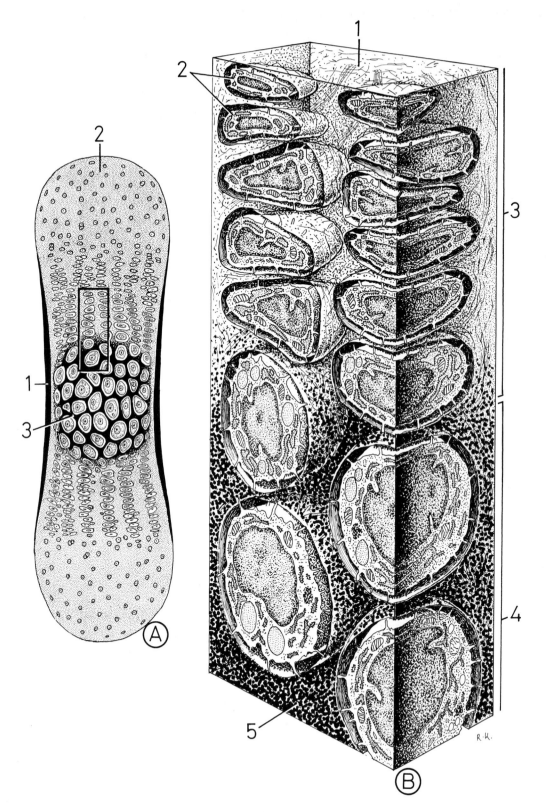

Plate 97. Bony Tissue. Indirect or Chondral Bone Formation. Continuation of Plate 96

At a later stage in the process, periosteal buds (Fig. A 2), made up of blood vessels and connective tissue, penetrate the diaphysis (Fig. A 1), which has been transformed into hypertrophic cartilage with a calcified cartilage matrix. First, however, the periosteal collar has to be perforated. The zone within the inset is enlarged in Fig. B to show the processes involved in greater clarity.

Some mesenchymal cells from the periosteal cambium (Fig. B 1) differentiate into polynuclear chondroclasts (Fig. B 2), which morphologically resemble osteoclasts (see Plate 111). An opening is created in the periosteal bony band (Fig. B 3) by the osteolytic enzymes of these cells. A capillary loop (Fig. B 4) then enters through this aperture. At the apex of the loop is a polynuclear structure (Fig. B 5) of endothelial origin, and it is believed that this structure also possesses a certain osteolytic effect, enhanced by the continual vascular pulsation. The blood vessels are accompanied by many actively dividing mesenchymal cells (Fig. B 6). Many of these cells will differentiate into the stem cells (Fig. B 7) of the primitive bone marrow, others into osteoblasts.

These processes take place in humans at about the 7th week of embryonic development, at a time therefore when red blood cells (Fig. B 8) are still nucleated.

Magnifications: Fig. A, ×65; Fig. B, ×1,000

REFERENCES

Ascenzi A, Bonucci E, Bocciarelli DS (1967) An electron microscope study on the primary periosteal bone. J Ultrastruct Res 18:605–618

Young RW (1962) Cell proliferation and specialization during endochondral osteogenesis in young rats. J Cell Biol 14:357–370

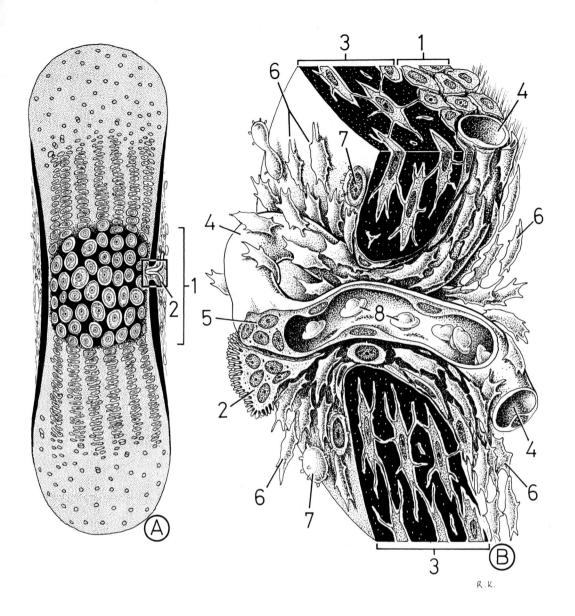

Plate 98. Bony Tissue. Indirect or Chondral Bone Formation. Continuation of Plate 97.

The cartilage-destroying effects of the chondroclasts cause a cavity to develop in the diaphysis, the primitive medullary or marrow cavity (Fig. **A 1**). A rich ramification of the blood vessels (Fig. **A 2**) that enter by the future foramen nutricium (Fig. **A 3**) takes place in this cavity. A three-dimensional illustration of the whole shaft is given in Fig. **B**.

The transverse section through the diaphysis shows the periosteal bony band (Fig. **B 1**) around the hyaline cartilage (Fig. **B 2**). The chondrocytes (Fig. **B 3**) have a columnar arrangement and are sectioned both vertically and horizontally. The periosteum (Fig. **B 4**) with numerous blood vessels (Fig. **B 5**) forms the outer coat of the periosteal bony band.

Through the opening in the periosteal bony band (foramen nutricium; Fig. **B 6**), blood vessels (Fig. **B 7**) and many mesenchymal cells (Fig. **B 8**) enter the primitive marrow cavity (Fig. **B 9**). The mesenchymal cells will soon differentiate into osteoblasts and gradually cover the entire medullary cavity in the form of an epithelioid layer.

The edges of the primary marrow cavity are very irregular due to the many dead chondrocytes (Fig. **B 10**) and their lacunae (Fig. **B 11**) destroyed by the activity of chondroclasts (Fig. **B 12**).

Magnifications: Fig. **A**, ×65; Fig. **B**, ×250

REFERENCES

Herring GM (1970) A review of recent advances in the chemistry of calcifying cartilage and bone matrix. Calcif Tissue Res 4 (Suppl 17)

Hohling HJ, Kreilos R, Neubauer G, Boyde A (1971) Electron microscopy and electron microscopical measurements of collagen mineralization in hard tissue. Z Zellforsch 122:36–52

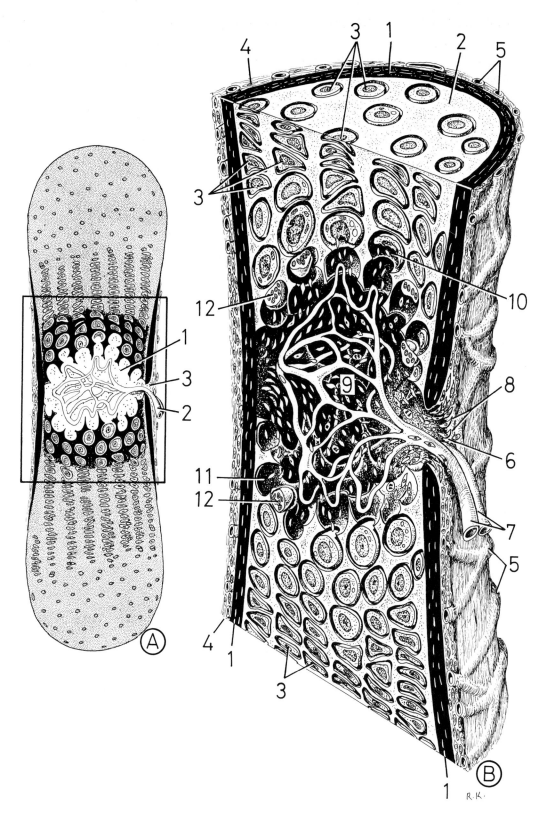

Plate 99. Bony Tissue. Indirect or Chondral Bone Formation. Continuation of Plate 98

At a somewhat later stage of endochondral ossification (in humans at about the 4th–5th month of fetal life) the following situation exists. The piece of hyaline cartilage (Fig. A) has in the meantime grown longer as a result of cartilage growth, and the limit of calcification (Fig. A 1) roughly presents a straight line. Distinct zones can now be seen between the epiphyseal cartilage and primitive medullary or marrow cavity. The area within the inset is enlarged in Fig. B.

The following zones can be observed, proceeding from top to bottom:

I. The reserve zone of hyaline cartilage not yet affected by ossification. The structure corresponds to that described in Plates 84 and 88.

II. The zone of proliferation or multiplication of chondrocytes (cartilage cell columns). As previously stated, cartilaginous epiphyses have no possibility of lateral extension. Thus, chondrocytes, which derive from a mother cell after vigorous mitoses (Fig. B 1), become stacked in a columnar fashion, which causes numerous isogenic cartilage cell groups to develop parallel to the axis of the future long bone.

III. The zone of hypertrophic cartilage and calcification. Here, the cartilage cells enlarge and accumulate cell organelles, alkaline phosphatase, and glycogen. Following advanced mineralization of the cartilage matrix, the cells gradually degenerate and/or become destroyed by the chondroclasts.

IV. The zone of cartilage erosion. Numerous chondroclasts (Fig. B 2), which destroy the thin uncalcified intercellular partitions (Fig. B 3) perpendicular to the bone axis, are characteristic of this zone.

V. The zone of endochondral ossification. In this zone, the activities of the chondroclasts are important in assisting the spreading of the capillaries and osteoblasts. However, many longitudinally oriented plate- or needlelike remains of calcified cartilage (Fig. B 4) persist in this zone and serve as attachment sites for vast numbers of osteoblasts (Fig. B 5). By means of the mechanism described in Plate 94, osteoblasts then form the osteoid (Fig. B 6), which becomes immature bone following mineralization. The first endochondral bony spicules (Fig. B 7) and osteocytes (Fig. B 8) develop in this way. Capillary loops (Fig. B 9) occur between the newly formed spicules. Directly and indirectly formed immature bones can be easily distinguished: Unlike membranous bone, bone that develops through indirect ossification – endochondral or cartilage bone – contains remnants of calcified hyaline cartilage. This differentiation can only be made prior to subsequent secondary bone formation.

In the epiphyses, as earlier in the diaphysis (see Plate 96), hypertrophy of the chondrocytes occurs with calcification of the cartilage matrix, and thus the epiphyseal ossification centers (Fig. A 2) develop. However, at the time of birth only two such centers are present – in the distal femur and in the proximal tibial epiphysis. (See Plates 175, 176 in KRSTIĆ 1979.)

Magnifications: Fig. A, ×65; Fig. B, ×500

REFERENCE

Dixon AD, Sarnat BD (eds) (1982) Factors and mechanisms influencing bone growth. Progress in clinical biological research, vol 101. Liss, New York

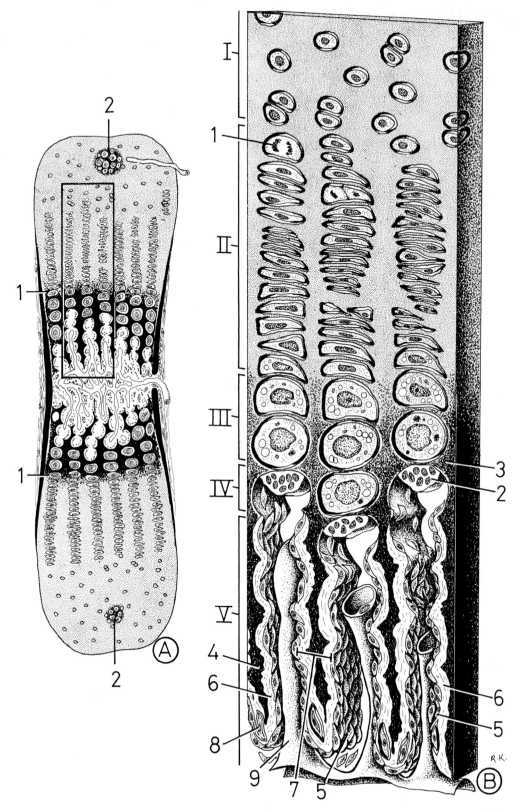

Plate 100. Bony Tissue. Indirect or Chondral Bone Formation. Continuation of Plate 99

The zones of hypertrophied and calcified cartilage, of erosion, and of endochondral ossification are depicted simultaneously in this plate.

Two very large hypertrophic chondrocytes (**1**) occur in the voluminous lacunae of hypertrophic cartilage. The intercellular substance has accordingly accumulated hydroxyapatite and therefore appears black.

In the zone of erosion, a thin, uncalcified transverse partition (**2**) begins to succumb to the action of a chondroclast (**3**). Another chondroclast (**4**) has already made its way into a lacuna, where it has largely destroyed the hypertrophic chondrocyte (**5**).

Since chondroclasts rapidly erode the noncalcified walls of lacunae, the lateral walls remain as mineralized cartilaginous debris (**6**). It is to this irregular surface that the osteoblasts (**7**) become attached and secrete osteoid (**8**). The endochondral i.e. cartilage bone develops following its mineralization.

In the transverse section, a capillary (**9**) and several osteoblasts (**7**) can be seen in the canal-like lacunae. The osteoblasts line the walls of the lacunae.

Magnification: × 1,800

REFERENCE

Rabinovitch LA, Anderson HC (1976) Biogenesis of matrix vesicles in cartilage growth plates. Fed Proc 35:112–116

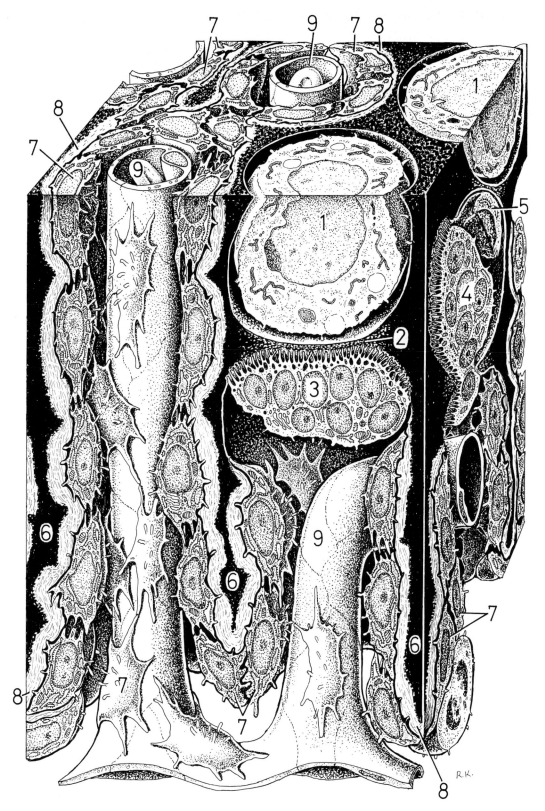

Plate 101. Bony Tissue. Indirect or Chondral Bone Formation. Epiphyseal Ossification Centers and Epiphyseal Plate. Continuation of Plate 99

As already mentioned in Plate 99 (Fig. A), the capillary buds with the mesenchymal cells penetrate the epiphyses, where subsequently each one forms an epiphyseal center of ossification. This plate depicts the process two- and three-dimensionally.

As bone formation of the diaphysis continues and the piece of skeleton becomes longer, cavities (Figs. A1, B1) develop in the epiphyseal cartilage, according to the same mechanism as seen earlier in the diaphysis (see Plates 96–98). Such epiphyseal ossification centers are visible to the naked eye as yellowish dots. Just as in the diaphysis, the blood vessels (Figs. A2, B2) that penetrate the epiphyseal cavities bring with them mesenchymal cells (Fig. B3), which differentiate into osteoblasts. Some vessels (Figs. A4, B4) from the diaphysis come into contact with vessels of the epiphyseal centers. The center of ossification consists of a zone of calcified hyaline cartilage (Figs. A5, B5), in which – like the diaphysis – vesicular, hypertrophied, degenerated cartilage cells (Figs. A6, B6) are found. The intercellular substance present between the chondrocytes becomes eroded by chondroclasts, and the calcified remnants serve as sites of attachment for osteoblasts. The activity of the osteoblasts and subsequent mineralization lead to the formation of endochondral bone (Fig. B7). Epiphyseal bone formation is a significantly longer process than diaphyseal ossification.

Ossification of the epiphyseal centers spreads in a centrifugal manner, though it never reaches the articular surface (Fig. B8) free of periosteum (Figs. A9, B9). Between the areas of diaphyseal and epiphyseal ossification, there is a noncalcified zone of hyaline cartilage, the epiphyseal plate or epiphyseal disc (Fig. B10), which can be clearly visualized in X-rays. Up until about the age of sexual maturity, the epiphyseal plate can compensate cartilage destruction caused by both the diaphyseal and epiphyseal processes of ossification by interstitial growth (proliferation of chondrocytes). Nevertheless, this zone becomes progressively thinner as a result of continuous loss of cartilage. The area within the inset in Fig. B is enlarged in Plate 102.

Magnifications: Fig. A, ×65; Fig. B, ×150

REFERENCES

Ali SY (1976) Analysis of matrix vesicles and their role in the calcification of epiphyseal cartilage. Fed Proc 35:135–142

Althoff J, Quint P, Krefting ER, Hoehling HJ (1982) Morphological studies on the epiphyseal growth plate combined with biochemical and X-ray microprobe analyses. Histochemistry 74:541–552

Barckhaus RH, Hoehling HJ (1978) Electron microscopical microprobe analysis of freeze dried and unstained mineralized epiphyseal cartilage. Cell Tissue Res 186:541–549

Laczko J, Levai G, Varga S (1976) Preliminary observations of the epiphyseal plate by scanning electron microscope. Jeol News 13:20–23

Schenk RK, Wiener J, Spiro D (1968) Fine structural aspects of vascular invasion of the tibial epiphyseal plate of growing rats. Acta Anat (Basel) 69:1–17

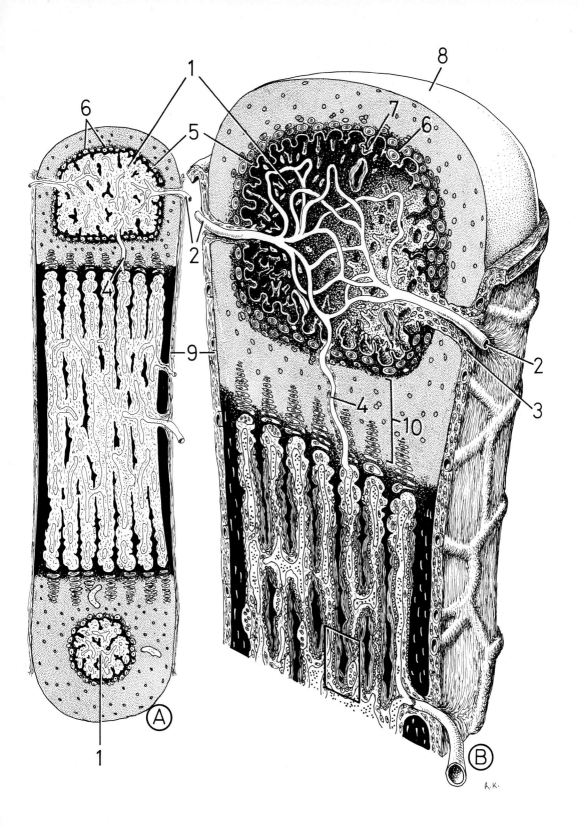

Plate 102. Bony Tissue. Indirect or Chondral Bone Formation. Endochondral or Cartilage Bone. Continuation of Plate 101

In the middle of a spicule (**1**) of endochondral bone is located a remnant of calcified hyaline cartilage (**2**), which has served as a site of attachment for the osteoblasts. The first osteoblasts to have arrived can be seen near the calcified cartilage as osteocytes (**3**) that have already become embedded in the bone matrix. On the surface of the spicule, there is a new layer of osteoblasts (**4**), which secrete osteoid (**5**), like the first generation of osteoblasts. This layer will also become enclosed in the bone matrix and differentiate into osteocytes. The initially thin spicule becomes thicker and transforms into a trabecula, owing to the continuous influx of new osteoblasts (**6**).

Ossification takes place parallel to a process of bone erosion effected by osteoclasts (**7**). The function of these polynuclear cells is to remodel the bone and provide it with a mechanically competent form.

The capillary (**8**) in the top left corner assures the transport of the substances necessary for synthesis of bone. (See Plates 5, 138, 139 in Krstić 1979.)

Magnification: ×2,000

REFERENCE
Urist MR (1976) Biochemistry of calcification. In: Bourne GH (ed) The biochemistry and physiology of bone, vol 4, 2nd edn. Academic, New York

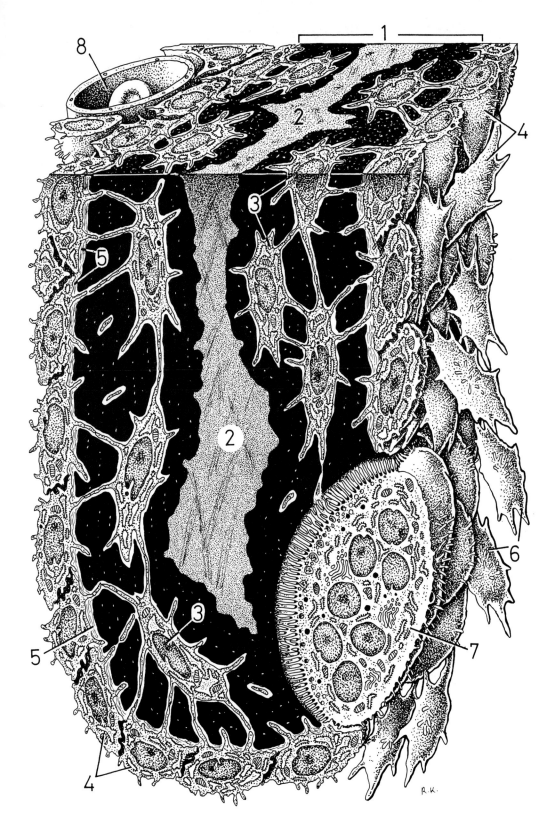

Plate 103. Bony Tissue. Indirect or Chondral Bone Formation. End of Longitudinal Growth and Articular Cartilage

When the supply of new cartilage cells through interstitial growth can no longer compete with the resorptive processes, the epiphyseal cartilage (Fig. **A 1**) is perforated and the epiphyseal cavities become continuous with the marrow cavity (Fig. **A 2**) of the diaphysis. It is clear that long bones can only increase in length if there are sufficient chondrocytes capable of proliferation in the epiphyseal cartilage. Longitudinal growth of the skeleton terminates with the disappearance of the epiphyseal plate.

After ossification has been completed, however, a 0.2- to 6-mm-thick layer of hyaline cartilage, articular cartilage (Figs. **A 3**, **B 3**), does persist at both ends of the long bone.

The area within the upper inset in Fig. **A** is viewed three-dimensionally in Fig. **B**.

The scanning electron microscope reveals a fine bas-relief (Fig. **B 4**) on the surface of the articular cartilage, corresponding to the underlying flattened chondrocytes (Fig. **B 5**). The articular surface appears macroscopically smooth and lustrous.

The uppermost zone consists of uncalcified hyaline cartilage (Fig. **B 6**). The collagen fibrils (Fig. **B 7**) near the surface are seen, by means of the polarizing microscope, to run initially parallel to the articular surface, the so-called tangential zone or gliding layer. They then curve down, penetrate an undulating well-calcified functional limit (Fig. **B 8**) between uncalcified and calcified cartilage (Fig. **B 9**), where they terminate. The border between articular cartilage and bone (Fig. **B 10**), the so-called histological limit, has an irregular course, which strengthens the junction of both tissues and prevents the articular cartilage from slipping. The collagen fibrils of articular cartilage do not penetrate the bone, as bone develops its own fibrillar system during endochondral ossification.

Articular cartilage is subject to great strain in the body and contains more chondroitin sulfates than other forms of hyaline cartilage that do not undergo compression. The isogenic groups and the functionally determined course of the collagen fibrils give a certain degree of elasticity to articular cartilage.

The area within the inset indicated in Fig. **A** by an arrow is presented three-dimensionally in the next plate to clarify diametric growth of bone.

Magnifications: Fig. **A**, × 65; Fig. **B**, × 500

REFERENCES

Ghadially FN, Moshurchak EM, Thomas I (1977) Humps on young human and rabbit articular cartilage. J Anat 124:425–435

Hough AJ, Banfield WG, Mottram FC, Sokoloff L (1974) The osteochondral junction of mammalian joints, an ultrastructural and microanalytic study. Lab Invest 31:685–695

Minns RJ, Steven FS (1977) The collagen fibril organisation in human articular cartilage. J Anat 123:437–457

Ratcliffe A, Fryer PR, Hardingham TE (1984) The distribution of aggregating proteoglycans in articular cartilage: Comparison of quantitative immunoelectron microscopy with radioimmunoassay and biochemical analysis. J Histochem Cytochem 32:193–201

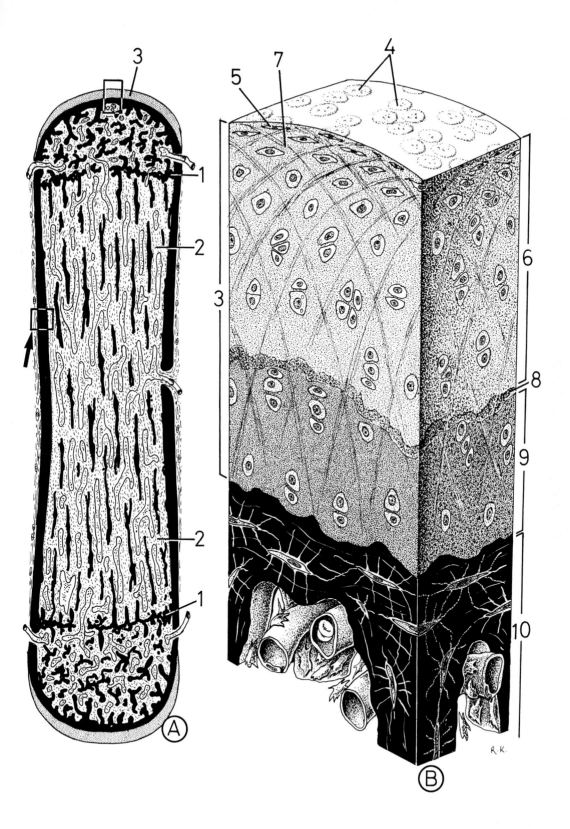

Plate 104. Bony Tissue. Diametric Growth. Continuation of Plate 103

The increase in bone diameter is brought about by subperiosteal appositional accumulation of new directly developed layers of bone with simultaneous bone resorption on the inner surface. For the sake of clarity, the section from Plate 103 (Fig. **A**, inset) has been drawn as viewed from the marrow cavity.

Two osteoclasts (**1**) and sporadic osteoblasts (**2**) are lying on the inner surface of the periosteal bony band (**3**). A few reticular fibers (**4**) and a capillary (**5**) can also be seen in this region. Other cells have been omitted from the drawing.

In the periosteal bony band, several osteocytes (**6**) are located and form a cell framework by means of their processes.

On the outer surface, several young cells (**7**), originating from the periosteal cambium, accumulate at the bone wall. These cells transform into osteoblasts (**8**) and form osteoid (**9**), i.e., new layers of bone, according to the mechanism of direct ossification described in Plates 91–94.

Since the majority of bone-forming cells occur on the periosteal side, new bone matrix increasingly accumulates here. To prevent the periosteal bony band (at a later stage, the substantia compacta) from becoming progressively wider, osteoclasts (**1**) resorb bone at the interior surface, thereby enlarging the medullary cavity. It will be evident from this description that diametric growth of long bones proceeds according to the same principle as the increase in skull volume (cf. Plates 92, 93).

Magnification: × 1,800

REFERENCES

Kallio DM, Garant PR, Minkin C (1972) Ultrastructural effects of calcitonin on osteoclasts in tissue culture. J Ultrastruct Res 39:205–216

Vaes G (1968) On the mechanism of bone resorption. The action of parathyroid hormone on the excretion and synthesis of lysosomal enzymes and on the extracellular release of acid by bone cells. J Cell Biol 39:676–697

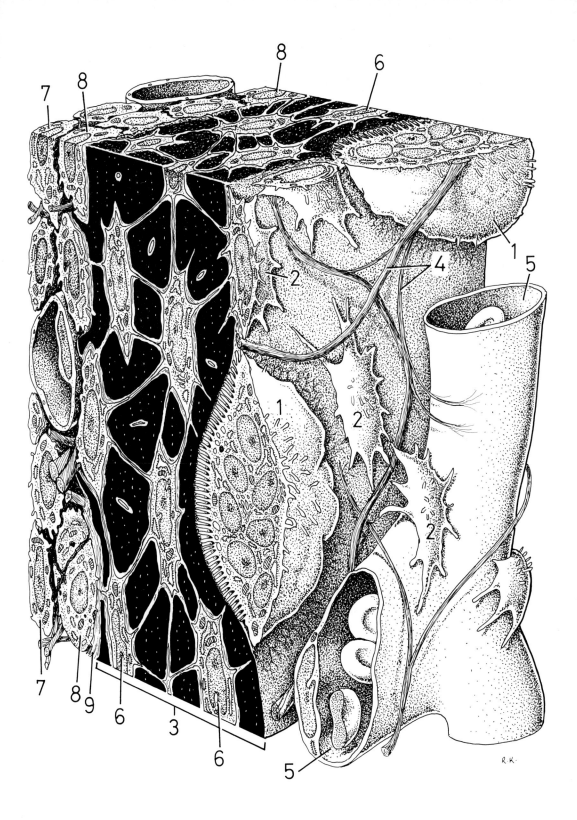

Plate 105. Bony Tissue. Secondary Bone Formation

The bundles of collagen fibrils in the bone matrix do not run in a functionally dependent manner during and immediately after direct and chondral ossification. This immature or woven bone is unable to withstand the high mechanical strains to which the organism is subject after birth. Therefore, this type of bone is replaced by structurally and dynamically highly differentiated mature or lamellar bone. The following three light-microscopic sections depict this process.

At the same time as immature bone (1) develops, osteoclasts (3) appear near the capillaries (2) and form longitudinal absorption cavities (4) in their immediate vicinity. The perivascular or Haversian spaces (5) thus formed are penetrated by osteoblasts (6) in addition to elements of primary bone marrow.

The bone-forming cells congregate at the periphery of the absorption cavities and deposit the first lamellae of bone (7). The concentric lamellae gradually constrict the perivascular spaces, which leads to the characteristic concentric appearance of an osteon (8) with its Haversian canal (9) in the center. As can be seen in the upper drawing, several processes of bone formation and erosion occur simultaneously. Osteons that have already formed can also become eroded. In the middle drawing, a new generation of osteons (10) can be seen which have partially replaced their predecessors. New perivascular absorption cavities (4) with osteoclasts (3) have formed in the woven bone (1).

The lower drawing corresponds to the appearance of a bone composed of several generations of osteons. The first concentric lamellar systems cannot be recognized as such, since they have been completely replaced by new generations of osteons (10). However, remnants of the earlier systems can be seen as interstitial lamellae (11). Secondary bone formation is continuous with the process of internal bone remodeling, through which bone constantly adapts, structurally and dynamically, to external and internal factors. Bone formation is subject to numerous hormonal influences, which reach the cells through the agency of cAMP. (See biochemistry, histology, and physiology texts for further information.)

Magnification: × 70

REFERENCES

Amprino R (1963) On the growth of cortical bone and the mechanism of osteon formation. Acta Anat (Basel) 52:177–187

Bordier PJ (1973) Aspects histologiques du remaniement osseux. Triangle 12:85–98

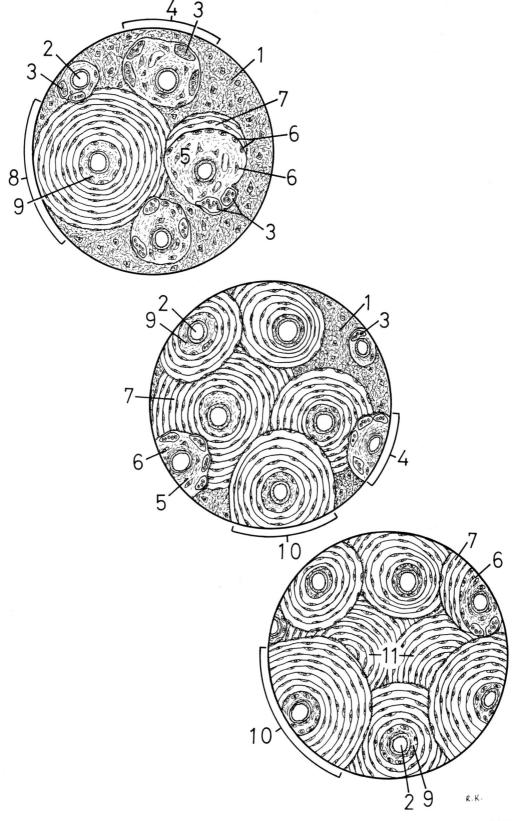

Plate 106. Bony Tissue. Structure

The structure of mature bone can best be studied in the thick compact bone or substantia compacta (Fig. A 1) of long bones. After sectioning a bone with a saw, it is possible to separate the periosteum (Fig. A 2) from the bone. A foramen nutricium (Fig. A 3) admits vessels, which are important for the vascularization of the blood-forming tissue, into the definitive medullary cavity (Fig. A 4).

The cancellous bone or substantia spongiosa (Fig. A 5) is significantly better developed in the epiphyses than in the diaphysis. The trabeculae of cancellous bone are composed of mature bone and their orientation is in accordance with structural and dynamic stress. The section of compact bone in the inset is enlarged in Fig. B and Plate 108.

A transverse section through the bone reveals first of all the well-vascularized periosteum (Figs. B 1, C 1, D 1), which is rich in collagen fibers. Beneath this layer, there is a lamellar system parallel with the surface, the outer circumferential lamellae (Figs. B 2, C 2, D 2). The osteons (Figs. B 3, C 3) characterize the middle layer of compact bone. The spaces between the osteons are filled with interstitial lamellae (Figs. B 4, C 4). As previously stated, these are remnants of earlier, partly eroded generations of osteons. The compact bone terminates at the medullary cavity (Fig. B 5) with an inconstant system of inner circumferential lamellae (Fig. B 6) continuous with cancellous bone. In Fig. C, corresponding to the inset in Fig. B, blood vessels penetrating the bone matrix through the transversely oriented Volkmann's canals (Figs. C 7, D 7) can be clearly seen.

In transverse section (Fig. C), the longitudinally oriented Haversian canals (Figs. C 8) appear as small openings surrounded by concentric bone lamellae. A Haversian canal with the surrounding system of concentric bone lamellae is termed an osteon (Fig. C). It can be observed that the Volkmann's canals (Figs. C 7, D 7) do not possess concentric lamellae. Even at relatively low magnification with appropriate staining, spiderlike cells, the osteocytes (Figs. C 9, D 9), are evident between the lamellae.

A longitudinal section (Fig. D) through the compact bone displays the periosteum (Fig. D 1) and the vascular system of the bone. The "perforating" transverse Volkmann's canals (Fig. D 7) connect the longitudinally directed Haversian canals (Fig. D 8) with one another and with the blood vessels of the periosteum. The compact bone is supplied with blood in this manner.

Magnifications: Fig. B, × 30; Figs. C, D, × 100

REFERENCE

Singh I (1978) The architecture of cancellous bone. J Anat 127:305–310

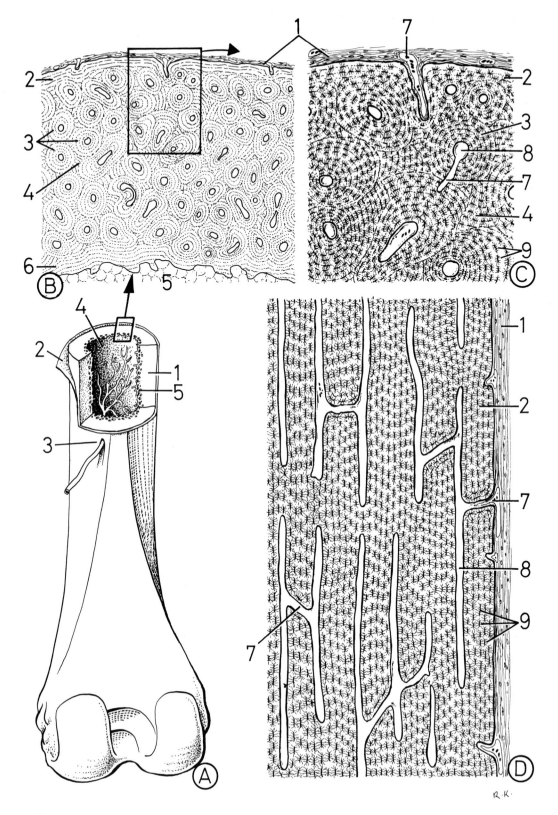

Plate 107. Bony Tissue. Osteons or Haversian Systems of Lamellae

An osteon (Fig. **A**) is a morphofunctional unit of the compact bone 100–500 μm in diameter and 0.5–2 cm long with a Haversian canal (Fig. **A1**) in the center. The canal is surrounded by about 3–15 concentrically arranged bony lamellae 10–20 μm thick with collagen fibrils running in a corkscrewlike manner. The orientation of the fibers varies from lamella to lamella. With increased light-microscopic magnification, punctate (Fig. **A2**) and striated (Fig. **A3**) lamellae are seen, according to the direction of the fibers.

Spindle-shaped bone cells, the osteocytes (Fig. **A4**), are found in or between the lamellae. A considerable number of long, only 1-μm-thick processes give these cells their characteristic spiderlike appearance. Osteocytes of the same and neighboring lamellae are connected by these cell processes, which run through minute canaliculi.

The osteocytes of the innermost lamella communicate with the Haversian canal. This is of great importance for the nutrition of the osteocytes, since diffusion cannot take place in bony tissue owing to the calcified intercellular substance. Thus, metabolic processes are carried out almost exclusively by means of the tissue fluid, which flows through the canaliculi from one osteocyte to another.

The osteons are separated from one another, and especially from the interstitial lamellae, by a particularly distinct cement line (Fig. **A5**). Only sporadic osteocytic processes manage to penetrate this boundary layer and make contact with the osteocytes of neighboring osteons (Fig. **A6**).

The course of the collagen fibers in the osteons becomes evident in polarized light. The punctate and striated lamellae comprising the osteons, as seen in Fig. **A**, appear as regularly alternating light and dark rings (Fig. **B1**) under the polarizing microscope. If an osteon appears dark in polarized light this usually signifies a vertical system of fibers (Fig. **B2**)

Magnifications: Fig. **A**, × 500; Fig. **B**, × 120

REFERENCES

Frasca P, Harper RA, Katz JL (1981) Scanning electron microscopy studies of collagen, mineral and ground substance in human cortical bone. Scan Electron Microsc 1981/III: 339–346

Schaffler, MB, Burr DB, Frederickson RG (1985) X-ray microprobe analysis of osteon cement lines. Anat Rec 211:168A–169A

Schenk RK (1974) Ultrastruktur des Knochens. Verh Dtsch Ges Pathol 58:72–83

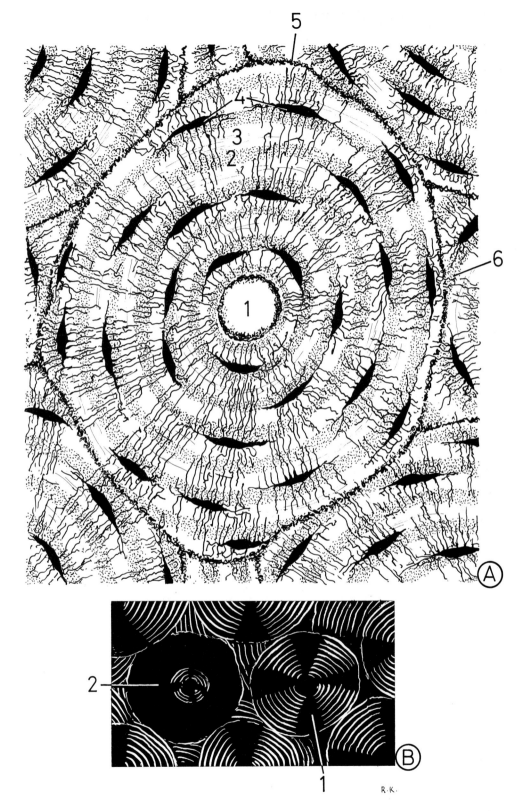

Plate 108. Bony Tissue. Structure of Mature Bone. Three-Dimensional View

The wall of a long bone, the substantia compacta or compact bone (1), is surrounded by periosteum (2), the vascular network (3) of which is responsible for vascularization of the bony tissue. A locally variable number of strong collagen fiber bundles, Sharpey's fibers (4), connect the bone substance with the periosteum.

The transversely oriented Volkmann's canals (5) with arterial branches perforate the outer circumferential lamellae (6) and open into the vessels of the Haversian canals (7). The compact bone is largely made up of densely packed osteons (8), which are in the form of narrow, 2-cm-long cylinders. Whereas the boundaries of the osteons can be easily recognized in transverse section because of the cement lines, these boundaries are not so distinct in longitudinal section. The collagen fibers of adjacent bone lamellae (9) run in different directions, as can be seen in one of the isolated osteons.

Several inner circumferential lamellae (10) are adjacent to the marrow cavity and continuous with the trabeculae (11) of the cancellous bone (12), which does not contain osteons and therefore receives nutrients by diffusion from the marrow cavity. The marrow surfaces of both the compact and cancellous bone are lined by a thin layer of flattened cells with potential bone-forming capacities, the endosteum (not shown).

In the diaphysis of a long bone, the cancellous bone is irregular and poorly developed. In the epiphyses, its trabeculae are oriented according to the structural and dynamic stress to which the bone is subjected (so-called trajectories).

Magnification: ×150

REFERENCES

Boyde A, Hobdell M (1969) Scanning electron microscopy of lamellar bone. Z Zellforsch 93:213–231

Horn V, Dvorak M (1974) Ultrastruktur der funktionellen Knochenkomponenten in der Raster- und Transmissionselektronenmikroskopie. Z Mikrosk Anat Forsch 88:836–848

Whitehouse WJ, Dyson ED (1974) Scanning electron microscope studies of trabecular bone in the proximal end of the human femur. J Anat 118:417–444

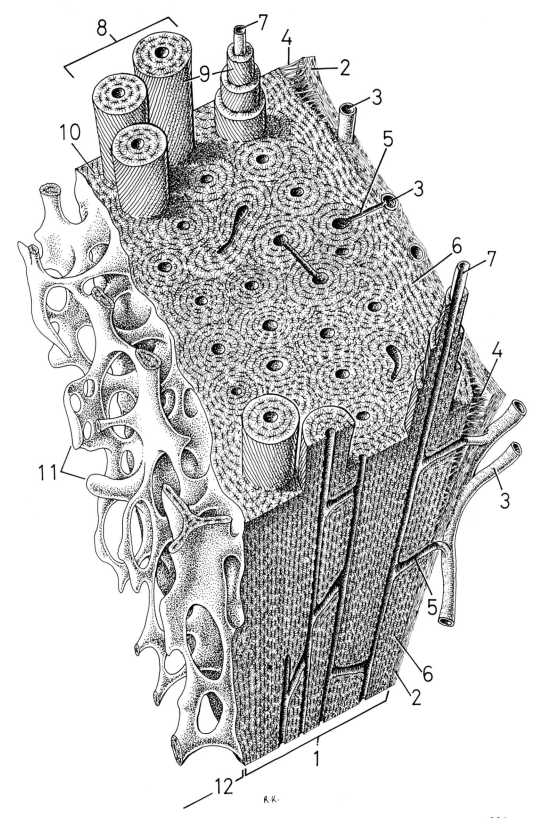

Plate 109. Bony Tissue. Central Part of an Osteon

Two concentric bone lamellae (1, 2) with interjacent osteocytes (3) surround a Haversian canal, through which runs a blood capillary (4). The perivascular space (5) contains osteoblasts (6; see Plate 100), which are less numerous in older than in younger osteons. Inactive osteoclasts (7), macrophages (8), nerve fibers (9), and lymphatic vessels may also occur in Haversian canals in addition to the blood vessels.

The bundles of collagen fibrils within bone lamellae do not run in an undulating manner, as in connective tissue, but are in the form of taut wires since they are already subject to tension, i.e., they are "prestretched." If a compressive load is applied and leads to a slight compression of the bone, the tensile stress of the collagen fibrils is reduced, but because of the prestretching the fibrils do not buckle.

The largely parallel collagen fibrils of a bone lamella are wound in spiral fashion, such that the fibrils of two adjacent lamellae cross more or less at right angles. This results in greater strength and plasticity of bone. Between neighboring lamellae, there is a complicated exchange of fibrils, which will not be examined here.

The presence of osteoblasts and osteoclasts in the perivascular spaces of connective tissue is of great importance in regeneration of bone (see Plate 112).

Magnification: ×1,800

REFERENCES

Hayek E (1967) Die Mineralsubstanz der Knochen. Klin Wochenschr 45:857–863

Hoehling HJ, Steffens H, Ashton BA, Nicholson WAP (1974) Molekularbiologie der Hartgewebebildung. Verh Dtsch Ges Pathol 58:54–71

Jande SS, Bélanger LF (1971) Electron microscopy of osteocytes and the pericellular matrix in rat trabecular bone. Calcif Tissue Res 6:280–289

Talmage RV (1970) Morphological and physiological considerations in a new concept of calcium transport in bone. Am J Anat 129:467–476

Thurston TJ (1982) Distribution of nerves in long bones as shown by silver impregnation. J Anat 134:719–728

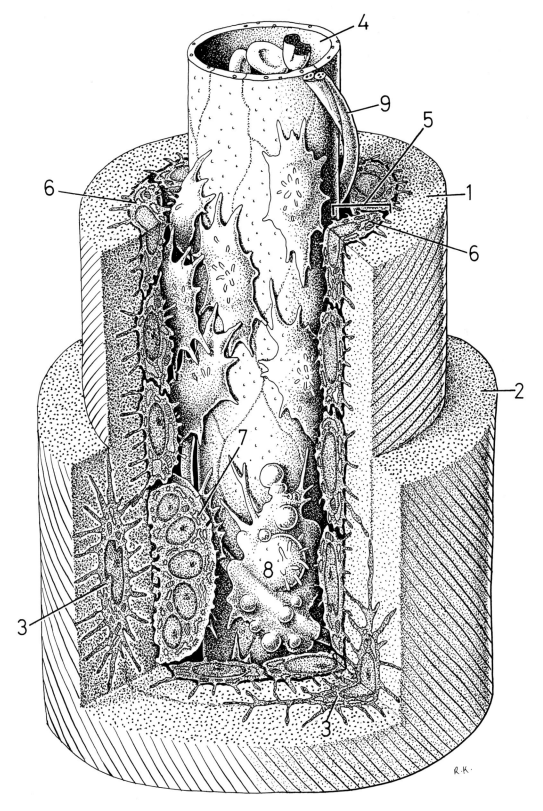

Plate 110. Bony Tissue. Osteocyte

Osteocytes (1) are stellate in transverse section (see Plate 107); in sections parallel to the Haversian canals they appear round or elliptical. The cell body contains a nucleus with a well-developed nucleolus. Within the cytoplasm are found mitochondria, some cisternae of rough endoplasmic reticulum, and a Golgi apparatus. The number and distribution of all cell components can vary considerably according to the histophysiological activities of the osteocytes (collagen synthesis). Young and active osteocytes naturally have better developed cell organelles.

Osteocytes are usually situated between, occasionally within, the bone lamellae (2). They lie in lacunae of calcified bone matrix (3), here drawn punctate or striated, and are surrounded by a 1- to 2-µm-wide zone of nonmineralized collagen microfibrils (4).

Numerous, very long, thin cytoplasmic processes (5), which run through the bone lamellae in narrow canaliculi (6), are characteristic of osteocytes; they are in contact with the processes of other osteocytes. A space, which is only 0.1 µm wide, separates the processes from the calcified bone matrix. Tissue fluid enters the canaliculi from the perivascular spaces of the Haversian canals and brings vital nutrients to the osteocytes.

The intersecting courses of the bundles of collagen fibrils (7) in the bone lamellae can also be seen in this plate. (See Plate 139 in KRSTIĆ 1979.)

Magnification: × 15,000

REFERENCES

Baud CG (1968) Submicroscopic structure and functional aspects of the osteocyte. J Clin Orthop 56:227–236

Ejiri S, Ozawa H (1982) Scanning electron microscopic observations of rat tibia using the HCl-collagenase method. Arch Histol Jpn 45:399–404

Jande SS (1971) Fine structural study of osteocytes and their surrounding bone matrix with respect to their age in young chicks. J Ultrastruct Res 37:279–300

Schulz A, Delling G (1976) Die Wirkung einer Calcitonin-Langzeittherapie auf Knochenzellen und Knochenmineralisation bei der Ratte. Virchows Arch [Pathol] 369:229–238

Wassermann F, Yaeger JA (1965) Fine structure of the osteocyte capsule and the wall of the lacunae in bone. Z Zellforsch 67:636–652

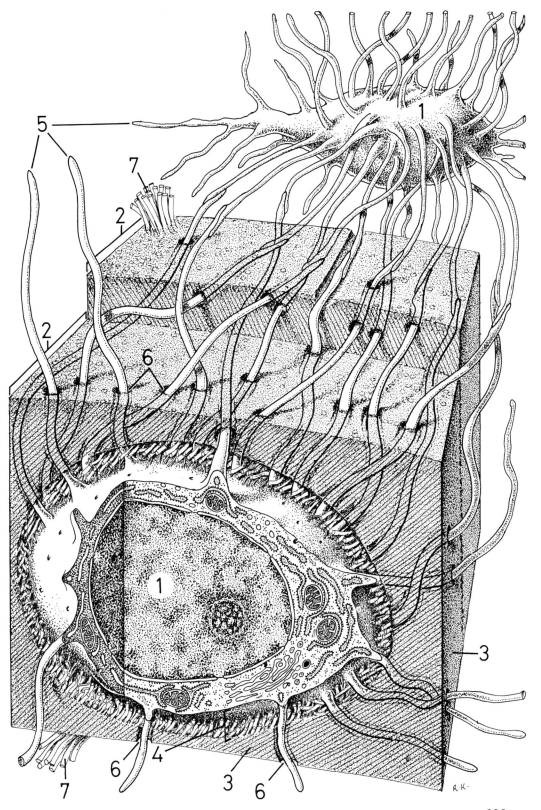

Plate 111. Bony Tissue. Osteoclast

Chondroclasts and osteoclasts differentiate from mesenchymal and possibly also endothelial cells. It is known that vascular endothelia can exert osteoclastic effects (cf. Plate 97, Fig. B, perforation of the periosteal collar by a capillary loop).

Chondroclasts and osteoclasts are giant polynuclear cells (20–100 µm) whose function is to erode cartilaginous and bony tissue, respectively; however, there are no structural differences between the two types of cell. In addition to multiple nuclei (5–50), the cytoplasm contains numerous mitochondria, cisternae of rough endoplasmic reticulum, multiple Golgi complexes, and many free ribosomes. A number of lysosomes, phagolysosomes, and residual bodies are also found in the cell.

Actively absorbing osteoclasts have a well-developed ruffled border made up of irregular, branching, microvillous cytoplasmic processes (**1**), producing an enormously increased area of contact with the bone matrix. Hydroxyapatite crystals (**2**) can be recognized between the microvilli. Following phagocytosis, the crystals make their way into phagolysosomes (**3**), where they are broken down by enzymes. Resorption of bone leads to the formation of hemispherical shallow cavities with an irregular polygonal relief, Howship's lacunae (**4**). The belt of cytoplasm (**5**) on the opposite side of the cell to the ruffled border is free of organelles, though still active. It is here that numerous irregular, predominantly microvillous processes (**6**) develop and resorptive processes of the surrounding area take place.

Osteoclasts are ameboid cells. Their morphology is highly variable according to cytophysiological demands. They can adopt both a very active form with the ruffled border and a rounded shape with only a slightly undulating plasmalemma (see Plate 109).

The effect of active osteoclasts is considerable: A single osteoclast is capable of resorbing what 100 osteoblasts have synthesized.

Osteoclasts are under the direct influence of parathyroid hormone. Increased secretion of this hormone leads to extensive absorption of bone (osteoporosis). (See Plate 5 in KRSTIĆ 1979.)

Magnification: × 4,500

REFERENCES

Addison WC (1979) Enzyme histochemical characteristics of human and kitten odontoclasts and kitten osteoclasts: a comparative study using whole cells. Histochem J 11:719–735

Hall BK (1975) The origin and fate of osteoclasts. Anat Rec 183:1–11

Jones JS, Boyde A (1977) Some morphological observations on osteoclasts. Cell Tissue Res 185:387–397

Kallio DM, Garant PR, Minkin C (1971) Evidence of coated membranes in the ruffled border of the osteoclast. J Ultrastruct Res 37:169–177

Knese KH (1972) Osteoklasten, Chondroklasten, Mineraloklasten, Kollagenoklasten. Acta Anat (Basel) 83:275–288

Stanka P, Bargsten G, Herrmann G (1981) Woher kommen die vielkernigen Osteoklasten? Verh Anat Ges 75:237–238

Takagi M, Parmley RT, Toda Y, Denys FR (1982) Extracellular and intracellular digestion of complex carbohydrates by osteoclasts. Lab Invest 46:288–297

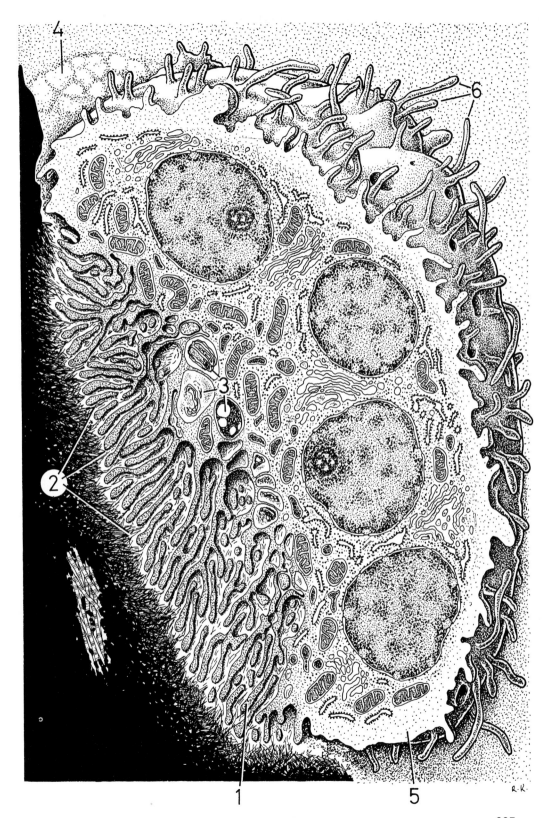

Plate 112. Bony Tissue. Regeneration

The main requirements for successful regeneration of bone are the presence of a histophysiologically active periosteum and excellent immobilization of the fragments.

After a bone has been fractured and following optimal surgical setting of the fragments by means of screws, wires, etc. (osteosynthesis Fig. A), young cells of the periosteal cambium and Haversian canals (see Plate 109) and some endosteal cells proliferate in the directions of the arrows and differentiate into osteoblasts, which synthesize new bone. The injured and dead bone is resorbed by osteoclasts formed in situ or activated osteoclasts. This primary healing is not accompanied by a callus formation and continues as secondary bone formation (see Plate 105).

In the case of nonoptimal fixation of the fragments (Fig. B), superficial cells of the periosteum differentiate under relatively avascular conditions into chondroblasts and chondrocytes, which produce a fibrocartilaginous callus (Figs. B1, C1). Cells of the inner periosteal layer, young cells of the Haversian canals, and cells of the endosteum differentiate into osteoblasts, which under good vascularization produce bone trabeculae (Fig. B2).

As in indirect bone formation, the fibrocartilaginous callus is gradually eroded and replaced by a bony callus (Fig. D), which naturally requires a good deal of time.

The type and location of fracture, age of the individual, and the therapeutic method applied affect the duration of fracture healing. It is obvious that osteosynthesis offers the greatest likelihood of achieving perfect regeneration. Poorly set bone fragments regenerate very slowly or sometimes not at all (so-called pseudoarthrosis).

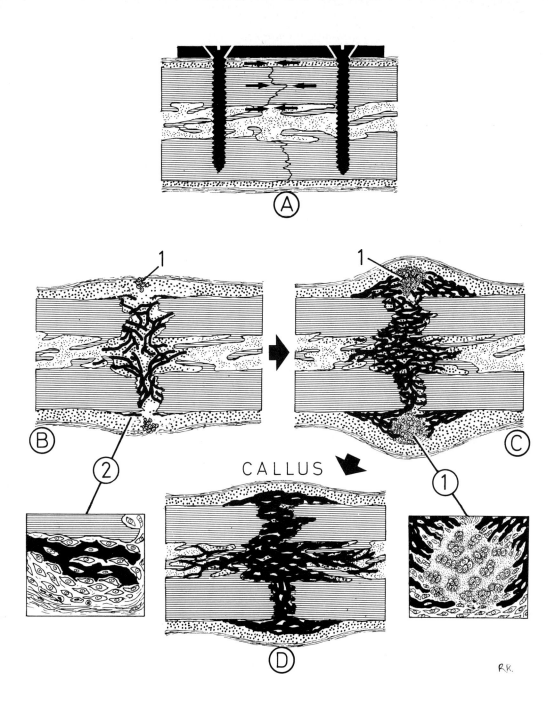

Plate 113. Dentin and Odontoblasts. Cementum

The two components of teeth, dentin and cementum, are closely allied to bony tissue.

A longitudinal ground preparation through a tooth reveals the pulp cavity (Fig. A 1), dentin (Fig. A 2), enamel (Fig. A 3), and cementum (Fig. A 4). The large inset in Fig. A corresponds to Fig. B.

In the interior of the pulp cavity (Fig. B 1), an epithelioid layer of tall, cylindrical cells, the odontoblasts (Figs. B 2, C), can be observed. The processes of these cells, Tomes' fibers (Fig. B 3), penetrate the mineralized dentin (Fig. B 4) and run in an undulating, S-shaped pattern as far as the enamel (Fig. B 5). A few odontoblasts are enlarged in Fig. C.

Despite their appearance, odontoblasts are not of epithelial origin but, like osteocytes, develop by differentiation from mesenchymal cells (see Plate 48). Unlike osteocytes, odontoblasts never become embedded in hard substance, since their product is only secreted at the apical cell pole.

Odontoblasts are columnar, distinctly polarized cells: The ellipsoidal nuclei, with large nucleoli, lie in the basal portion of the cell; the apical pole narrows to form the Tomes' fibers (Fig. C 1). The cytoplasm of odontoblasts contains a considerable number of cisternae of rough endoplasmic reticulum, many mitochondria, free ribosomes, and a well-developed Golgi apparatus. Vacuoles with striated contents (Fig. C 2) occur in the Golgi zone. A short distance above these appear vacuoles with dense, coarse-granular contents (Fig. C 3), which move to the apical pole and reach the cytoplasmic processes as highly osmiophilic secretory granules (Fig. C 4). All three types of inclusion are secretory products of the odontoblasts – collagen and dentinal matrix proteoglycans at various stages of maturity. Dentinal matrix vesicles (Fig. C 5) cause the organic material secreted by the cell to become calcified. This does not take place in the immediate vicinity of the cell but at a distance of a few micrometers, and thus it is possible to differentiate the slightly mineralized predentin (Fig. C 6) from the dentin (Fig. C 7), which is completely impregnated with calcium salts.

The Tomes' fibers extend as far as the enamel in narrow channels, the dentinal tubules (Fig. C 8).

Evidence that odontoblasts are not epithelial cells is supplied by the presence of capillaries (Fig. C 9) between the cells.

Dental cementum (Figs. A 4, D) occurs at the root of a tooth; the structure of the cementum is similar to that of bony tissue. Between the irregular, calcified lamellae (Fig. D 1) of cementum are scattered the cementocytes (Fig. D 2), which with their cytoplasmic processes are very similar to osteocytes. Sharpey's fibers (Fig. D 3) penetrate the exterior surface of the cementum, whose interior border contacts dentin (Fig. D 4).

In the region of the neck of the tooth (Fig. A, arrow) the cementum becomes thinner and acellular. (See Plate 99 in KRSTIĆ 1979.)

Magnifications: Fig. A, × 5;
Fig. B, × 40; Fig. C, × 5,000;
Fig. D, × 250

REFERENCES

Almuddaris MF, Dougherty WJ (1979) The association of amorphous mineral deposits with the plasma membrane of pre- and young odontoblasts and their relationship to the origin of dentinal matrix vesicles in rat incisor teeth. Am J Anat 155:223–244

Goldberg M, Escaig F (1981) Odontoblastes: collagène dans la prédentine et la dentine de l'incisive de rat. Etude par cryofracture. Biol Cell 40:203–216

Katchburian E, Severs NJ (1982) Origin of matrix vesicles of developing dentine investigated by freeze-fracture. J Anat 134:615

Weinstock M (1981) Gap junctions in the odontoblasts of the rat incisor teeth. Anat Rec 199:270A

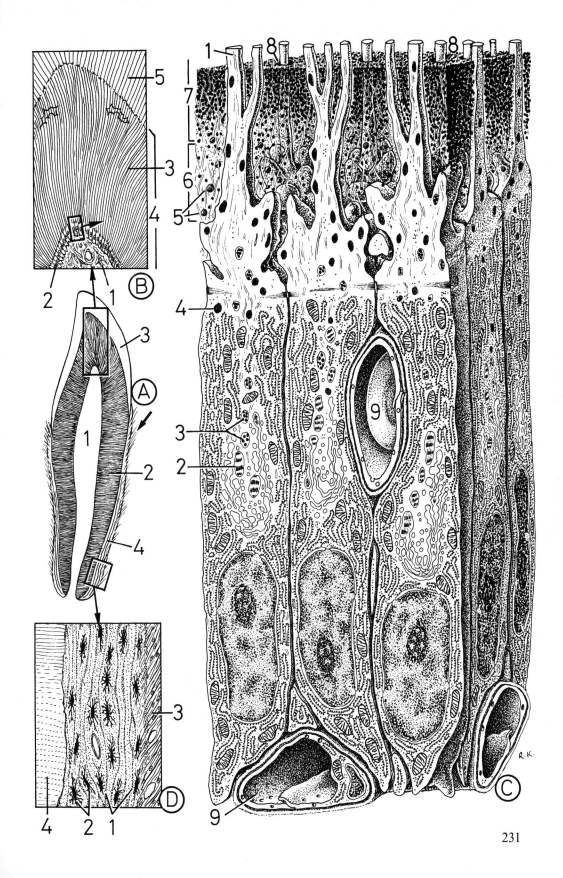

231

Plate 114. Possibilities of Transplanting Connective and Supporting Tissues

In principle, connective and supporting tissues can be grafted successfully, particularly in autografting. This plate shows a few examples.

Underdevelopment of the mammary gland can be corrected by cosmetic surgery by transferring adipose tissue from the buttocks to the breasts of the same patient (Figs. **A–C**).

If it is necessary to replace dura mater (Fig. **D1**) or articular capsules (Fig. **D2**), the dense connective tissue of the fascia lata (Fig. **D3**) is employed.

The cornea of the eye can be successfully homografted. The material for such operations (keratoplasty) derives from corpses. Figures **E–H** show some of the commonest forms of corneal transplantation.

Since tendons (Fig. **I**) possess a high degree of regenerative power, they can be autografted relatively easily. However, there are some problems associated with grafting tendons with synovial vaginae.

Cartilaginous tissue (Fig. **J**), provided that the perichondrium is intact, can be successfully grafted. The transplant is taken from the costal cartilage of the same patient.

Autoplastic transplantation of bone can also be accomplished successfully. Thin periosteum-containing platelets of bone (Fig. **K2**) are fixed between the specially prepared bone fragments (Figs. **K1, L1**). These platelets do not become integrated into the structure at the damaged site, instead they stimulate and direct bone formation (Fig. **L2**; Plate 112). These bone chips are gradually resorbed upon release of the material necessary for ossification (Fig. **L**).

(Surgery texts provide details about further possibilities of transplantation.)

REFERENCES

Fernando NV, Movat HZ (1963) Fibrillogenesis in regenerating tendon. Lab Invest 12:214–229

Larsson Å (1973) Studies on dentinogenesis in the rat. Z Anat Entwickl Gesch 142:103–115

Weinstock M (1977) Centrosymmetrical crossbanded structures in the matrix of rat incisor predentin and dentin. J Ultrastruct Res 61:218–229

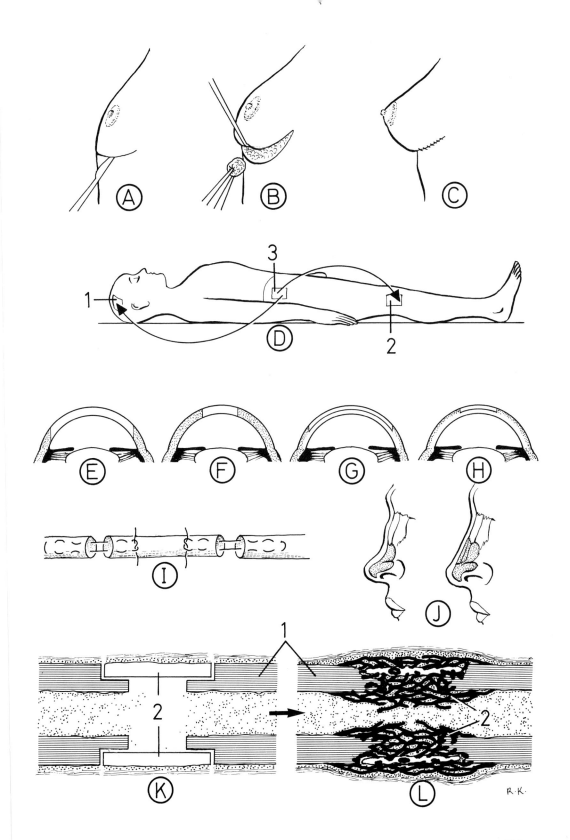

IV Muscular Tissue

MUSCULAR TISSUE

Plate 116. Three Types of Muscular Tissue

In the body of mammals, three types of muscular tissue occur. Since they all contain myofilaments in their cells or fibers, they possess the ability to contract.

Striated, skeletal voluntary musculature (Fig. **A**) is the most widely distributed type in the body. Striated cardiac musculature (Fig. **B**) is independent of the will. Smooth musculature (Fig. **C**) is largely found in the walls of hollow organs and, like cardiac musculature, is involuntary.

Muscular tissue can be summarized as follows (after BUCHER 1977):

Muscular tissue develops almost exclusively from mesoderm. Myoepithelial cells, which derive from ectoderm, are described with the smooth muscle cells on account of their ability to contract. The two iris muscles are of neuroectodermal origin and thus constitute another exception.

Magnifications: Figs. **A–C**, × 300

REFERENCE
Bucher O (1973) Diagnostic et diagnostic differentiel en cytologie et en histologie normales. Masson, Paris; Huber, Berne

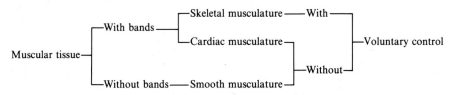

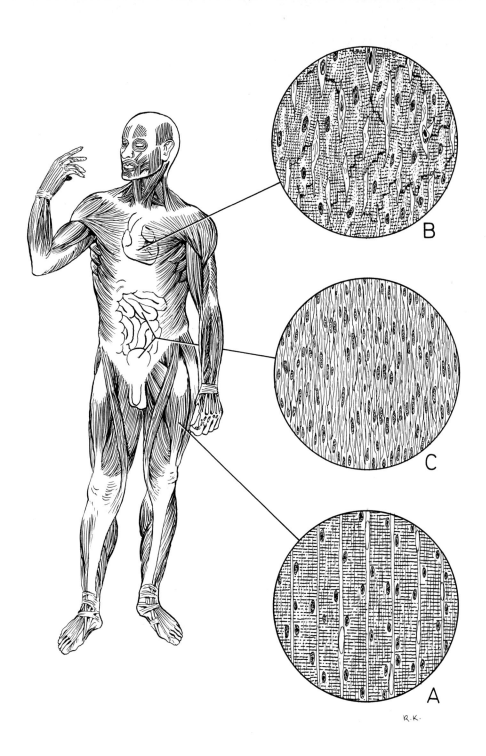
239

Plate 117. Occurrence of Smooth Musculature

Smooth musculature is made up of spindle-shaped muscle cells (**1**). It forms, in particular, the muscular coats (tunicae musculares) of the hollow organs whose inner diameters can be altered. This plate gives, without aiming to be complete, a general impression of the occurrence of smooth muscular tissue.

Smooth musculature is found from the middle of the esophagus (**2**) to the anus, in the trachea (**3**), bronchi (**4**), and bronchioli (**5**), in blood and lymphatic vessels (**6**), in certain areas under the endocardium (**7**) of the heart, in the gallbladder (**8**), ureter (**9**) and urinary bladder, ductus deferens (**10**), stroma of the prostate (**11**), in trabeculae between cavernous sinuses of the corpus cavernosum penis (**12**), in the ductus epididymidis (**13**), etc.

In the female sex organs, smooth musculature occurs in the vagina (**14**), uterus (**15**), and oviduct (**16**).

REFERENCES

Barr L, Berger W, Dewey MM (1968) Electrical transmission at the nexus between smooth muscle cells. J Gen Physiol 51:347–368

Coob JLS, Bennett T (1969) A study of nexuses in visceral smooth muscle. J Cell Biol 41:287–297

Cooke PH, Fay FS (1971) Correlation between fiber length, ultrastructure, and the length-tension relationship of mammalian smooth muscle. J Cell Biol 52:105–116

Sawada H (1981) Three-dimensional observation on muscular tissues. Scan Electron Microsc 4:7–15

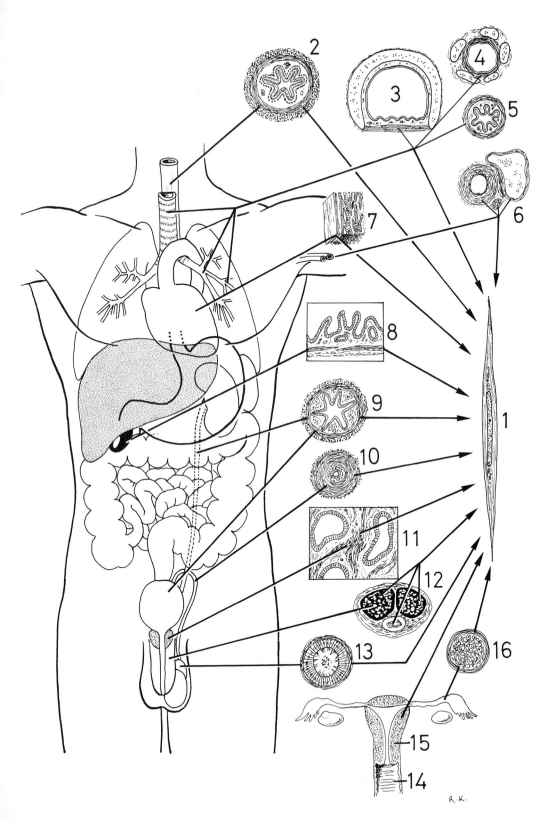

241

Plate 118. Smooth Musculature. Smooth Muscle Cells from the Tunica Muscularis of the Rat Small Intestine. Light-Microscopic Appearance

Smooth musculature occurs in hollow organs whose contents have to be transported by rhythmic contractions of the walls.

In a section from the small intestine, the intestinal villi (Fig. **A 1**) are most conspicuous. The intestinal crypts (Fig. **A 2**) open between the villi and are the sites of a high level of mitotic activity (see Plate 47). Beneath a thin layer of smooth muscle cells, the lamina muscularis mucosae (Fig. **A 3**) and tela submucosa (Fig. **A 4**), composed of loose connective tissue, is situated the tunica muscularis (Fig. **A 5**). The outer surface of the small intestine is covered with peritoneal mesothelium (Fig. **A 6**).

The tunica muscularis comprises two layers (an inner circular and an outer longitudinal) of smooth muscle cells and thus can be observed in both longitudinal and transverse sections (Fig. **B**). Under the light microscope, the spindle-shaped muscle cells appear "smooth," i.e., without striation. Most smooth muscle cells are 50–200 µm long and 5–15 µm wide. The nuclei of smooth muscle cells are rod-shaped and 10–25 µm long. In the contracted state they adopt a screwlike appearance. During pregnancy, some hypertrophied smooth muscle cells attain a length of 800 µm.

Smooth muscle cells are rarely branched at both ends. They arrange themselves such that their broad nucleus-containing portions (Fig. **B 1**) come to be located between the tapered, anuclear parts of neighboring cells. This gives rise to seemingly anuclear muscle cells in transverse sections (Fig. **B 2**). With increased magnification, a fine longitudinal striation (Fig. **B 3**) corresponding to the myofibrils becomes evident in the smooth muscle cells.

Bundles of muscle cells surrounded by loose fibrous connective tissue are termed "smooth muscle fibers" (see Plate 180).

Magnifications: Fig. **A**, × 100; Fig. **B**, × 400

REFERENCES

Bo WJ, Odor DL, Rothrock ML (1969) Ultrastructure of uterine smooth muscle following progesterone or progesterone-estrogen treatment. Anat Rec 163:121–132

Bülbring E, Branding AF, Jones AW, Tomita T (eds) (1981) Smooth muscle: An assessment of current knowledge. Arnold, London

Campbell GR, Chamley JH (1975) Thick filaments in vertebrate smooth muscle. Cell Tissue Res 156:201–216

Devine CE, Somlyo AP (1971) Thick filaments in vascular smooth muscle. J Cell Biol 49:636–649

Hinek A, Thyberg J (1977) Electron microscopic observations on the formation of elastic fibers in primary cultures of aortic smooth muscle cells. J Ultrastruct Res 60:12–20

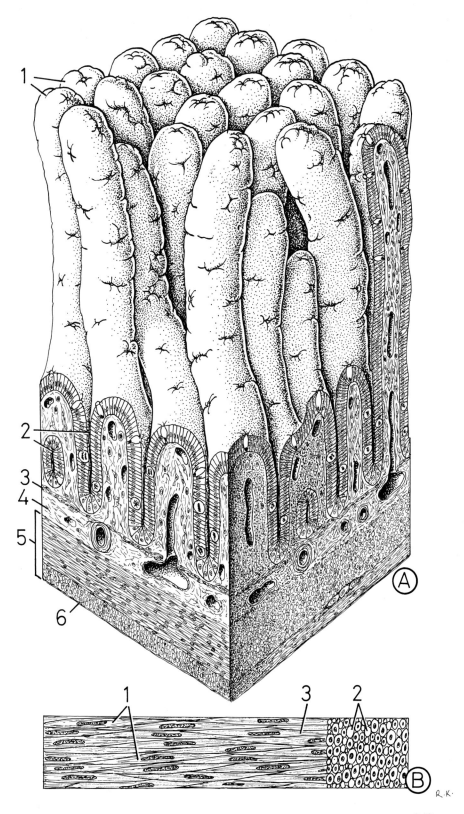

Plate 119. Smooth Musculature. Smooth Muscle Cells. Section from a Smooth Muscle Fiber

In the spindle-shaped body of a smooth muscle cell, there is a cylindrical, partially invaginated nucleus (Fig. **A 1**) with a distinct nucleolus. The endoplasm (Fig. **A 2**) is a myofilament-free zone at both poles of the nucleus and contains the mitochondria, rough and smooth endoplasmic reticulum, Golgi apparatus, lysosomes, lipofuscin granules, and glycogen particles. Countless myofilaments (Fig. **A 3**) run through the rest of the cytoplasm.

All muscle cells are surrounded by a basal lamina (Fig. **A 4**) and collagen and reticular microfibrils (Fig. **A 5**). In addition, occasional elastic fibers (Fig. **A 6**) branch between the cells. At the nexus (Fig. **A 7**), muscle cells form junctions with one another without an interposing basal lamina. The zones in the rectangular and circular insets are enlarged in Figs. **B** and **C**. Beneath the plasmalemma, here termed sarcolemma, (Figs. **B 1**, **C 1**), of muscle cells is located a thin belt of cytoplasm, here termed sarcoplasm (Figs. **B 2**, **C 2**), with ribosomes and smooth-walled tubules of endoplasmic reticulum (Figs. **B 3**, **C 3**). It is possible that these perform a similar function to the sarcoplasmic reticulum of skeletal musculature in storing Ca^{2+} ions during relaxation. This peripheral sarcoplasmic zone contains a large number of micropinocytotic vesicles (Figs. **B 4**, **C 4**). The basal lamina (Figs. **B 5**, **C 5**) occurs outside the sarcolemma. In the sarcoplasm and at the sarcolemma, osmiophilic, fine-granular dense areas appear, the so-called dense bodies (Figs. **B 6**, **C 6**).

Recent studies have shown that there are three types of microfilament in the cytoplasm of muscle cells:
- The thin (about 5 nm) actin myofilaments (Figs. **B 7**, **C 7**) are anchored to the dense bodies, which can thus be considered equivalent to Z-lines in striated muscle fibers.
- The roughly 10-nm-thick intermediate desmin microfilaments (Figs. **B 8**, **C 8**) run more or less as bundles from one dense body to another.
- The thick, short myosin myofilaments (Figs. **B 9**, **C 9**) have a diameter of about 15 nm. They can be identified considerably easier in contracted than in relaxed muscle cells. In transverse section, they resemble free ribosomes. (See Plates 6, 72, 73, 85, 113 in KRSTIĆ 1979.)

Magnifications: Fig. **A**, ×6,000; Figs. **B**, **C**, ×55,000

REFERENCES

Devine CE, Somlyo AV, Somlyo AP (1972) Sarcoplasmic reticulum and excitation-contraction in mammalian smooth muscles. J Cell Biol 52:690–718

Fay FS, Cooke PH (1973) Reversible disaggregation of myofilaments in vertebrate smooth muscle. J Cell Biol 56:399–411

Gabella G, Blundell D (1979) Nexuses between the smooth muscle cells of the guinea-pig ileum. J Cell Biol 82:239–247

Kelly RE, Rice RV (1968) Localization of myosin filaments in smooth muscle. J Cell Biol 37:105–116

Somlyo AP, Devine CE, Somlyo AV, North SR (1971) Sarcoplasmic reticulum and the temperature dependent contraction of smooth muscle in calcium free solutions. J Cell Biol 51:722–741

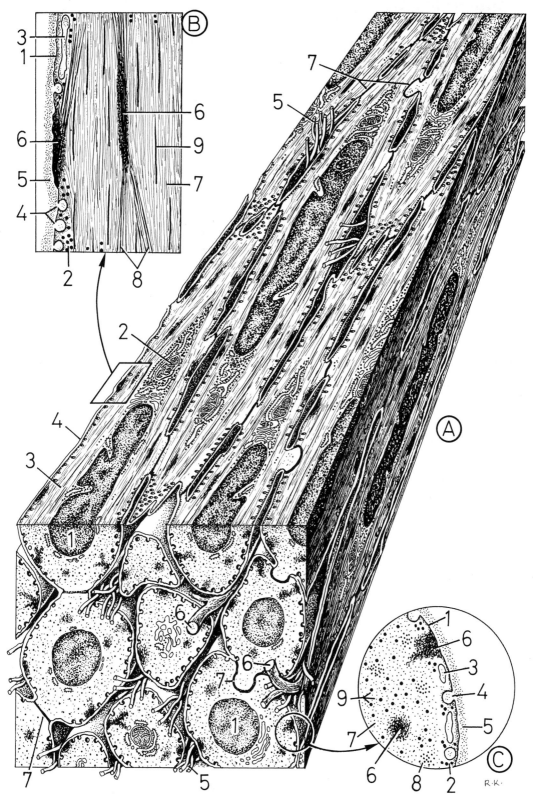

245

Plate 120. Smooth Musculature. Smooth Muscle Cells

This three-dimensional figure illustrates the plane of contact between the longitudinally and circularly coursing muscle layers of the tunica muscularis of the intestine. In addition to the nuclei (1) and endoplasm (2) with its organelles, the muscle cells can be seen to contain longitudinally oriented microfilaments (3), dense bodies (4), a peripheral zone of cytoplasm with micropinocytotic vesicles (5), and tubules of smooth endoplasmic reticulum (6).

If the basal lamina (7) is lifted back, the fairly regular arrangement of the micropinocytotic vesicles (5) comes into view. These vesicles occur everywhere in the plasmalemma except where the spindle-shaped dense bodies (4) contact the sarcolemma. Micropinocytotic vesicles and tubules of smooth endoplasmic reticulum (6) are particularly conspicuous in the tapered region of the muscle cell (arrow). This part of the cell usually has two or three short processes which, by means of the reticular microfibrils (8) anchored in the basal lamina, are connected with the collagen microfibrils (9) and elastic (10) fibers.

As stated in the previous plate, smooth muscle cells contact one another in the region of the nexus (11), the surface of which is characterized by hexagonally arranged, 8- to 9-nm-wide hemispherical particles (12).

Two kinds of nerve ending exist in the smooth musculature. The first type (13) occurs in large numbers, scattered between the cells, and does not form any contacts with muscle cells. The other type (14) is less frequent; it penetrates the basal lamina and contacts the sarcolemma. There is, however, no morphological differentiation in terms of a neuromuscular synapse. Both kinds of ending contain synaptic vesicles with osmiophilic core (see Plate 180).

Muscle cells can receive stimuli over their entire surface. Thus, for example, the neurotransmitter released by the free nerve endings induces contractions. Conductance of the bioelectrical impulses across the nexus regions, which can therefore be regarded as nonvesicular electrotonic synapses, enables all the muscle cells to contract synchronously.

Smooth muscle cells also have the ability to form reticular and collagen microfibrils and elastic fibers. Smooth musculature has no regenerative capacity. (See Plates 131, 151 in KRSTIĆ 1979.)

Magnification: × 15,000

REFERENCES

Nonomura Y (1968) Myofilaments in smooth muscle of guinea pig taenia coli. J Cell Biol 39:741–745
Oakes BW, Batty AC, Handley CJ, Sandberg LB (1982) The synthesis of elastin, collagen, and glycosaminoglycans by high density primary cultures of neonatal rat aortic smooth muscle. An ultrastructural and biochemical study. Eur J Cell Biol 27:34–46
Sawada H (1981) Scanning electron microscopy of guinea pig taenia coli. Biomed Res [Suppl] 2:153–158
Watanabe H, Yamamoto TY (1974) Freeze-etch study of smooth cells from vas deferens and taenia coli. J Anat 117:553–564
Yamamoto M (1977) Electron microscopic studies on the innervation of the smooth muscle and the interstitial cell of Cajal in the small intestine of the mouse and rat. Arch Histol Jpn 40:171–201

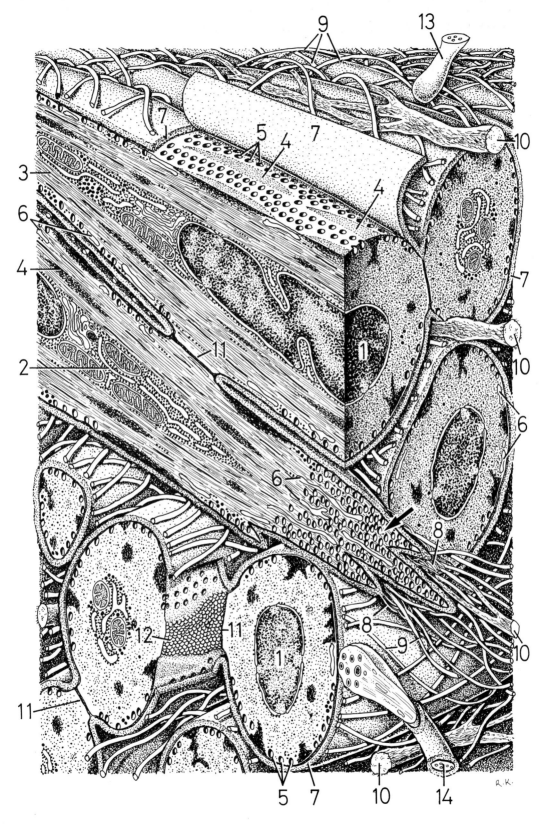

247

Plate 121. Smooth Musculature. Contraction of Smooth Muscle Cells

It has already been indicated that three types of microfibril occur in the sarcoplasm of muscle cells. Whereas actin (**1**) and myosin (**2**) myofilaments are responsible for shortening of the cell, the intermediate desmin microfilaments (**3**) and dense bodies (**4**) form an intracellular network which ensures the simultaneous contraction of the whole muscle cell. It should not be forgotten here that the actin myofilaments are also attached to the dense bodies.

Figure **A** shows a relaxed muscle cell with the rod-shaped nucleus. The contracted state is displayed in Fig. **B**. The force created by the interaction of actin and myosin myofilaments brings about an alteration in the position of the dense bodies within the cell. This change is transferred by the intermediate microfilaments to the neighboring dense bodies; muscle cells shorten their longitudinal axis and adopt a jagged outline. Depending on the strength of contraction, the nuclei of smooth muscle cells can become short and thick or, as here, take on a corkscrewlike form.

Recent investigations have indicated that the numerous micropinocytotic vesicles figure in eliciting the contraction of muscle cells by reducing the electrical resistance of the cell membrane.

REFERENCES

Gabella G (1976) Structural changes in smooth muscle cells during isotonic contraction. Cell Tissue Res 170:187–201

Kelly RE, Rice RV (1969) Ultrastructural studies on the contractile mechanism of smooth muscle. J Cell Biol 42:683–694

Rice RV, Moses JA, McManus GM, Brady AC, Blasik LM (1970) The organization of contractile filaments in mammalian smooth muscle. J Cell Biol 47:183–196

Small JV, Sobieszek A (1980) The contractile apparatus of smooth muscle. Int Rev Cytol 64:241–306

Stephens NL (ed) (1984) Smooth muscle contraction. Dekker, New York

Thaemert JC (1966) Ultrastructural interrelationships of nerve processes and smooth muscle cells in three dimensions. J Cell Biol 28:37–49

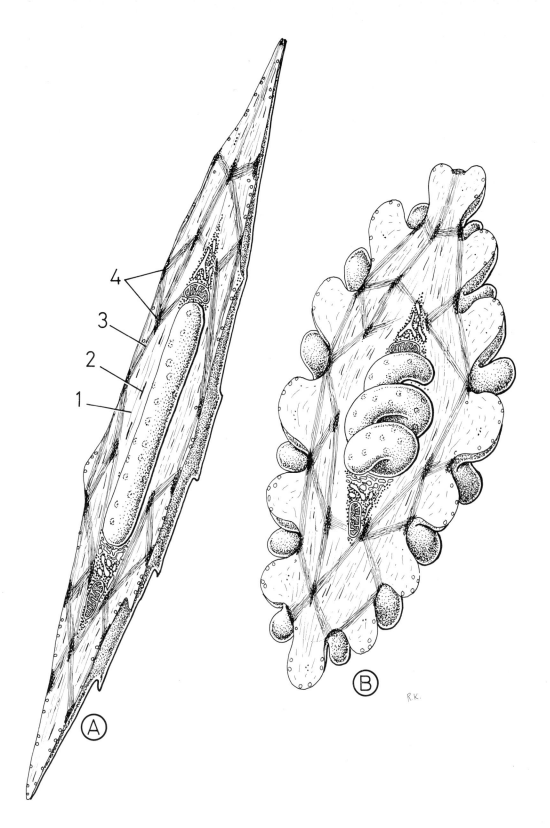

Plate 122. Smooth Musculature. Hypothetical Model of the Contraction Mechanism (After PANNER and HONIG 1967)

The contraction mechanism and the interaction between actin and myosin filaments in smooth muscle cells are far from being clear. A possible explanation is given in the following scheme. In Fig. **A**, the peripheral zone of the sarcoplasm of a muscle cell has been cut and removed so that the sarcolemma (Fig. **A 1**) and the associated dense body (Fig. **A 2**) can be seen. Further to the left, another ellipsoidal dense body (Fig. **A 3**) is situated free in the sarcoplasm. The two bodies are connected by means of actin myofilaments (Fig. **A 4**) and intermediate desmin microfilaments (here omitted). All dense bodies are composed of a dense, fine-granular, amorphous material, the contractile protein actinin, and are arranged parallel to the axis of the cell. They have a diameter of 0.2–0.5 µm and are up to 3 µm long and about 1–3 µm apart. It seems that they are able to contract actively.

The actin filaments anchored in the dense bodies or those that pass straight through have short, lateral 1.5- to 2.5-nm-thick branchings (Fig. **A 5**), which are only visible in negatively contrasted sections. These structures serve as junctions between the actin filaments and between the latter and the material of the dense bodies.

In a relaxed muscle cell, only sporadic, short myosin myofilaments (Figs. **B 2**, **C 2**) occur between the actin myofilaments (Figs. **B 1**, **C 1**). The dense bodies (Fig. **B 3**) are relatively far apart. During contraction, the actin filaments glide toward one another, with the assistance of the myosin units, such that the dense bodies draw closer together (Fig. **C 3**). As a result, the cell membrane becomes deformed, and the longitudinal axis (arrow) of the cell shortens. The movement of the dense bodies is transferred by means of intermediate desmin microfilaments (not shown; see Plate 121) to other points of fixation, which leads to a synchronous contraction of the muscle cell.

The fact that myosin filaments are found more easily and in larger numbers in the contracted than in the relaxed state supports the inference that the myosin units are broken down during muscular relaxation. For this reason, more myosin elements have been illustrated in Fig. **C** than in Fig. **B**.

REFERENCES

Bond M, Somlyo AV (1982) Dense bodies and actin polarity in vertebrate smooth muscle. J Cell Biol 95:403–413

Ebashi S, Maruyama K, Endo M (eds) (1980) Muscle contraction. Its regulatory mechanisms. Springer, Berlin Heidelberg New York

Panner BJ, Honig CR (1967) Filament ultrastructure and organization in vertebrate smooth muscle. Contraction hypothesis based on localization of actin and myosin. J Cell Biol 35:303–321

Panner BJ, Honig CR (1970) Locus and state of aggregation of myosin in tissue sections of vertebrate smooth muscle. J Cell Biol 44:52–61

Tsukita S, Tsukita S, Ishikawa H (1983) Association of actin and 10 nm filaments with the dense body in smooth muscle cells of the chicken gizzard. Cell Tissue Res 229:233–242

Uehara Y, Campbell GR, Burnstock G (1971) Cytoplasmic filaments in developing and adult vertebrate smooth muscle. J Cell Biol 50:484–497

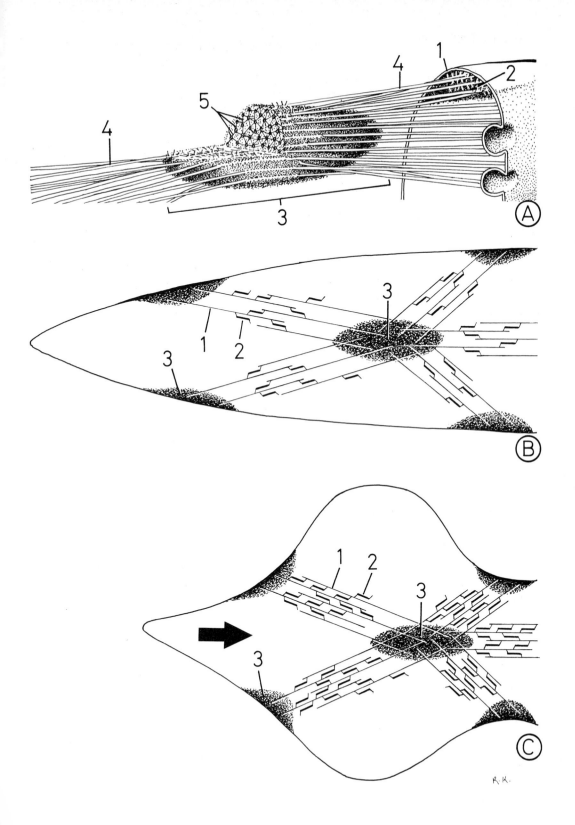

251

MUSCULAR TISSUE

Plate 123. Myoepithelial Cells from a Human Sweat Gland

As indicated in Plate 116, myoepithelial cells do not stem from the mesenchyme, but from the ectoderm. They are specially differentiated epithelial cells that resemble smooth muscle cells because of their faculty of contraction. For this reason, these cells are described here in the muscular tissue section.

In the form of stellate or parallel spindle-shaped cells, myoepithelial cells assist through their contraction in the drainage of secretory products in salivary, mammary, lacrimal, ceruminous, and sweat glands. A sweat gland is chosen to illustrate this type of cell here.

A few spindle-shaped myoepithelial cells (**1**) surround the terminal secretory portion of a sweat gland, which is made up of several cuboidal epithelial cells (**2**). Like smooth muscle cells, myoepithelial cells contain a rod-shaped nucleus and an endoplasm (**3**) with organelles. Myofilaments (**4**) and dense bodies (**5**) are conspicuous in the cytoplasm.

Myoepithelial cells always have a common basal lamina (**6**) with the secretory epithelium. The spicular cell processes (**7**) are attached by means of reticular microfibrils (**8**) to the surrounding collagen microfibrils (**9**). In this way, stable points of insertion are created for contraction of the myoepithelial cells, which helps to bring about expulsion of the secretory product. (See Plate 156 in KRSTIĆ 1979.)

Magnification: × 5,500

REFERENCES

Ellis RA (1965) Fine structure of the myoepithelium of the eccrine sweat glands of man. J Cell Biol 27:551–563

Murakami M, Sugita A, Abe J, Hamasaki M, Shimada T (1981) SEM observation of some exocrine glands with special reference to configuration of the associated myoepithelial cells. Biomed Res [Suppl] 2:99–102

Richardson KC (1962) The fine structure of autonomic nerve endings in smooth muscle of the rat vas deferens. J Anat 96:427–442

Ross R (1971) The smooth muscle cell: II. Growth of smooth muscle in culture and formation of elastic fibers. J Cell Biol 50:172–185

Tandler B (1965) Ultrastructure of the human submaxillary gland: III. Myoepithelium. Z Zellforsch 68:852–863

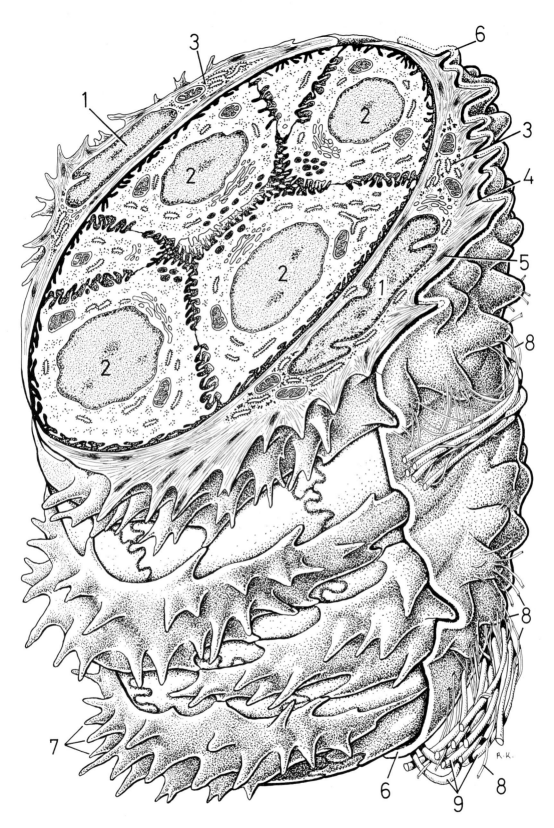

Plate 124. Histogenesis of Skeletal Musculature

At an early stage of development (in humans at about day 20), somites (Fig. A 1) appear lateral to the neural groove (Fig. A 2). The somite cells proliferate and some migrate to the notochord (Fig. B 1), where they form the sclerotome (Fig. B 2). Subsequently, a new cell layer, the myotome (Fig. C 1), develops from the somite wall.

The myotome cells become spindle-shaped and form the vigorously proliferating presumptive myoblasts (arrows; M and S). Some of these differentiate gradually into real myoblasts (Fig. D). These cells arrange themselves end to end (Fig. E) and the points of contact of their cell membranes fuse (Fig. F). In this manner, the syncytial muscular tubules or myotubes (Fig. F) develop, to which further myoblasts can attach (Fig. F 1). Other presumptive myoblasts remain undifferentiated and migrate to the myotubes, where they become satellite cells (S).

Differentiated myotubes (Fig. G) are long, cylindrical structures with the nuclei (Fig. G 1) of earlier myoblasts arranged in a row in the axial zone. Myotubes and satellite cells (Figs. G–I, S) are surrounded by the same basal lamina (Figs. G 2, H 2, I 2).

Some satellite cells (Fig. G 3) continue to divide mitotically and furnish nuclei for the myotubes (Fig. H 1). The number of nuclei in the myotubes thus increases. At the same time, further differentiation and myofibrillogenesis take place, following which the nuclei are displaced to the sarcolemma (Fig. I). Skeletal muscle fibers develop in this way. A certain number of satellite cells remain in an undifferentiated state with the muscle fibers as a reserve of cells capable of division.

Myofibrillogenesis signifies the formation of myofibrils in the differentiation of myoblasts, via myotubes, to form mature skeletal muscle fibers. To clarify this process, sections from each of Figs. D, F, H, and I are magnified in Figs. O–R.

Myoblasts contain a vacuolar rough endoplasmic reticulum (Fig. O 1). Myofilaments (Fig. O 2) are synthesized on the helicoidal polyribosomes (Fig. O 3) and are initially randomly oriented in the cytoplasm.

Fingerlike smooth-walled processes (Fig. P 1) sprout from the rough endoplasmic reticulum (Fig. P 2) of young, newly formed myotubes, while the two types of myofilament (Fig. P 3) arrange themselves longitudinally. The cell membrane forms a slight invagination (Fig. P 4).

In further phases of differentiation, the rough endoplasmic reticulum disappears, and the smooth fingerlike endoplasmic tubules come into contact with one another. The basis for the longitudinal L-system of the sarcoplasmic reticulum (Fig. Q 1) is thus established. As the myofilaments (Fig. Q 2) adopt a regular pattern, the invagination in the sarcolemma (Fig. Q 3) becomes deeper.

Finally, the triad (Fig. R 1) is formed and the differentiation of striated muscle fibers is complete.

Magnifications: Figs. O–R, × 10,000

REFERENCES

Goldspink DF (ed) (1980) Development and specialization of skeletal muscle. Cambridge University Press, Cambridge

Isobe Y, Shimada Y (1983) Myofibrillogenesis in vitro as seen with the scanning electron microscope. Cell Tissue Res 231:481–494

Nag AC, Foster JD (1981) Myogenesis in adult mammalian skeletal muscle in vitro. J Anat 132:1–8

Pearson ML, Epstein HF (eds) (1982) Muscle development: Molecular and cellular control. Cold Spring Harbor Laboratory, New York

Russell RG, Oteruelo FT (1981) An ultrastructural study of the differentiation of skeletal muscle in the bovine fetus. Anat Embryol 162:403–417

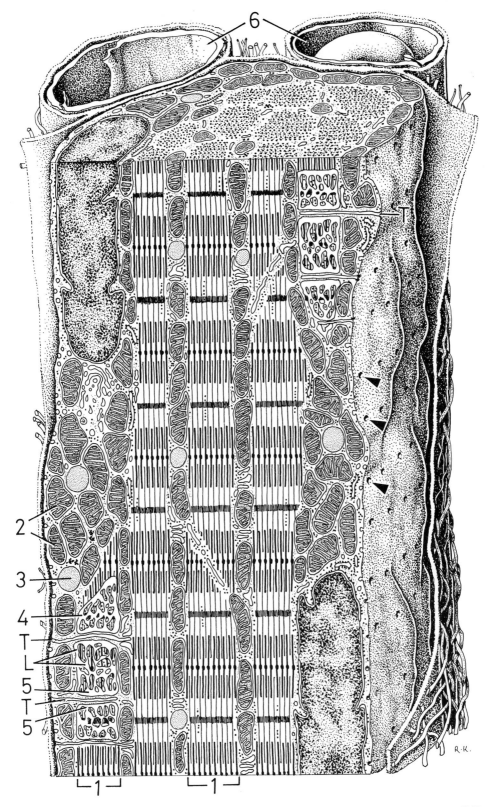

Plate 129. Skeletal Musculature. Three-Dimensional View of the Internal Structure of a Muscle Fiber

Every myofibril (**1**) is surrounded by segmented sarcoplasmic reticulum (**2**) at the level of the A- and I-bands. In these areas, the tubules of the sarcoplasmic reticulum run longitudinally, i.e., parallel to the myofibrils, for which reason this arrangement of the tubules is termed the L-system. At the boundary between the A- and I-bands, all the tubules of a segment join to form the terminal cisternae (**3**), which are perpendicular to the direction of the myofibrils.

Neighboring terminal cisternae are not in contact with one another since they are always separated by a tubular invagination of the sarcolemma, i.e., the T-tubule (**T**), running transversely to the myofibrils. The T-tubules, which only anastomose transversely, form a T-system. The sarcoplasmic reticulum around the first myofibril on the right has, like the T-tubules, been tangentially cut. The two terminal cisternae and the interjacent T-tubules are transversely sectioned in the middle myofibril. This tripartite structure – two terminal cisternae with a T-tubule in the middle – is termed a triad (**4**).

In the lower part of the muscle fiber, it can be seen that the T-tubules (**5**) only anastomose transversely and are in direct communication with the intercellular space. The openings (arrowheads) of the T-tubules appear on the surface of the muscle fiber at regular intervals. Outside the sarcolemma (**6**) lie the basal lamina (**7**) and a dense framework of reticular and collagen microfibrils (**8**). Sarcoplasmic reticulum is important in muscular contraction: During contraction, the Ca^{2+} ions leave the reticulum and accumulate there again during relaxation. (See biochemistry and physiology texts for further information.)

Magnification: $\times 24{,}000$

REFERENCES

Dowben RM, Shay JW (eds) (1983) Cell and muscle motility, vol 4. Plenum, New York

Kelly DE (1969) The fine structure of skeletal muscle triad junctions. J Ultrastruct Res 29:37–49

Rowe RW (1971) Ultrastructure of the Z line of skeletal muscle fibers. J Cell Biol 51:674–685

Walker SM, Schrodt GR, Currier GJ, Turner VE (1975) Relationship of the sarcoplasmic reticulum to fibril and triadic junction development in skeletal muscle fibers of fetal monkeys and humans. J Morphol 146:97–128

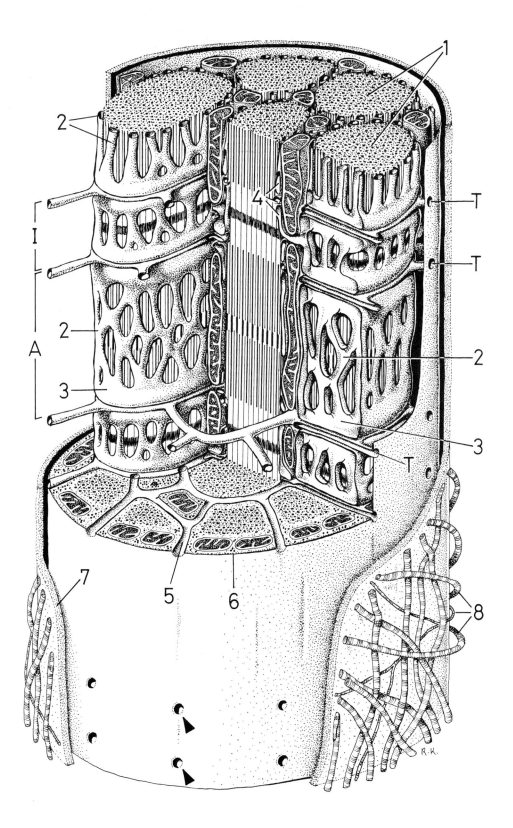

MUSCULAR TISSUE

Plate 130. Skeletal Musculature. Scheme of a Relaxed and a Contracted Striated Myofibril

In a very relaxed myofibril (Fig. **A**), the actin (Figs. **A 1**, **B 1**) and myosin (Figs. **A 2**, **B 2**) myofilaments only interlock to a minor extent.

It is now generally accepted that during contraction (Fig. **B**) the thin actin myofilaments slide between the thick myosin filaments toward the M-line, thereby shortening the myofibrils. However, the two types of filament are kept apart by the movable myosin heads (Figs. **A 3**, **B 3**), described in detail in the following plate. In this way the diameter of a myofibril, and thus the whole muscle, increases. This sliding of the myofilaments into one another causes the Z-lines to come closer together, and the I- and H-bands become narrower or disappear completely, leading to anisotropy of the contracted muscle under polarized light.

Transverse sections reveal different patterns of the interlocking myosin and actin myofilaments depending on the level. Figures **C–G** show the appearance of various cross sections of a myofibril.

The section at the level of the I-band only contains the hexagonally arranged actin myofilaments (Fig. **C**). Where the myosin and actin myofilaments interlock (Fig. **D**), one myosin myofilament is surrounded by six actin myofilaments. A hexagonal arrangement of the thicker parts of the myosin filaments can be seen in a transverse section through the M-line (Fig. **E**). A similar pattern is created by the thinner parts of the same myofilaments at the level of the H-band (Fig. **F**).

As the actin myofilaments run through the I-band toward the Z-line, they change their course by about 10 nm, producing the quadratic pattern in Fig. **G**. Short portions from relaxed and contracted myofibrils (asterisks) are shown in the following plate. (See Plate 71 Krstić 1979.)

REFERENCES

Bagshaw CR (1982) Muscle contraction. Chapman and Hall, New York

Franzini-Armstrong C (1970) Studies on the triad: I. Structure of the junction in frog twitch fibers. J Cell Biol 47:488–499

Franzini-Armstrong C, Porter KR (1964) The Z disc of skeletal muscle fibrils. Z Zellforsch 61:661–672

Kelly DE (1967) Models of muscle Z-band fine structure based on a looping filament configuration. J Cell Biol 34:827–840

Kelly DE, Cahill MA (1972) Filamentous and matrix components of skeletal muscle Z-discs. Anat Rec 172:623–642

Wang K, Ramirez-Mitchell R (1983) A network of transverse and longitudinal intermediate filaments is associated with sarcomeres of adult vertebrate skeletal muscle. J Cell Biol 96:562–570

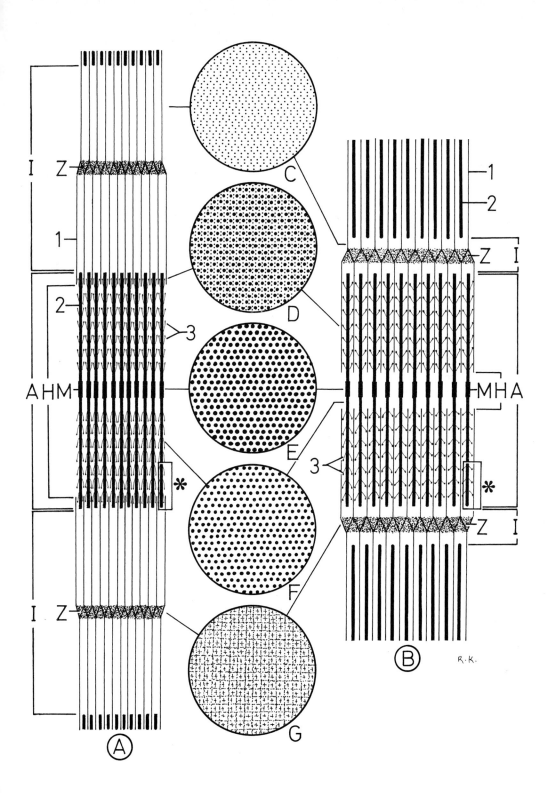

Plate 131. Partly Hypothetical Representation of the Interaction Between Actin and Myosin Myofilaments in Relaxed and Contracted Myofibrils
(Modified after HAM 1974 and MOREL and PINSET-HÄRSTRÖM 1975)

The thick myosin myofilaments (Figs. **A 1**, **B 1**) are composed of bipartite, threadlike, approximately 150-nm-long myosin molecules. Light meromyosin (Figs. **A 2**, **B 2**), which is approximately 80 nm long, makes up the actual backbone of myosin myofilaments, whereas heavy meromyosin (Figs. **A 3**, **B 3**) forms shorter sections and terminates in a globular head (Figs. **A 4**, **B 4**).

X-ray diffraction studies have shown that the globular heads (Figs. **A 4**, **B 4**) occur in pairs located on opposite sides of the myosin filament. There is an angle of 120° between each pair and the next, and thus, six rows of these globular units appear along the length of the myosin filaments. In the longitudinal direction, the globular pairs are 14.3 nm apart, therefore meromyosin heads of the same row are separated by a distance of 42.9 nm.

The thin actin myofilaments (Figs. **A 5**, **B 5**) are composed of two helicoidally entwined fibrillar chains of F-actin, which is the result of longitudinal aggregation of the 5.6-nm-wide globular molecules of G-actin. Thin tropomyosin molecules (Figs. **A 6**, **B 6**), which are approximately 40 nm long, occur along the chains of F-actin. Troponin molecules (Figs. **A 7**, **B 7**) are bound to the tropomyosin and are located along the actin filaments at intervals of about 40 nm.

During relaxation, interaction between the meromyosin heads (Fig. **A 4**) and actin myofilaments (Fig. **A 5**) is prevented by interpolation of the tropomyosin molecules (Fig. **A 6**).

At the beginning of contraction, however, troponin modifies tropomyosin, under the influence of Ca^{2+} ions, (released from the sarcoplasmic reticulum), such that tropomyosin comes to lie in the groove between the globular actin molecules. This movement of the tropomyosin brings the actin and myosin elements into contact. At the same time, heavy fragments of meromyosin spread out (broken arrow, Fig. **B**), whereby the actin myofilaments are displaced upward and laterally (unshaded arrow). At a later stage, the movements of the meromyosin heads (in the direction of the short arrows, Fig. **B**) are also applied to the actin myofilaments. It is even possible that the meromyosin heads are able to move over the globular actin molecule in a caterpillar action.

The energy for contraction is provided by the splitting of adenosine triphosphate (ATP). (See biochemistry and physiology texts for further information.)

REFERENCES

Eaton BL (1976) Tropomyosin binding to F-actin induced by myosin heads. Science 192:1337–1338

Ham AW (1974) Histology, 7th edn. Lippincott, Philadelphia

Huxley GE (1969) The mechanism of muscular contraction. Science 164:1356–1366

Huxley HE (1983) Molecular basis of contraction in cross-striated muscles and relevance to motile mechanisms in other cells. In: Stracher A (ed) Muscle and nonmuscle motility, vol 1. Academic, New York

Morel JE, Pinset-Härström I (1975) Ultrastructure of the contractile system of striated skeletal muscle and the processes of muscular contraction: I. Ultrastructure of the myofibril and source of energy. Biomedicine 22:88–96

Pepe FA (1983) Macromolecular assembly of myosin. In: Stracher A (ed) Muscle and nonmuscle motility, vol 1. Academic, New York

Toma M, Berl S (1982) Structural characterization of myosin from bovine brain. Eur J Cell Biol 28:122–129

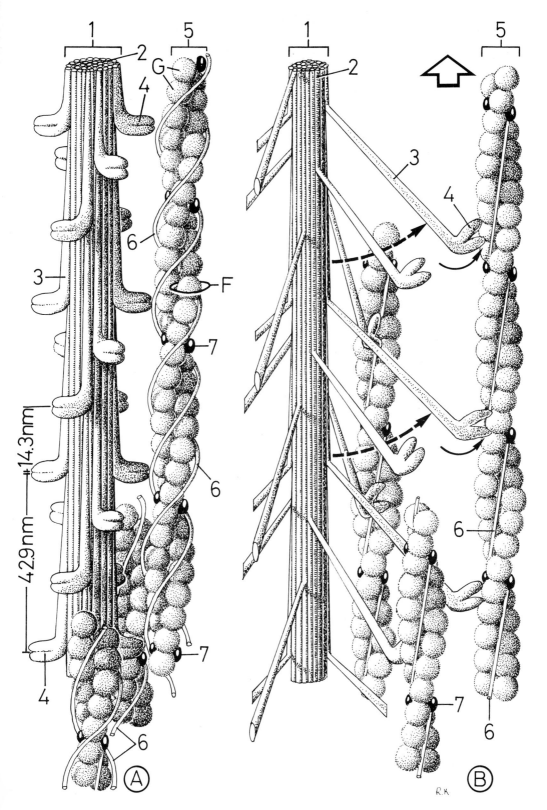

Plate 132. Skeletal Musculature. Myotendinal Junction

As can be seen in the light-microscopic image in Fig. A, the connective tissue (Fig. A 1) surrounding the muscle fibers (Fig. A 2) gradually continues into the tendon fibers (Fig. A 3). The boundary formed by the conical ends of the muscle fibers against the tendon indicates that there is no true continuity between the contractile substance and the network of collagen fibers. The area within the inset in Fig. A is enlarged in Fig. B to elucidate structural details of the myotendinal junction.

At its end, the muscle fiber gradually tapers and forms longitudinal depressions (Fig. B 1) and tubular invaginations (Fig. B 2), lined with basal lamina (Fig. B 3), which are penetrated by tendon microfibrils (Fig. B 4). The area of contact between the muscle fiber and tendon microfibrils is considerably augmented by these infoldings, and this naturally consolidates the junction between tendon and muscle fiber. Collagen tendon microfibrils, however, are incapable of directly anchoring to the basal lamina; they require the assistance of reticular microfibrils (Fig. B 5).

This relationship is shown at greater magnification in Fig. C (after HANAK and BÖCK 1971).

On the external layer of the sarcolemma (Fig. C 1) of the muscle fiber are located numerous loops of reticular microfibrils (Fig. C 2), which penetrate the basal lamina (Fig. C 3). The free ends of these microfibrils, outside the basal lamina, intertwine with the tendon microfibrils (Fig. C 4), giving rise to a somewhat elastic, though firm, junction between the sarcolemma and the bundles of tendon fibers.

Magnifications: Fig. A, × 300; Fig. B, × 20,000; Fig. C, × 85,000

REFERENCES

Hanak H, Böck P (1971) Die Feinstruktur der Muskel-Sehneverbindung von Skelet- und Herzmuskel. J Ultrastruct Res 36:68–85

Mackay B, Harrop TJ, Muir AR (1969) The fine structure of the muscle-tendon junction in the rat. Acta Anat (Basel) 73:588–604

Trotter JA, Corbett K, Avner BP (1981) Structure and function of the murine muscle-tendon junction. Anat Rec 201:293–302

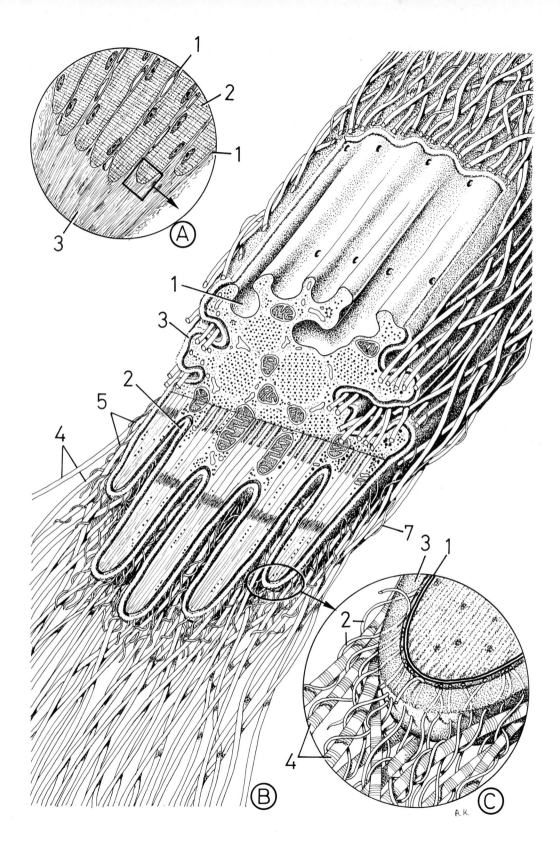

Plate 133. Skeletal Musculature. Regeneration of Skeletal Muscle Fibers

Despite very high morphofunctional differentiation, skeletal muscle fibers are capable of regeneration. The most important conditions for this to take place are an uninterrupted basal lamina and that the distance to undamaged muscle fibers is not too great. Muscle fibers regenerate:
I. Continuously
II. Discontinuously

I. The first possibility can be summarized in simplified form as follows. Macrophages (Fig. A 1) enter the damaged zone of muscle fiber (Fig. A 2) via the basal lamina (Fig. A 3) and phagocytize the debris (Fig. A 4). About 3–4 days after the necrosis, muscle fiber nuclei (Fig. B 5) become larger in the undamaged ends of the fibers. They arrange themselves in a line and move into the newly formed buds (Fig. B 6), which grow toward one another (arrow) and fuse.

In the re-formed muscle fiber, the nuclei initially occupy a central position (Fig. C), just as in the myotubes, however they soon become displaced toward the sarcolemma (Fig. D) as a result of myofibril formation. With this displacement, continuous regeneration terminates, approximately 2 weeks after damage to the muscle fibers.

II. Discontinuous regeneration requires the assistance of satellite cells. In this process also, macrophages (Fig. E 1) first clear the debris (Fig. E 2) from the damaged section of muscle fiber (Fig. E 3) surrounded by the basal lamina (Fig. E 4). The satellite cells (Fig. E 5) show no signs of mitosis at this stage.

About 3–4 days after injury, the satellite cells begin to divide (Fig. F 5) and approach one another in the direction of the arrows. Shortly after, the necrotic zone is bridged by young cells, similar to myoblasts (Fig. G 6), which subsequently fuse. In this manner, the syncytium of the muscular tube is once again established with initially centrally located nuclei (Fig. H). These nuclei are displaced toward the periphery of the sarcoplasmic cylinder upon completion of myofibrillogenesis (Fig. I).

With respect to Fig. F, it should here be stated that mitoses of the satellite cells in reality occur much less dramatically than depicted here. The peripheral mitotic activity in regenerating muscle fibers has long been established by light-microscopic observation, though for some time it was ascribed to the muscle fiber nuclei.

REFERENCES

Carlson BM (1973) The regeneration of skeletal muscle – a review. Am J Anat 137:119–150

Carlson BM, Gutmann E (1975) Regeneration in free grafts of normal and denervated muscles in the rat: morphology and histochemistry. Anat Rec 183:47–61

Hall-Craggs ECB (1974) The regeneration of skeletal muscle fibers per continuum. J Anat 117:171–178

Jakubiec-Puka A, Kulesza-Lipka D, Kordowska J (1982) The contractile apparatus of striated muscle in the course of atrophy and regeneration: II. Myosin and actin filaments in mature rat soleus muscle regenerating after reinnervation. Cell Tissue Res 227:641–650

Ontell M, Hughes D, Bourke D (1982) Secondary myogenesis of normal muscle produces abnormal myotubes. Anat Rec 204:199–207

Snow MH (1977) Monogenic cell formation in regenerating rat skeletal muscle injured by mincing: I. A fine structural study; II. An autoradiographic study. Anat Rec 188:181–218

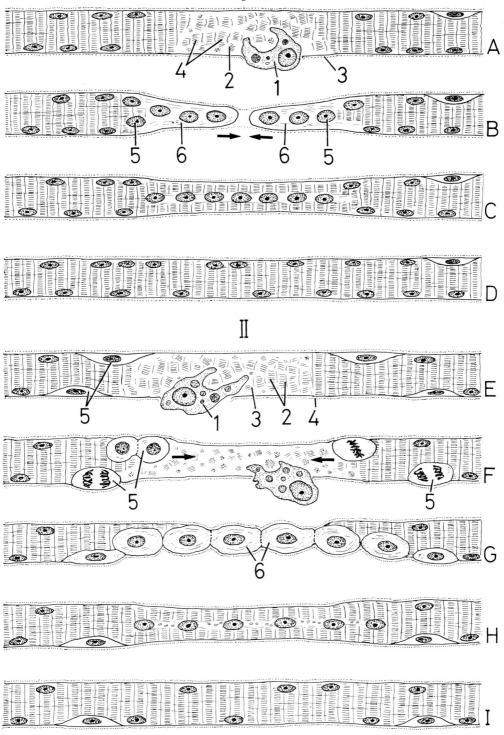

MUSCULAR TISSUE

Plate 134. Cardiac Musculature. Light-Microscopic Appearance

Cardiac musculature constitutes the main mass of the heart, the myocardium (Fig. A). Part of the ventricular wall is shown in transverse section in Fig. B and longitudinal section in Fig. C.

A cross section through the working cardiac musculature reveals the great variation in the appearance of the fibers (Fig. B1); no two are identical in shape and diameter. The nuclei (Fig. B2) are located in the middle of the sarcoplasm and surrounded by radial Cohnheim's fields of myofibrils (Fig. B3). A well-developed endomysium (Fig. B4) with a large number of blood capillaries (Fig. B5) surrounds the fibers.

On the longitudinal section through the cardiac musculature, it can be seen that the cardiac muscle fibers branch, bifurcate, and anastomose quite freely – a phenomenon found in neither smooth nor skeletal musculature. The nuclei (Fig. C1) are surrounded by myofibrils. At both nuclear poles, therefore, just as in smooth muscle cells, small conelike areas of sarcoplasm develop, which here are also termed endoplasm (Fig. C2). The intercalated discs (Fig. C3) are visible under the light microscope and characteristic of cardiac muscles. These discs traverse the muscle fibers in a steplike fashion and divide the fibers into 50- to 120-μm-long mono- or binuclear segments, i.e., into cardiac cells (Fig. C4).

Cardiac musculature is thus made up of cellular elements, like smooth musculature. Unlike skeletal muscle fibers, cardiac musculature is not a syncytium.

A well-developed, capillary-rich endomysium (Fig. C5) is found between the heart muscle fibers.

Magnifications: Figs. B, C, ×600

REFERENCES

Fawcett DW, McNutt NS (1969) The ultrastructure of the cat myocardium: I. Ventricular papillary muscle. J Cell Biol 42:1–45

Grzycki S, Tochman A (1975) Some observations on the intercalated disc of a rat cardiac muscle. Z Mikrosk Anat Forsch 89:219–230

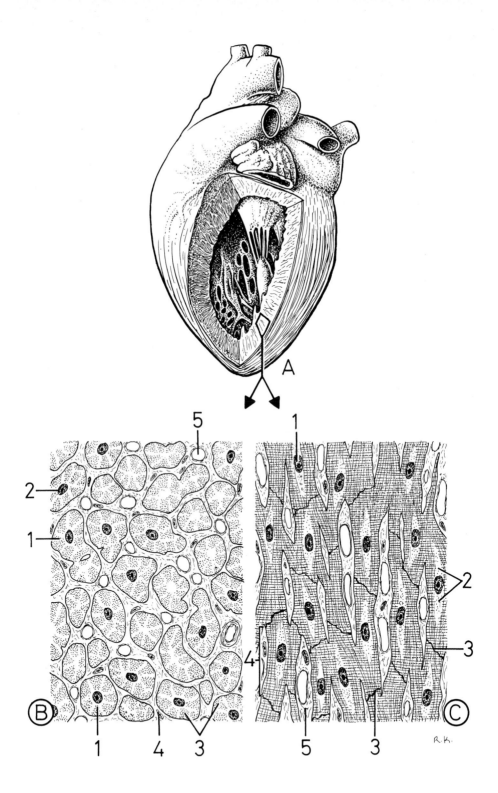

Plate 135. Cardiac Musculature. Three-Dimensional View of Several Fibers

Cardiac muscle fibers (1) form, as stated above, a netlike structure, whose interstices contain numerous capillaries (2) and the loose fibrous connective tissue of the endomysium (3).

Transverse sections clearly reveal the irregular outlines of the cardiac muscle fibers with their central nuclei. A large number of mitochondria (4) are evident both in transverse and longitudinal sections. The intercalated discs (5) and the division of the fibers into mono- and binuclear cardiac muscle cells, however, are only apparent in longitudinal sections.

A mononuclear cardiac muscle cell (arrow) is presented three-dimensionally in Plate 137.

Magnification: ×2,200

REFERENCES

McNutts NS, Fawcett DW (1974) Myocardial ultrastructure. In: Mammalian myocardium. Wiley, New York

Page E, McCallister LP (1973) Studies on the intercalated disk of rat left ventricular myocardial cells. J Ultrastruct Res 43:388–411

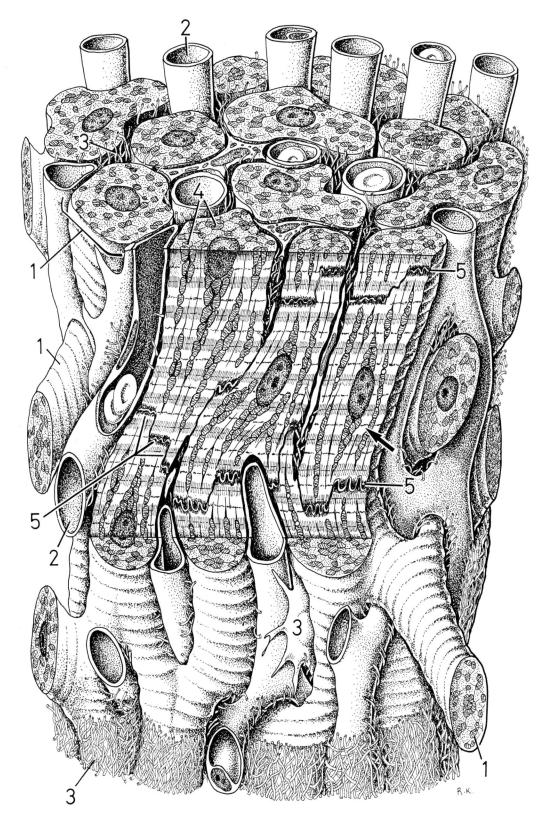

277

Plate 136. Cardiac Musculature. Electron-Microscopic View of Transverse and Longitudinal Sections Through a Cardiac Muscle Cell

A transverse section through a cardiac muscle cell (Fig. A) reveals the centrally located nucleus and numerous mitochondria to be dominant features. The myofibrils (Fig. A 1) here are much more poorly defined than in skeletal musculature, though they show the same arrangement of myosin and actin myofilaments. According to the level of transverse section through cardiac muscle cells, it is possible to distinguish all the patterns seen in Plate 130.

In the peripheral sarcoplasmic zone and between the mitochondria, many glycogen granules (Fig. A 2) and a few tubules of smooth endoplasmic reticulum (Fig. A 3) can be recognized. A fine connective tissue envelope made up of reticular and collagen microfibrils (Fig. A 4) and a basal lamina (Fig. A 5) surrounds the cardiac muscle cell. Numerous blood capillaries (Fig. A 6), which are essential for the high level of metabolism, cross the interspaces.

The large mitochondria with their many cristae are conspicuous in longitudinal sections. The triads (Fig. B 1), whose structure is not as distinct here as in skeletal musculature, are found at the level of the Z-lines. Numerous glycogen granules (Fig. B 2) appear between the myofilaments and in the interfibrillar spaces.

From the morphological point of view, an intercalated disc is an intercellular borderline, consisting of a highly indented portion running transversely through the fiber and a longitudinal smooth section. In the fasciae adherentes (Fig. C 1) of the transverse elements are attached the actin filaments (Fig. C 2). Intermediate microfilaments, about 10 nm thick (not shown), are anchored in the desmosomes (Fig. C 3). In the region of the fascia adherens and in the desmosomes, the two cell membranes are 20–30 nm apart and bound to one another by a glycoproteinaceous material.

The longitudinal cell boundaries form a nexus (Fig. C 4). The circular inset on the left of Fig. C shows a cleft (Fig. C 6), approximately 2 nm wide, between the two cell membranes (Fig. C 5). After freeze-etching, this cleft can be seen to correspond to a surface made up of hexagonally arranged membranous particles, called connexons (circular inset, lower right).

The highly interdigitating transverse portions of the intercalated discs ensure cohesion between the cells, whereas bioelectrical impulses for contraction of cardiac muscle are spread through the nexus zones – just as in smooth muscle cells. (See Plate 15 in KRSTIĆ 1979).

Magnifications: Fig. A, × 17,000;
Figs. B, C, × 35,000;
circular insets, × 85,000

REFERENCES
Barr L, Dewey MM, Berger W (1965) Propagation of action potentials and the structure of the nexus in cardiac muscle. J Gen Physiol 48:797–824
McNutt NS, Weinstein RS (1970) The ultrastructure of the nexus. A correlated thin-section and freeze-cleavage study. J Cell Biol 47:666–688
Robinson TF (1980) Lateral connections between heart muscle cells as revealed by conventional and high voltage transmission electron microscopy. Cell Tissue Res 211:353–359

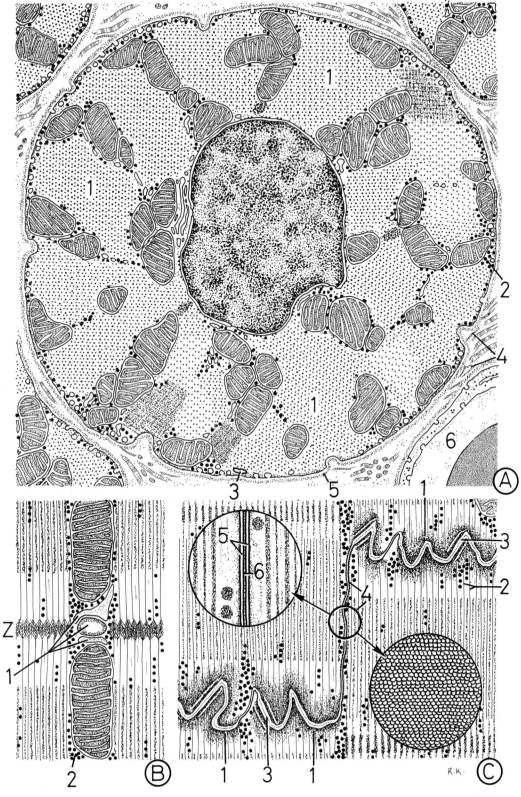

Plate 137. Cardiac Musculature.
Three-Dimensional View of an Isolated Cardiac Muscle Cell

The nucleus (Fig. A 1), with its conspicuous pores, is located in the center of the cell body. At both nuclear poles between the diverging and converging myofibrils (Fig. A 2), sarcoplasm-rich endoplasm (Fig. A 3) develops with organelles, glycogen granules, lipid droplets, and – in the musculature of the atria – specific atrial granules (arrow, Fig. B 1). The latter are highly osmiophilic, membrane-bound granules, 0.3–0.4 µm across, and contain hormone cardiodilatin. Since cardiac muscle cells belong to the group of cells that are incapable of expelling residual bodies or lipofuscin granules, these accumulate in the course of life, for which reason large numbers of lipofuscin granules (Fig. A 4) can be found in the endoplasm of older individuals.

As in skeletal musculature, the transverse striation due to the presence of two types of myofilament is characteristic of cardiac muscle cells.

At the lower right of Fig. A at the level of a Z-line, a T-tubule can be seen as well as the openings of other T-tubules (arrowheads) beneath the basal lamina (Fig. A 5). The irregularly branching profiles of smooth endoplasmic reticulum (Fig. A 6) are in close contact with the T-tubule.

The transversely striated portion of the intercalated discs bears processes (Fig. A 7) of varying length, which interlock with those of other cells. The surface of the nexus (Fig. A 8), parallel with the cell axis, is distinguished by the presence of numerous connexons.

Outside the basal lamina, there are a few collagen and reticular microfibrils (Fig. A 9) of the endomysium. (See Plates 59, 113, 114 in KRSTIĆ 1979.)

Magnifications: Fig. A, × 6,000;
Fig. B, × 30,000

REFERENCES

Forssmann WG, Birr C, Carlquist M, Christmann M, Finke R, Henschen A, Hock D, Kirchheim H, Kreye V, Lottspeich F, Metz J, Mutt V, Reinecke M (1984) The auricular myocardiocytes of the heart constitute an endocrine organ. Cell Tissue Res 238:425–430

Hibbs RG, Ferrans VJ (1969) An ultrastructural and histochemical study of rat atrial myocardium. Am J Anat 124:251–280

Jamieson JD, Palade GE (1964) Specific granules in atrial muscle cells. J Cell Biol 23:151–172

McNutt NS, Fawcett DW (1969) The ultrastructure of the cat myocardium: II. Atrial muscle. J Cell Biol 42:46–67

Saetersdal T, Rotevatn S, Myklebust R, Ødegården S (1980) Development of specific granules in embryonic cardiac myocytes of the human auricular wall. Anat Embryol 160:1–9

Segretain D, Rambourg A, Clermont Y (1981) Three dimensional arrangement of mitochondria and endoplasmic reticulum in the heart muscle fiber of the rat. Anat Rec 200:139–151

Seiden D (1979) Specific granules of the rat atrial muscle cell. Anat Rec 194:587–602

Yunge L, Benchimol S, Cantin M (1980) Ultrastructural cytochemistry of atrial muscle cells: VIII. Radioautographic study of synthesis and migration of proteins. Cell Tissue Res. 207:1–11

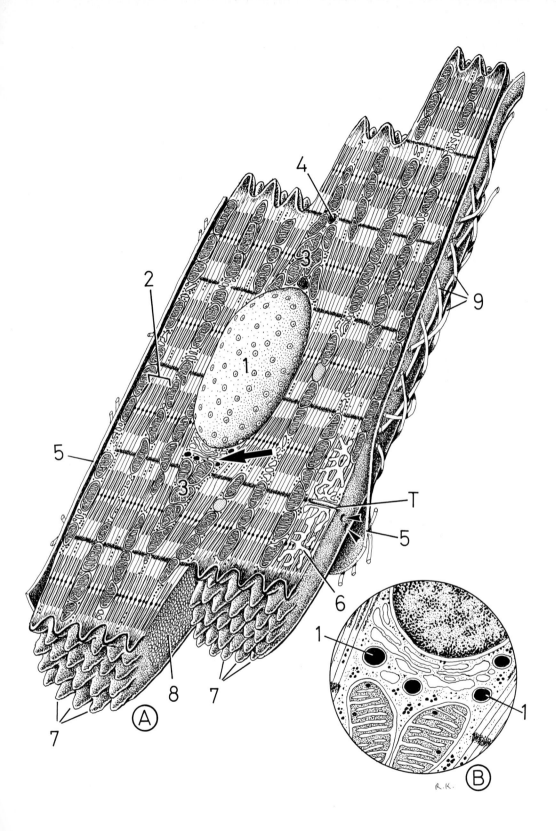

Plate 138. Cardiac Musculature. Three-Dimensional View of the Internal Structure of a Cardiac Muscle Fiber

A cardiac muscle fiber has been cut at the level of the Z-line to show the intracellular extension of a T-tubule, which runs through the quadratic pattern of the Z-line. The T-tubule has a glycosaminoglycan-containing lining, which can be regarded as a continuation of the basal lamina (**1**). For this reason, the T-tubules in the cardiac musculature have a considerably greater diameter (100–200 nm) than in skeletal musculature (50 nm).

Since the T-tubules are located at the level of the Z-line, only one segment of the sarcoplasmic reticulum (**2**) is found in every sarcomere. As in skeletal musculature, this sarcoplasmic reticulum surrounds every myofibril, however in cardicac muscle fibers it has neither a regular system of longitudinally anastomosing tubules (L-system) nor well-defined terminal cisternae. Thus, the saclike terminations of the sarcoplasmic reticulum are only partially in contact with the T-tubules, and dyads (**3**) are more frequent then triads at the points of contact.

As in skeletal musculature, the sarcoplasmic reticulum and T-system play an important part in the contraction of cardiac muscle. Here, the Ca^{2+} ions are supported by the action of cAMP, which is not the case in skeletal musculature (see biochemistry and physiology texts for further information).

Large numbers of mitochondria and glycogen granules (**4**) are evident in the sarcoplasm. In the lower part of the fiber, its continuity is interrupted by an intercalated disc (**5**).

A network of reticular and collagen microfibrils (**6**) envelops the cardiac muscle fiber.

Magnification: ×24,000

REFERENCES

Forssmann WG, Girardier L (1970) A study of the T-system in rat heart. J Cell Biol 44:1–19

Scales DJ (1983) Three-dimensional electron microscopy of mammalian cardiac sarcoplasmic reticulum at 80 kV. J Ultrastruct Res 83:1–9

Simpson FO, Rayns DG (1968) The relationship between the transverse tubular system and other tubules at the Z disc levels of myocardial cells in the ferret. Am J Anat 122:193–208

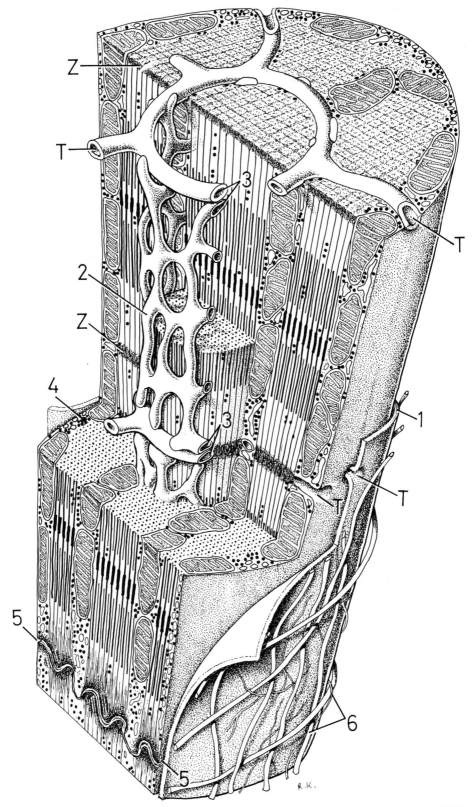

Plate 139. Cardiac Musculature. Vascularization, Necrosis, Transplantation

The uninterrupted contractions of cardiac muscle demand exceptional amounts of energy and, therefore, an intensive blood supply. Figure A illustrates the ramifications of the larger arteries (Fig. **A 1**) and the development of a dense capillary network in the ventricular wall of a rat heart. It has been calculated that virtually every muscle cell is in direct contact with a capillary.

Sudden obstruction of the larger branches of the coronary arteries (Fig. **B 1**) leads to infarction (Fig. **B 2**). This is a local anemic necrotic focus which, depending on size and localization, can be fatal. The cardiac muscular tissue dies in such areas and becomes replaced by a connective tissue scar, since the muscle is incapable of regeneration. This scar tissue cannot adapt to the hemodynamic pressure and thus is constantly liable to rupture.

Very severe heart conditions can now be treated by homografting of cardiac musculature. A possible surgical intervention is shown in Figs. **C** and **D** (after NETTER 1971). Following ligature of the large blood vessels, the ventricles (Fig. **C 1**) of the diseased heart are removed, such that the atria (Fig. **C 2**), or parts thereof, remain in the body. Subsequently, the appropriately prepared ventricle of the donor is sewn onto the atria (Fig. **D**). After all the vessels have been connected, the transplanted heart is made to contract by electric shock and take over the function of the diseased heart. The same conditions and mechanisms obtain here with regard to graft rejection as described in Plate 115. (See surgery and immunology texts for further information.)

Magnification: Fig. **A**, × 30

REFERENCES

Forbes RDC, Kuramochi T, Guttmann RD, Klassen J, Knaack J (1975) A controlled segmental morphologic study of hyperacute cardiac allograft rejection in the rat. Lab Invest 33:280–288

Netter FH (1971) The Ciba collection of medical illustration, vol 5, Heart. Colopress, New York

Polezhaev LV (1975) Myocardial regeneration. Ontogenez 6:154–162 (in Russian)

Rumyantsev P (1977) Interrelations of the proliferation and differentiation processes during cardiac myogenesis and regeneration. Int Rev Cytol 51:187–273

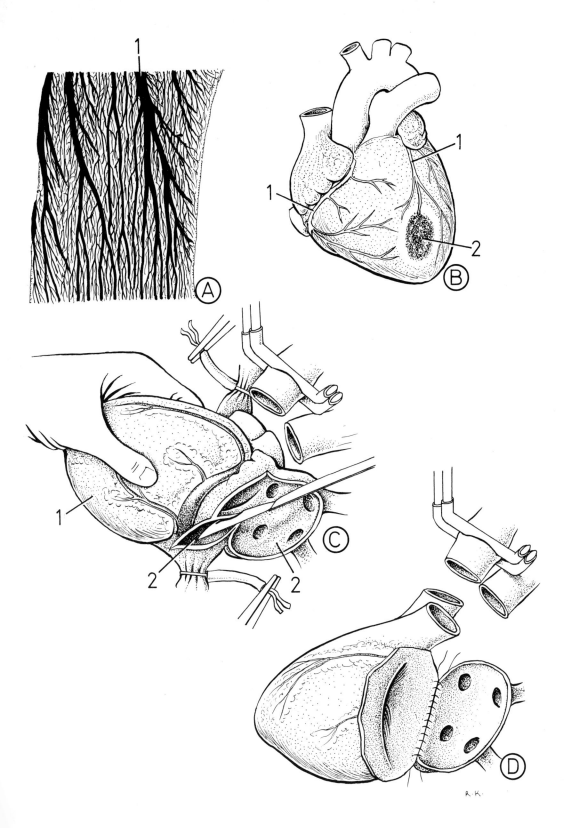

Plate 140. Cardiac Musculature. Impulse-Conducting System

A notable feature of cardiac musculature is the specially differentiated cardiac muscle tissue responsible for the regulation and propagation of the rhythmic heart contractions – the impulse-conducting system.

The sinus node (not shown), where the cardiac rhythm originates, is also termed the pacemaker of the heart. The impulses produced here reach the atrioventricular node (Fig. **A 1**), spread through the bundle of His (Fig. **A 2**), and finally extend to the latter's ramifications in the cardiac musculature, the Purkinje fibers (for further details here see anatomy texts).

The light-microscopic appearance of Purkinje fibers in the cardiac musculature is presented in Figs. **B** and **C**. The fibers were drawn from a sheep heart preparation, which is particularly suitable for demonstration of the impulse-conducting system. Purkinje fibers (Fig. **B 1**) are made up of large, often binuclear Purkinje cells (Figs. **B 2**, **C 2**), which are about 100 μm long, 50 μm wide, and arranged in chain-like rows. Isolated, mainly longitudinally oriented myofibrils (Figs. **B 3**, **C 3**) run through the glycogen-rich sarcoplasm. Some myofibrils, however, pursue a spiral course through the cell body and appear in sections as hypolemmal circular myofibrils (Fig. **C 4**).

Purkinje fibers are separated from the neighboring working cardiac musculature (Fig. **B 4**) by a connective tissue envelope (Figs. **B 5**, **C 5**).

Magnifications: Fig. **B**, × 150;
Fig. **C**, × 800

REFERENCES

Davies MJ, Anderson RH, Becker AE (1982) The conductive system of the heart. Huber, Bern

Defelice LJ, Challice CE (1969) Anatomical and ultrastructural study of the electrophysiological atrioventricular node of the rabbit. Circ Res 24:457–474

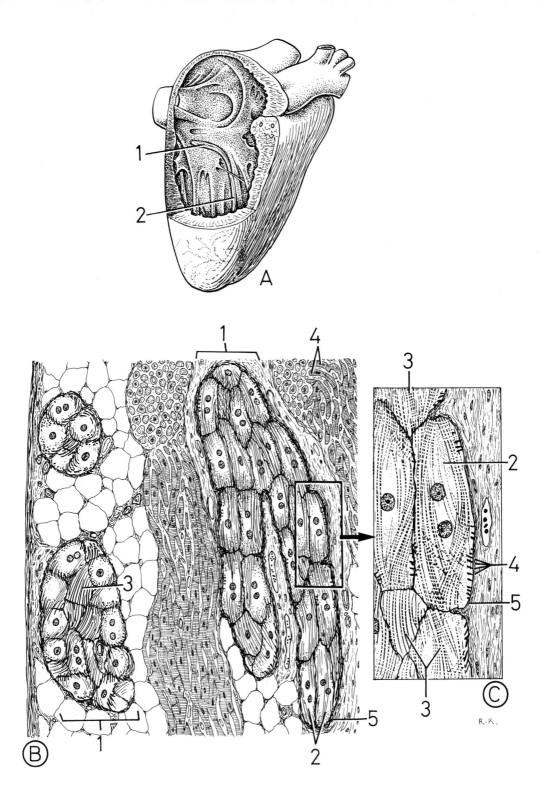

287

Plate 141. Cardiac Musculature.
Three-Dimensional View of a Purkinje Fiber

In addition to two nuclei, conspicuous features of the sarcoplasm of Purkinje cells include portions of the longitudinally (Fig. **A 1**) and transversely (Fig. **A 2**) cut myofibrils, small mitochondria (Fig. **A 3**), and accumulations of glycogen (Fig. **A 4**). In the vicinity of the nucleus are located more or less well-developed Golgi complexes. Sarcoplasmic reticulum is present to a minor extent, and the T-system is wholly absent.

The cell boundaries are irregular, and the cells interdigitate in a complex manner. In Fig. **B**, which shows an enlargement of the longitudinally oriented section of the cell boundary in the small central inset, can be recognized a desmosome (Fig. **B 1**) and a nexus (Fig. **B 2**).

The transverse cell boundaries in the impulse-conducting system roughly correspond to those of the intercalated discs of cardiac muscle fibers, which is attested by the presence of desmosomes (Fig. **C 1**), occasional fasciae adherentes (Fig. **C 2**), and sometimes limited nexus (Fig. **C 3**).

Each Purkinje fiber is surrounded by a basal lamina (Fig. **A 5**) and a connective network (Fig. **A 6**).

In this plate, the diameter of the cells of the impulse-conducting system can be compared with that of working cardiac muscle fibers (Fig. **A 7**). The basal lamina of the latter has been excluded for the sake of clarity.

Magnifications: Fig. **A**, ×2,200;
Figs. **B**, **C**, ×35,000

REFERENCES

Canale E, Campbell GR, Uehara Y, Fujiwara T, Smolich JJ (1983) Sheep cardiac Purkinje fibers: Configurational change during the cardiac cycle. Cell Tissue Res 232:97–110

Forsgren S, Strehler E, Thornell LE (1982) Differentiation of Purkinje fibres and ordinary ventricular and atrial myocytes in the bovine heart: an immuno- and enzyme histochemical study. Histochem J 14:929–942

Kim S, Baba N (1971) Atrioventricular node and Purkinje fibers in the guinea pig heart. Am J Anat 132:339–354

Shimada T, Nakamura M, Notohara A (1984) The Purkinje fiber-myocardial cell region in the goat heart as studied by combined scanning electron microscopy and chemical digestion. Experientia 40:849–850

Truex RC, Copenhaver WM (1947) Histology of the moderator band in man and other mammals, with special reference to the conduction system. Am J Anat 80:173–201

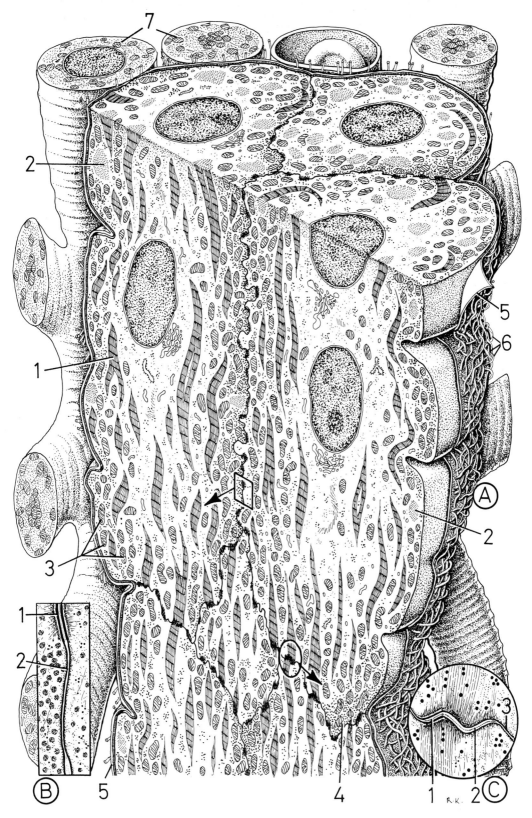

Plate 142. Cardiac Musculature. Three-Dimensional View of a Cell of the Impulse-Conducting System

The presence of myofibrils (**1**) entails that Purkinje cells (**2**) are capable of contraction. This is of great significance, since the last cell of the impulse-conducting system contacts the first working cardiac cell (**3**), which performs the contractile rhythm that has been transmitted to it.

The Purkinje cell in this plate has been drawn transparent, and the pattern of the myofibrils (**1**) can be seen in the sarcoplasm. Though the majority of fibrils are longitudinally oriented, the hypolemmal fibrils run in spirals through the cell body. Fasciae adherentes (**4**) develop at the points of insertion of the myofibrils. The small, rounded structures are mitochondria (**5**).

There are differences between cells of the working cardiac musculature and Purkinje cells with respect to the presence of enzymes. Whereas the former contain enzymes for oxidative breakdown of glucose (cytochrome oxidase, succinic dehydrogenase), the metabolism of Purkinje cells is equipped for anaerobic glycolysis with, among others, lactate dehydrogenase. The slight oxygen requirements of Purkinje cells explains the relative paucity of mitochondria.

Upon breakdown or severe disturbance of the regulatory action of the sinus and atrioventricular nodes, the cells of the Purkinje fibers become free of the influence of both nodes and independently produce impulses for cardiac contractions, a situation that can have fatal implications. This danger can be overcome by implanting a minute electric generator (pacemaker, **6**) under the abdominal skin. The electric impulses produced in this apparatus chemically or by nuclear power are conducted to the cardiac musculature by fine, implanted electrodes (**7**).

Magnification: × 4,000

REFERENCES

Bogusch G (1979) Electron microscopic investigations on the differentiation of Purkinje cells in the ontogenetic development of the chicken heart. Anat Embryol 155:259–271

Nunez-Duran H (1980) Sarcoplasmic reticulum in Purkinje cells of the goat heart. A three-dimensional reconstruction. Acta Anat 107:177–187

Osculati F, Franceschini F, Cinti S, Gazzanelli G, Amati S (1980) Subcellular aspects of Purkinje fibres in the pig heart. J Submicrosc Cytol 12:73–84

Sommer JR, Johnson EA (1968) Cardiac muscle. A comparative study of Purkinje fibers and ventricular fibers. J Cell Biol 36:497–526

Thaemert JC (1973) Fine structure of the atrioventricular node as viewed in serial sections. Am J Anat 136:43–66

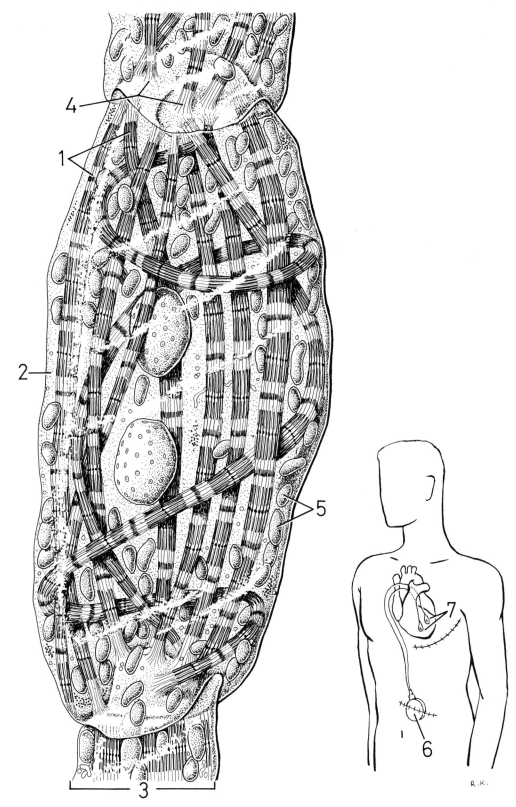

V Nervous Tissue

Plate 143. Histogenesis of Nervous Tissue

The entire nervous tissue develops from the ectoderm (see Plate 1). Initially, it is a simple columnar epithelium (Figs. a, A), beneath which the notochord is located (Figs. a–e, Ch). The notochord induces the formation of the medullary or neural plate (Fig. b1), the wall of which becomes a thick pseudostratified epithelium as a result of mitotic divisions (Fig. B). Subsequently, the neural plate deepens, forming the neural groove (Fig. c1). The cells of the neural crest (Fig. c2) temporarily appear at the two lateral folds, and the epithelium of the neural groove continues to thicken through high mitotic activity (Fig. C).

At a somewhat later stage, the neural folds fuse and give rise to the medullary or neural tube (Fig. d). Simultaneously, the neural crest cells separate from the folds and form a continuous plate (Fig. d1) under the surface ectoderm (Fig. d2). After the neural tube has formed, the neuroepithelial cells continue to divide (Fig. D) and the wall of the tube becomes even thicker. This pseudostratified epithelium is termed neuroepithelium. Once the neural tube is formed (Fig. e), a new type of cell differentiates from the neuroepithelial cells, the so-called primitive nerve cells or neuroblasts, whose number soon increases as a result of vigorous division. They form an easily recognizable mantle zone (Figs. e1, E1).

When the short genesis of the neuroblasts is over, primitive support cells, glioblasts, develop from the neuroepithelial cells and also migrate into the mantle zone. Neurons, i.e., the definitive nerve cells complete with processes, can only differentiate from neuroblasts. Glial cells develop by differentiation of the dividing glioblasts. The mantle zone subsequently becomes the gray matter of the spinal cord. The processes of the neuroblasts form a marginal zone (Figs. e2, E2), from which the white matter of the spinal cord originates.

Only one layer (formerly termed the matrix) of cuboidal cells with basal processes, ependymoblasts (Fig. e3, E3), persists from the neuroectoderm, and these cells differentiate into the ependymal cells that line the central canal (Fig. e4) and cerebral ventricles. The cells of this layer – called epithelium – also serve as a primordium for the development of other components of the central nervous system (see Plate 144).

At the stage of the neural tube, the neural crest (Fig. e5) is divided into rounded segments, which supply various types of cell (see Plate 144). (See embryology texts for further information.)

Magnifications: Figs. A–E, × 1,000

REFERENCES

Bourne GH (ed) (1972) Structure and function of nervous tissue, vols 1–5. Academic, New York

Meller K, Tetzlaff W (1975) Neuronal migration during the early development of the cerebral cortex. Cell Tissue Res 163:313–325

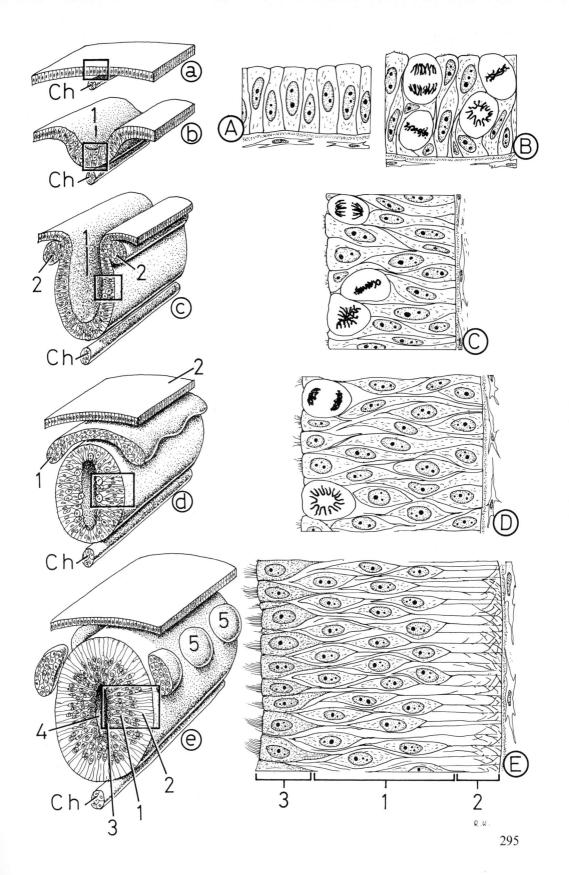

Plate 144. Cells Originating from the Neural Tube and Neural Crest

This plate gives a general view of the cells that derive from the neural crest (**A**), mantle (**B**), and epithelium (**C**).

A. From the neural crest develop:
1. Neuroblasts, i.e., nerve cells of the spinal ganglia
2. Glioblasts of the peripheral glia, from which develop the satellite cells (**2a**) of the spinal ganglia cells and Schwann's cells (**2b**)
3. Sympathoblasts, which develop into both nerve cells of the sympathetic ganglia (**3a**) and medulloblasts, i.e., the chromaffin cells of the adrenal medulla (**3b**) and part of the paraganglia
4. Melanoblasts, i.e., melanocytes (**4a**)
5. Microglia cells and part of the mesenchyme in the head region ("mesectoderm")
6. C cells of the thyroid and possibly other endocrine cells of the amine precursor uptake and decarboxylation (APUD) system
7. Cells of the leptomeninges.

B. From the mantle zone of the neural tube develop:
8. Glioblasts, i.e., glia cells of the central nervous system: protoplasmic (**8a**) and fibrillar (**8b**) astrocytes and oligodendrocytes (**8c**)
9. Neuroblasts, i.e., nerve cells (**9a**) of the central nervous system and some sensory cells (**9b**).

C. From the epithelium develop:
10. Ependymoblasts, i.e., ependymocytes (**10a**) and so-called ependymal differentiations
11. Epithelial cells of the choroid plexus
12. Pituicytes, i.e., cells of the neurohypophysis
13. Pinealocytes, i.e., cells of the pineal body

Although there is still some debate with respect to some of these details (see embryology texts for further information), this summary indicates the pluripotentiality of the neural tube and crest.

REFERENCES

Blakemore WF, Jolly RD (1972) The subependymal plate and associated ependyma in the dog. An ultrastructural study. J Neurocytol 1:69–84

Bruni JE, Clattenburg RE, Montemurro DG (1974) Ependymal tanycytes of the rabbit third ventricle: a scanning electron microscopic study. Brain Res 73:145–150

Leonhardt H, Lindemann B (1973) Über ein supraependymales Nervenzell-, Axon- und Gliazellsystem. Eine raster- und transmissionselektronenmikroskopische Untersuchung am IV. Ventrikel des Kaninchengehirns. Z Zellforsch 139:285–302

Millhouse EO (1972) Light and electron microscopic studies of the ventricular wall. Z Zellforsch 149:149–174

Schmitt FO, Worden FG, Adelman G, Dennis SG (eds) (1981) The organization of the cerebral cortex. MIT Cambridge

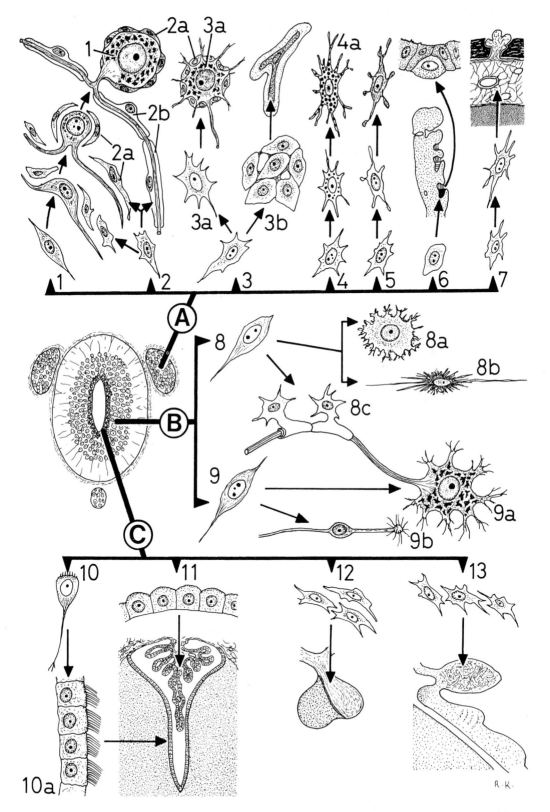

Plate 145. General Distribution of Nervous Tissue. Central and Peripheral Nervous Systems

The central nervous system consists of the cerebrum (**1**) with brain stem, cerebellum (**2**), and spinal cord (**3**).

Spinal ganglia (**4**), peripheral nerves (**5**), efferent and afferent nerve endings (**6**), and autonomic ganglia (**7**) form the peripheral nervous system. (See anatomy texts for further information.)

REFERENCES

Braak H (1980) Architectonics of the human telencephalic cortex. Springer, Berlin Heidelberg New York

Leonhardt H (1976) Die Liquorkontaktfortsätze im Zentralkanal des Rückenmarks. Z Mikrosk Anat Forsch 90:1–15

Lindemann B, Leonhardt H (1973) Supraependymale Neuriten, Gliazellen und Mitochondrienkolben im caudalen Abschnitt des Bodens der Rautengrube. Z Zellforsch 140:401–412

Mestres P, Breipohl W (1976) Morphology and distribution of supraependymal cells in the third ventricle of the albino rat. Cell Tissue Res 168:303–314

Noack W, Dumitrescu L, Schweichel JU (1972) Scanning and electron microscopical investigation of the surface structures of the lateral ventricles in the cat. Brain Res 46:121–129

Oksche A, Vollrath L (eds) 1978) Handbuch der mikroskopischen Anatomie des Menschen, vol 4, part 1, Nervengewebe. Das peripherische Nervensystem. Das Zentralnervensystem. Springer, Berlin Heidelberg New York

Vigh-Teichmann I, Vigh B (1975) The infundibular cerebrospinal fluid contacting neurons. Adv Anat Embryol Cell Biol 50/2:1–91

Yamadori T, Yagihashi S (1975) A scanning and transmission electron microscopic observation of the fourth ventricular floor in the mouse. Arch Histol Jpn 37:415–432

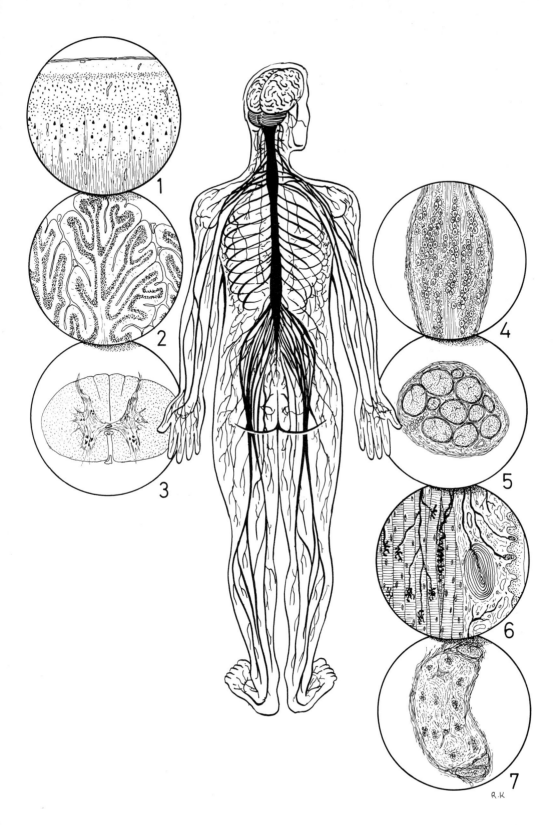

Plate 146. Simplified Scheme of the Relationship Between Neurons and Glia Cells in the Central and Peripheral Nervous Systems (Modified from Rhodin 1974)

The rectangle depicts cells of the central nervous system (CNS); the structures outside this rectangle belong to the peripheral nervous system. The direction of nervous impulses is indicated by arrows.

The nervous system is made up of nerve cells (**1**) and glia cells. Each nerve cell has a body, the soma or perikaryon, and usually several processes. In the cell body is found the nucleus (**2**), surrounded by cytoplasm, here termed neuroplasm. A nerve cell basically has only one cellulifugal process, the axon (**3**), and several cellulipetal processes, the dendrites (**4**).

A nerve cell with its processes constitutes a structural and functional unit termed a neuron. At certain sites, termed synapses (**5**), neurons come into contact with one another. In this manner, neuronal switching circuits develop, which are essential for the function of the nervous system.

Since the nervous tissue is very highly differentiated it possesses its own mechanically supporting tissue, the neuroglia, or glia for short. The glia cells are located between the blood capillaries (**6**) and neurons and constitute an extremely important cell system involved in the metabolism of the neurons.

In the gray matter of the CNS, protoplasmic astrocytes (**7**) are situated between the neurons and capillaries. With their processes (**7a**), these astrocytes delimit the CNS from the peripheral nervous system (PNS) and from adjacent tissues. In the white matter, the nerve fibers are supported by fibrous astrocytes (**8**). Axons in the region of the CNS are surrounded by a myelin sheath, produced by oligodendrocytes (**9**). These segmented envelopes are separated from one another by the nodes of Ranvier (**10**). Microglia or Hortega cells (**11**) represent the phagocytes of the CNS.

In the PNS, there are two types of neuron – one type occurs in spinal (**12**) and the other in autonomic (**13**) ganglia. Here also, the glia, though this time peripheral, surrounds the nerve cells. Amphicytes or satellite cells (**14**) envelop the neurons of the peripheral ganglia (**12, 13**). Axons in the PNS are invested by Schwann's cells (**15**): In myelinated nerve fibers (**16**) only one axon per cell is found, whereas several axons are enclosed within one Schwann's cell in unmyelinated nerve fibers (**17**). Nodes of Ranvier (**10**) occur where Schwann's cells of myelinated nerve fibers are contiguous.

The nerve fibers that derive from the autonomic ganglia (**13**) innervate the inner organs (**18**). Myelinated nerve fibers (**16**) originating from the CNS conduct the efferent nerve impulses to the skeletal musculature (**19**). Information from the periphery (e.g., from an epithelium; **20**) reaches the CNS via spinal ganglion cells (**12**), whose fibers are thus said to be afferent. There is no histological difference between efferent and afferent fibers.

All the peripheral nerve fibers referred to are components of spinal nerves (**21**)

REFERENCE

Rhodin JAG (1974) Histology. A Text and Atlas. Oxford University Press, New York London Toronto

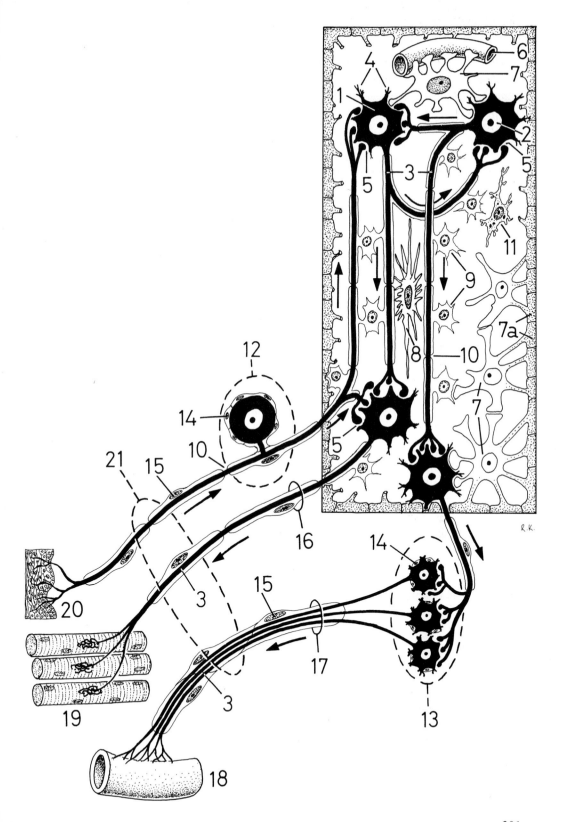

Plate 147. Glia of the Central Nervous System. Ependymal Cells

The neuroglia performs several important functions in the nervous system. In addition to its role in mechanical support, isolation, and metabolism of nerve cells, it is involved in synthesis of the myelin sheaths, phagocytosis, and scar formation. The neuroglia is classified as follows:

A. Glia of the central nervous system
1. Ependymal cells or ependymocytes
2. Astrocytes (macroglia)
3. Oligodendrocytes
4. Microglia (Hortega cells or mesoglia)
5. Epithelial cells of the choroid plexus (see Plate 12)

B. Glia of the peripheral nervous system
1. Schwann's cells
2. Amphicytes (mantle or satellite cells)

The ependyma forms a layer of simple cuboidal or columnar cells which lines the cavities of the brain and spinal cord. In the embryo, a branching process, the ependymal fiber (Fig. A 1), grows from the base of the ependyma; this fiber usually recedes (Fig. A 2) in adults. Thus, e. g., in the mature organism, the posterior median septum (Fig. A 3) of the spinal cord develops from the interwoven ependymal fibers. In all fetal ependymal cells, the free surface still bears cilia (Fig. A 4).

Differentiated ependymal cells have a round, basally or centrally located nucleus (Fig. B 1) with a nucleolus. Mitochondria are somewhat more numerous in the basal half of the cell than in the supranuclear cytoplasm. The Golgi apparatus consists of only a few flattened cisternae and vacuoles. The profiles of rough endoplasmic reticulum are short and narrow. Large numbers of free ribosomes and a few smooth vesicles, lysosomes, residual bodies, and microfilaments are found scattered in the cell body.

The majority of differentiated ependymal cells bear microvilli and bunches of cilia (Fig. B 2) on the free surface.

Many ependymocytes have processes (Fig. B 3) which lead from the basal pole, extend between the subependymal cells, and come into contact with the blood capillaries (Fig. B 4). Ependymal cells with such processes are termed tanycytes.

It has recently been discovered that the surface of various ependymal regions is covered with a network of long cytoplasmic extensions. Further studies have shown these structures to be largely supraependymal unmyelinated axons with expanded mitochondria-rich areas (Fig. B 5). They form synapselike junctions (Fig. B 6) with the ependymal cells. In addition to the axons, nerve cells (Fig. B 7), from which these axons presumably derive, are found on the ependymal surface. The function of these nerve cells still remains to be clarified; they are possibly a type of receptor.

One or more cell-poor subependymal glia cell layers (Fig. B 8), with predominantly horizontal glia cell processes (Fig. B 9), are located beneath the ependymal cells. This zone of undifferentiated glia elements is important in adults in the regeneration of glia cells.

Between the ependymal cells, which contact one another by means of terminal bars, are interposed cilia-bearing processes of cerebrospinal fluid-contacting neurons (Fig. B 10), the precise function of which has not been identified (possibly osmoreceptors). (See Plates 170, 171 in KRSTIĆ 1979.)

Magnification: Fig. B, ×7,000

REFERENCES

Bruni JE, Clattenburg RE, Millar E (1983) Tanycyte ependymal cells in the third ventricle of young and adult rats: a Golgi study. Anat Anz 153:53–68

Coates PW, Davis SL (1982) Tanycytes in long-term ovariectomized ewes treated with estrogen exhibit ultrastructural features associated with increased cellular activity. Anat Rec 203:179–187

Leonhardt H (1980) Ependym und circumventriculäre Organe. In: Oksche A, Vollrath L (eds) Handbuch der mikroskopischen Anatomie des Menschen, vol 4, part 10. Nervensystem. Springer, Berlin Heidelberg New York

Vigh B, Vigh-Teichmann I (1981) The CSF contacting neurosecretory cell: A protoneuron. In: Farner DS, Lederis K (eds) Proceedings 8th International Symposium on Neurosecretion. Plenum, New York

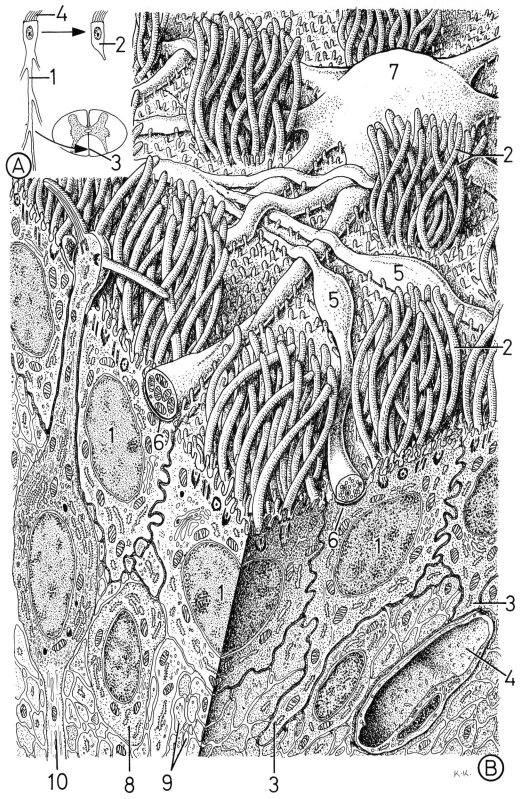

Plate 148. Glia of the Central Nervous System. Light-Microscopic Appearance of Astrocytes

Sections from different parts of the central nervous system – cerebrum (Fig. **A 1**), spinal cord (Fig. **A 2**), and cerebellum (Fig. **A 3**) – are shown in Figs. **B–D** as they appear under the light microscope after routine staining. In each section, a rectangular inset marks the gray matter, and a circular inset the white. The structures in the rectangular insets are summarized in Fig. **E**.

Protoplasmic astrocytes (Fig. **E 1**), following Golgi staining, appear as stellate cells which contact both nerve cells (Fig. **E 2**) and blood capillaries (Fig. **E 3**) with their processes. The connections formed between astrocytes lead to the formation of a glial supporting framework with neurons located in the interstices.

The astrocyte processes that are attached to capillaries terminate in footlike expansions, termed perivascular feet, which cover most of the capillary surface. Thus, a membrana limitans gliae perivascularis develops, which in the central nervous system replaces the connective tissue envelope and is involved in the formation of the blood-brain barrier. Other astrocyte processes give rise to the membrana limitans gliae superficialis at the outer surface of the central nervous system (see Plate 153).

The circular insets in Figs. **B–D** display an enlargement of a fibrous astrocyte (Fig. **F 1**) within Fig. **E**. Since these cells occur in the white matter, their processes follow, i.e., are largely parallel to, the nerve fibers (Fig. **F 2**).

The smaller cells adjacent to the nerve cells (Fig. **E 2**) and between the nerve fibers are oligodendrocytes (Figs. **E 4**, **F 3**).

A further, special form of astrocyte (Fig. **G 1**) is found in the cerebellum. These cells have winglike (velate) extensions and appear in the region of the cerebellar glomeruli (Fig. **G 2**) of the granular layer (see Plate 152).

Magnifications: Fig. **B**, ×20;
Fig. **D**, ×10, Figs. **E**, **F**, ×400;
Fig. **G**, ×700

REFERENCES

Allt G (1980) Astrocytes, filaments, and microtubules. Trends Neurosci 3:72–73

Phillips DE (1973) An electron microscopic study of macroglia and microglia in the lateral funiculus of the developing spinal cord in the fetal monkey. Z Zellforsch 140:145–167

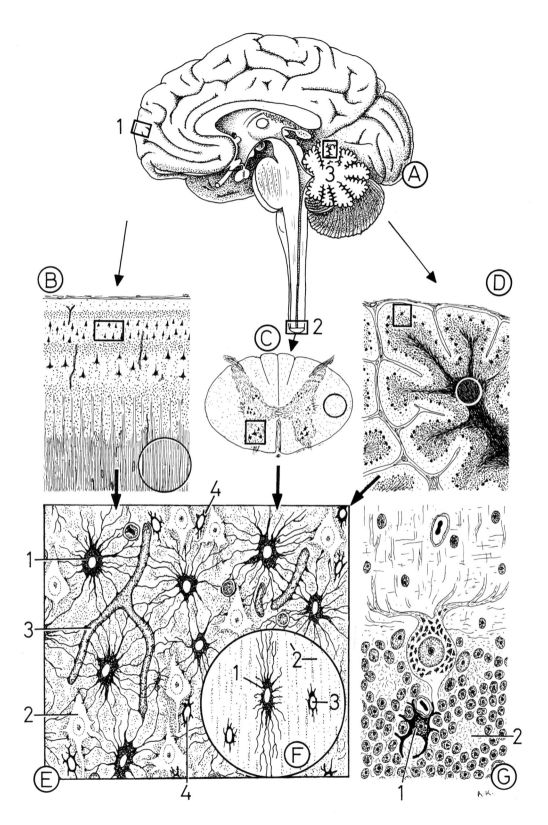

Plate 149. Glia of the Central Nervous System.
Membrana Limitans Gliae Perivascularis. Three-Dimensional View

This plate provides a clearer impression of the relationship between a nerve cell (**1**) and protoplasmic astrocytes (**2**), an oligodendrocyte (**3**), and blood capillaries (**4**).

A nerve cell normally has several branching dendrites (**5**) and an axon (**6**). Short, cut boutons terminaux (**7**) of other nerve cells are located both on dendrites and on the axon, forming synapses with the depicted neuron.

Near the nerve cells are located two astrocytes (**2**) with their processes (**8**). These processes are of variable length and have footlike or platelike terminations (**9**), which are in contact both with the nerve cell (**1**) and, as perivascular feet, with the capillaries (**4**). The astrocytes contact one another at the site indicated by an arrow. The perivascular feet of astrocytes, as the membrana limitans gliae perivascularis, cover about 80% of the capillary surface. Areas not covered by the membrane (**10**), where only the capillary basal lamina is present, are uncommon.

Short, rounded protrusions (**11**) also form on the body of astrocytes and serve as supports for the numerous processes of the various nerve cells.

An oligodendrocyte (**3**) surrounds an axonal segment with its myelin sheath (**12**). Below the node of Ranvier (**13**), another oligodendrocyte supplies the neuron with a myelin sheath.

To avoid misinterpretation of this plate, it should be noted that great numbers of many different kinds of cell processes occur between the cells and capillaries – the so-called neuropil. These processes have been omitted here for the sake of clarity. The actual appearance of a section through the close cell union of the nervous tissue has already been shown in Plate 2.

Magnification: × 2,700

REFERENCES

Kings JS (1968) A light and electron microscopic study of perineuronal glial cells and processes in the rabbit neocortex. Anat Rec 161:111–124

Lierse W (1968) Die Hirncapillaren und ihre Glia. Acta Neuropathol (Berl) Suppl 4:40–52

Ling EA, Patterson JA, Privat A, Mori S, Leblond CP (1973) Identification of glial cells in the brain of young rats. J Comp Neurol 149:43–72

Oksche A (1980) Neuroglia I. In: Oksche A, Vollrath L (eds) Handbuch der mikroskopischen Anatomie des Menschen, vol 4, part 10, Nervensystem. Springer, Berlin Heidelberg New York

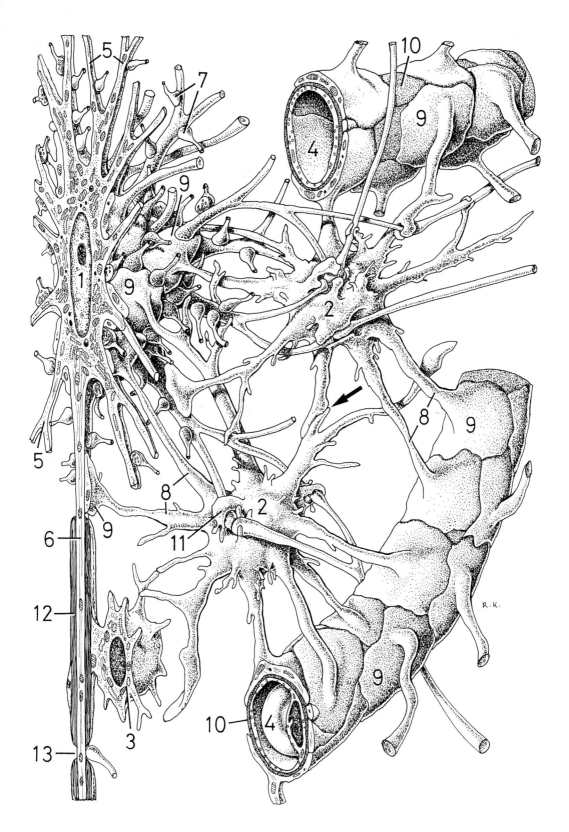

Plate 150. Glia of the Central Nervous System. Protoplasmic Astrocyte

A protoplasmic astrocyte (**1**) is a relatively large, stellate cell that predominantly occurs in the gray matter of the brain and spinal cord. The cell body contains a large, spherical nucleus with very finely dispersed chromatin, which facilitates identification of astrocytes in the electron microscope. Around the nucleus, several short cisternae of rough endoplasmic reticulum and a medium-sized Golgi apparatus are found. In addition to a small number of mitochondria, glycogen particles and centrioles occur in the slightly osmiophilic cytoplasm, since astrocytes are capable of division under certain circumstances.

Gliofibrils (**2**) are particularly characteristic of astrocytes; they are seen is sections to run transversely and longitudinally, largely in the vicinity of the nucleus, whence they extend into the processes. Gliofibrils, which probably have a mechanical function, are composed of 5- to 10-nm-thick gliofilaments consisting of glia fibrillary acidic protein.

The short processes of a protoplasmic astrocyte extend on one side to the nerve cells (**3**) and on the other to the blood capillaries (**4**). Whereas capillaries and perivascular feet (**5**) are separated by a 30- to 50-nm-thick capillary basal lamina (**6**), the cytoplasmic end plates (**7**) approach the nerve cells at a distance of less than 20 nm. Thus, some axosomatic synapses (**8**) become covered by astrocyte extensions.

Many unmyelinated (**9**) and myelinated nerve fibers (**10**) as well as nerve endings (**11**) are found around the body of the astrocyte at a distance of only 20 nm. They indent the surface of the cell body somewhat, giving rise to many small platelike cytoplasmic processes. It is, therefore, supposed that an astrocyte is also able to embrace various synapses (e.g., axodendritic synapses, **12**; axospinous synapses, **13**) in addition to surrounding nerve fibers. A few dendritic spines (**14**) can also be observed outside the astrocyte.

The presence of glycogen in the cytoplasm of astrocytes and the high phosphorylase activity indicate that these cells are of major importance in the metabolism of carbohydrates in the central nervous system. In contrast to nerve cells, glial elements do not shown any acetylcholinesterase activity, though they do have an increased concentration of adenosine triphosphate (ATP) bound to the cell membrane. As a result of their energetic involvement in the sodium pump, astrocytes play an important role in ionic regulation of the CNS. Hitherto, there has been little evidence of phagocytotic activity of astrocytes.

Magnification: × 10,000

REFERENCES

Duncan D, Morales R (1973) Fine structure of astrocyte mitochondria in the spinal cord of the dog, cat, and monkey. Anat Rec 175:519–528

Frederickson RG, Low FN (1969) Blood vessels and tissue space associated with the brain of the rat. Am J Anat 125:123–146

Ling EA, Paterson JA, Privat A, Mori S, Leblond CP (1973) Investigation of glial cells in semithin sections: I. Identification of glial cells in the brain of young rats. J Comp Neurol 149:43–72

Phillips DE (1973) An electron microscopic study of macroglia and microglia in the lateral funiculus of the developing spinal cord in the fetal monkey. Z Zellforsch 140:145–167

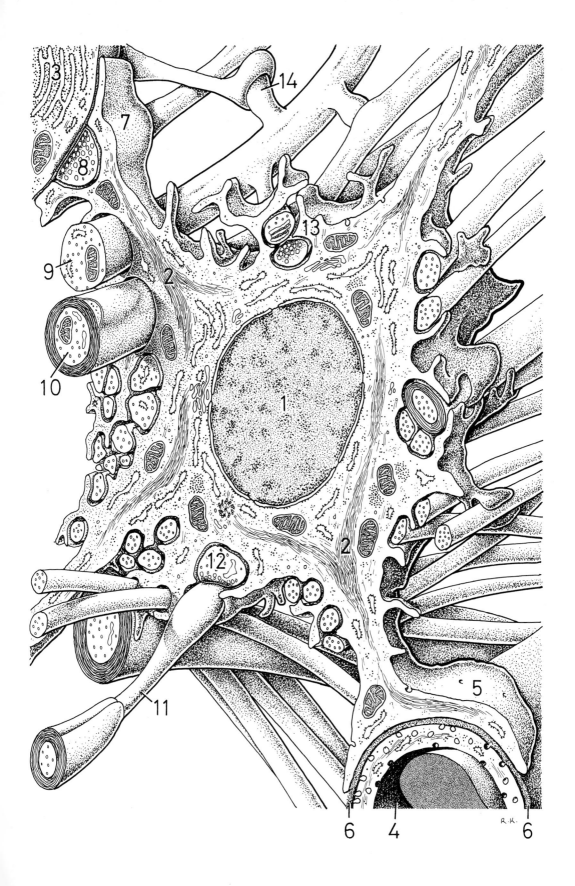

Plate 151. Glia of the Central Nervous System. Fibrous Astrocyte

The 8- to 10-μm-wide fibrous astrocytes occur, as stated in Plate 148, predominantly in the white matter of the central nervous system. These cells are characterized by 20–40 wirelike, occasionally branched processes (**1**) of various lengths (10–50 μm) that are inserted between myelinated (**2**) and unmyelinated (**3**) nerve fibers. The longer processes run parallel with the nerve fibers, whereas many of the transverse processes (arrow) contact blood capillaries (**4**).

The astrocyte body contains a nucleus with homogeneously dispersed heterochromatin. Moderately developed organelles, a few lysosomes, and small clusters of glycogen particles (**5**) occur in the cytoplasm.

As in protoplasmic astrocytes, gliofibrils (**6**), made up of gliofilaments, represent specific differentiations in fibrous astrocytes. Gliofilaments form a mechanically resistant skeleton in the perikaryon and its processes.

Fibrous astrocytes fulfill the same histophysiological functions in the white matter of the central nervous system as the protoplasmic astrocytes do in the gray. It has been proved that the two are variations of the same cell type which have adapted to different environments.

Following loss of nervous tissue, it is largely the function of fibrous astrocytes to replace the damaged or affected area. Though the ability of glia cells to divide is slight after birth, it is possible in the case of damage for mitoses to take place. Scar formation then occurs in the CNS and is termed gliosis.

Two nodes of Ranvier (**7**) can be recognized on myelinated nerve fibers. (See Plate 68 in KRSTIĆ 1979.)

Magnification: × 8,000

REFERENCES

Hirano A, Dembitzer HM (1967) A structural analysis of the myelin sheath in the central nervous system. J Cell Biol 34:555–567

Ribadeau Dumas J-L, Poirier J (1972) L'astrocyte. Nouv Presse Med 1:1091–1096

Vaughn JE, Peters A (1967) Electron microscopy of the early postnatal development of fibrous astrocytes. Am J Anat 121:131–152

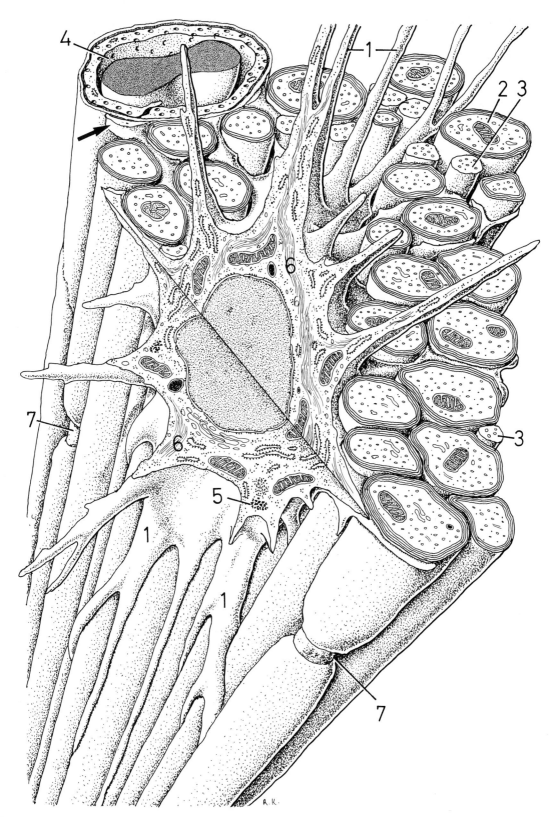

311

Plate 152. Glia of the Central Nervous System. Velate Astrocyte (Reconstructed from electron micrographs of CHAN-PALAY and PALAY 1972)

In the two previous plates, it was seen how protoplasmic and fibrous astrocytes affix nerve cells and nerve fibers by their processes. All these elements, together with the microglia cells and capillaries, form the neuropil – a densely packed cellular conglomerate which fills spaces between the perikarya of the nerve cells of the CNS. This figure shows that astrocytes also play a mechanical and metabolic role in the extensive synaptic regions.

The light-microscopic morphology of velate astrocytes in the granular layer of the cerebellum has already been seen in Plate 148. The cells appear stellate following special staining techniques, though the number of their processes is relatively small.

Electron-microscopic studies have shown that these cells are a specially differentiated type of astrocyte which have winglike processes, similar to cabbage leaves in appearance (arrows), instead of cylindrical processes. With these cytoplasmic sheets, the cells envelop capillaries (**1**) and nerve cells (**2**, here a small granule cell), as well as parts of the cerebellar glomeruli (**3**), which are characteristic of the granular layer of the cerebellar cortex. Under the light microscope, these cerebellar glomeruli appear as pale anuclear areas (see Plate 148); electron-microscopically, it has been determined that in these structures, synaptic contacts exist between mossy fibers (**4**) and dendrites (**5**) of the small granule cells. These complex structures are thus more or less bounded by the astrocyte extensions. It is assumed that some dendrites and/or axons (**6**) of nerve cells may pass through the astrocyte processes.

The cytoplasmic sheets appear free in this plate though they do of course surround other nerve cells or cerebellar glomeruli, which have been omitted for the sake of clarity.

From the behavior of the processes of this type of astrocyte, it is evident that these cells are not solely involved in the metabolism of nerve cells, but also supply the synapses with energy.

The basal lamina (**7**) can be observed between the astrocyte processes and the non-fenestrated blood capillary.

Magnification: × 5,000

REFERENCES

Chan-Palay V, Palay LS (1972) The form of velate astrocytes in the cerebellar cortex of monkey and rat: High voltage electron microscopy of rapid Golgi preparations. Z Anat Entwickl Gesch 138:1–19

Reese TS, Karnovsky MJ (1967) Fine structural localization of a blood-brain barrier to exogenous peroxidase. J Cell Biol 34:207–218

Vaughn JE, Peters A (1968) A third neuroglial cell type. An electron microscopic study. J Comp Neurol 133:269–288

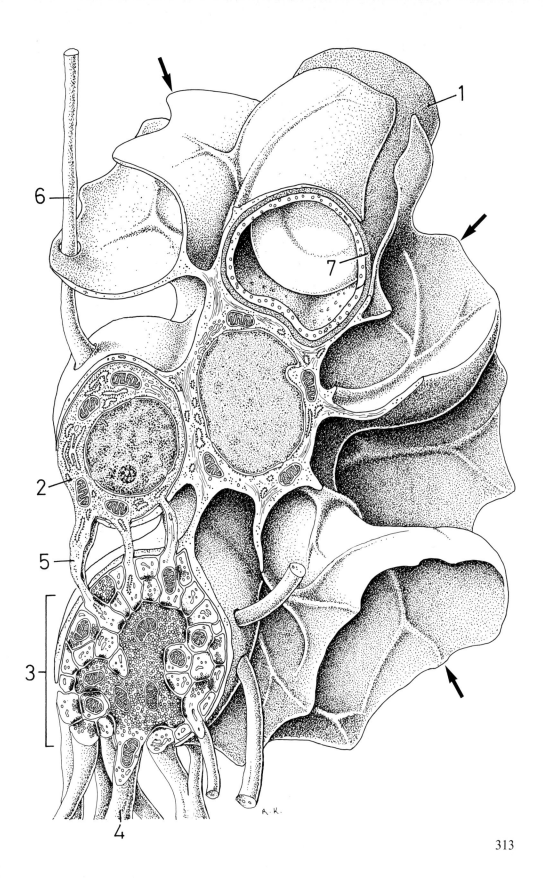

Plate 153. Glia of the Central Nervous System.
Membrana Limitans Gliae Superficialis and Perivascularis

In addition to other functions, astrocytes provide a boundary to the tissue of the CNS against the blood vessels that supply it and other adjacent tissues. The structure of these boundaries is as follows.

With their processes, protoplasmic astrocytes (**1**) form a three-dimensional network, the interstices of which contain neurons (not shown). All footlike astrocyte processes that extend to the surface of the brain are arranged adjacent to one another such that a membrana limitans gliae superficialis (**2**) forms. The innermost of the leptomeninges, the pia mater (**3**), is closely apposed to this membrane. Above the pia mater is found a connective tissue layer poor in blood vessels, the arachnoid membrane (**4**), connected to the pia mater by a loose trabecular framework (**5**). Thus, between the two leptomeninges there is a space containing numerous blood vessels (**6**), termed the subarachnoid space (**7**). Between the arachnoid and the dura mater (**8**), which is composed of dense connective tissue, occurs the slitlike subdural space (**9**).

Extensions of the subarachnoid space, the Virchow-Robin spaces (**10**), accompany the penetrating blood vessels (**11**) to a certain depth. These spaces contain collagen and reticular fibers, which gradually diminish with the reduction in blood vessel diameter until eventually the capillaries (**12**) completely lack a connective tissue envelope. In its place, as already mentioned in Plates 149 and 150, the astrocyte perivascular feet form a membrana limitans gliae perivascularis (**13**).

The two glial membranes provide the essential specific environment and isolation for the nervous tissue of the CNS. Here, the membrana limitans gliae perivascularis is important since it forms part of the blood-brain barrier.

The structures in insets **A** und **B** are enlarged in the corresponding figures of Plate 154.

Magnification: × 1,000

REFERENCES

Bondareff W, McLone DG (1973) The external glial limiting membrane in Macaca: ultrastructure of a laminated glioepithelium. Am J Anat 136:277–296

Jones EG (1970) On the mode of entry of blood vessels into the cerebral cortex. J Anat 106:507–520

Lierse W (1969) Was wissen wir über die Blut-Hirn-Schranke? Wissenschaft Dienst "Roche" 37:9–16

Morse DE, Low FN (1972) The fine structure of the pia mater of the rat. Am J Anat 133:349–368

Plate 154. Glia of the Central Nervous System.
Membrana Limitans Gliae Superficialis and Perivascularis.
Blood-Brain Barrier. Barrier-Free Regions of the Brain.
Continuation of Plate 153

The end feet (Fig. A 1) of the astrocytes contact one another by means of a nexus (Fig. A 2) and form the membrana limitans gliae superficialis. The glial processes are separated from the pia mater (Fig. A 3) and its collagen microfibrils (Fig. A 4), fibrocytic processes (Fig. A 5), and blood capillaries (Fig. A 6) by a basal lamina (Fig. A 7). Beneath the surface glial membrane are located numerous processes (Fig. A 8) of nerve and glia cells belonging to the neuropil.

The membrana limitans gliae perivascularis is, as already mentioned, made up of the perivascular feet (Fig. B 1) of astrocytes. The perivascular glial layer is separated from a nonfenestrated brain capillary (Fig. B 2) and its pericyte (Fig. B 3) by a 20- to 50-nm-thick basal lamina (Fig. B 4). Outside the membrana limitans gliae perivascularis is the neuropil (Fig. B 5). Calculations have shown that 25%–80% of the neuropil volume of the rat consists of glia cell processes. The inset in Fig. B corresponds to Fig. C.

It has been known for some time that vital stains, e. g., trypan blue, which is capable of staining certain cells blue, are unable to penetrate the brain, with the exception of a few restricted areas. The same applies to some substances circulating in the blood plasma. This selectivity of the brain, which affects metabolic processes, is termed the blood-brain barrier (BBB). Morphologically, the BBB consists of zonulae occludentes (Fig. C 1) joining nonfenestrated endothelial cells (Fig. C 2), a basal lamina (Fig. C 3), and the perivascular feet (Fig. C 4) of astrocytes. The absence of pericapillary spaces is very characteristic of the BBB.

The extracellular spaces in the CNS according to a recent estimation comprise 17%–20% of the total volume of the brain. Investigations have shown that these spaces hold a proteoglycan-containing ground substance which could hinder free ionic diffusion. In addition, endothelial cells and astrocytes are able to regulate the selectivity of the BBB. It is assumed that some endothelial cells exhibit polar differentiation and transport certain substances only in one direction. (See physiology texts for further information.)

There are also, however, areas of the brain which lack a BBB. Here, the capillaries are fenestrated and surrounded by broad pericapillary spaces; the membrana gliae perivascularis is only partially present. Such structural features appear in the choroid plexus (Fig. D 1), pineal body (Fig. D 2), subcommissural organ (Fig. D 3), area postrema (Fig. D 4), neurohypophysis (Fig. D 5), median eminence (Fig. D 6), and infundibulum.

The BBB is in practice of great importance in the medical treatment of cerebral diseases, since it can prevent the passage of drugs into nervous tissue. (See Plates 148, 150 in KRSTIĆ 1979.)

Magnifications: Figs. **A, B,** × 10,000;
Fig. **C,** × 40,000

REFERENCES

Cragg B (1979) Brain extracellular space fixed for electron microscopy. Neurosci Lett 15:301–306

Hajos F, Basco E (1984) Surface-contact glia. Adv Anat Embryol Cell Biol 84:1–100

Wagner H-J, Barthel J, Pilgrim C (1983) Permeability of the external glial limiting membrane of rat parietal cortex. Anat Embryol 166:427–437

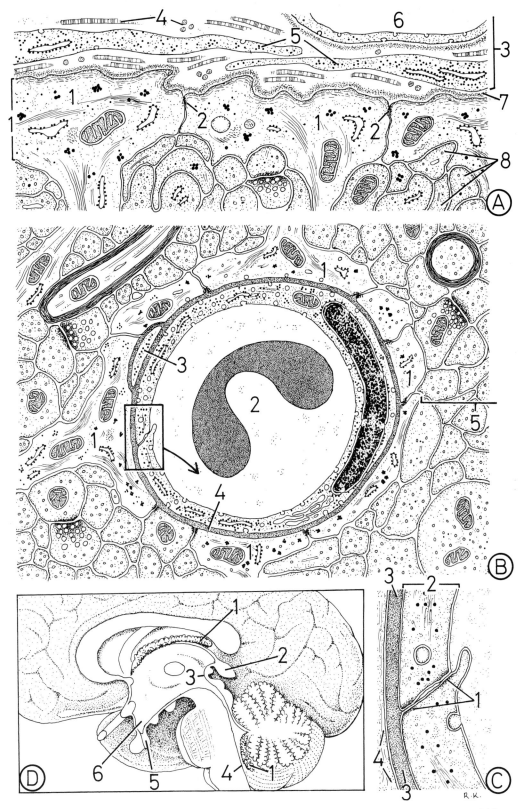

Plate 155. Glia of the Central Nervous System. Oligodendrocytes

The 6- to 8-μm-large oligodendrocytes are found in the direct vicinity of nerve cells and their processes and blood capillaries. The light-microscopic appearance of these cells has already been shown in Plate 148.

Oligodendrocytes (Fig. A 1) usually appear in the electron microscope as ovoid or ellipsoidal cells with a large heterochromatin-rich nucleus. A few mitochondria and relatively voluminous Golgi complexes are present in the narrow belt of cytoplasm. Owing to the large numbers of cisternae of rough endoplasmic reticulum and free ribosomes, the cytoplasm of oligodendrocytes appears significantly darker than that of astrocytes. Oligodendrocytes bear only a few relatively short conical and platelike processes. Functionally, these processes are of particular importance in forming the myelin sheaths in the CNS. It is assumed that the leaflike processes with their tonguelike cytoplasmic thickenings (Fig. A 2) roll around the axons (Fig. A 3), though the exact manner is unknown. One oligodendrocyte can provide a segment or internode of the myelin sheath (Fig. A 4) for several axons. The nodes of Ranvier (arrows) are sites where the myelin sheath is lacking. At these places, the axons are in direct contact with their environment, e.g., with boutons terminaux (Fig. A 5). Schmidt-Lanterman incisures are almost never present in the myelin sheaths of the CNS. In the upper right of the drawing, a cutaway section of a nerve cell (Fig. A 6) and the axon hillock (Fig. A 7) can be seen.

Several boutons terminaux (Fig. A 8) are in contact with the neuron. Figure **B** shows the development of the myelin sheath as currently understood, which is based on the fact that the innermost myelin lamella is always the shortest. It is supposed that the lamellar oligodendrocyte processes (Fig. B 1) have a trapezoid form with the broader base (Fig. B 2) toward the cell. Following attachment of the shorter side (Fig. B 3) to the axon (Fig. B 4), the cell process then probably rotates about the axon, thus forming the myelin sheath.

The peculiar contractions of oligodendrocytes observed in tissue culture for the main part remain to be clarified. (See Plate 89 in Krstić 1979.)

Magnification: Fig. **A**, × 14,000

REFERENCES

Gonatas NK, Hirayama M, Stieber A, Silberberg DH (1982) The ultrastructure of isolated rat oligodendroglial cell cultures. J Neurocytol 11:997–1008

Hirano A (1968) A confirmation of the oligodendroglial origin of myelin in the adult rat. J Cell Biol 38:637–640

Knobler RL, Stempak JG, Laurencin M (1976) Nonuniformity of the oligodendroglial ensheathment of axons during myelination in the developing rat central nervous system. J Ultrastruct Res 55:417–432

Kruger L, Maxwell DS (1966) Electron microscopy of oligodendrocytes in normal rat cerebrum. Am J Anat 118:411–436

Norton WT (ed) (1984) Oligodendroglia. Plenum, New York

Sturrock RR (1982) Gliogenesis in the prenatal rabbit spinal cord. J Anat 134:771–793

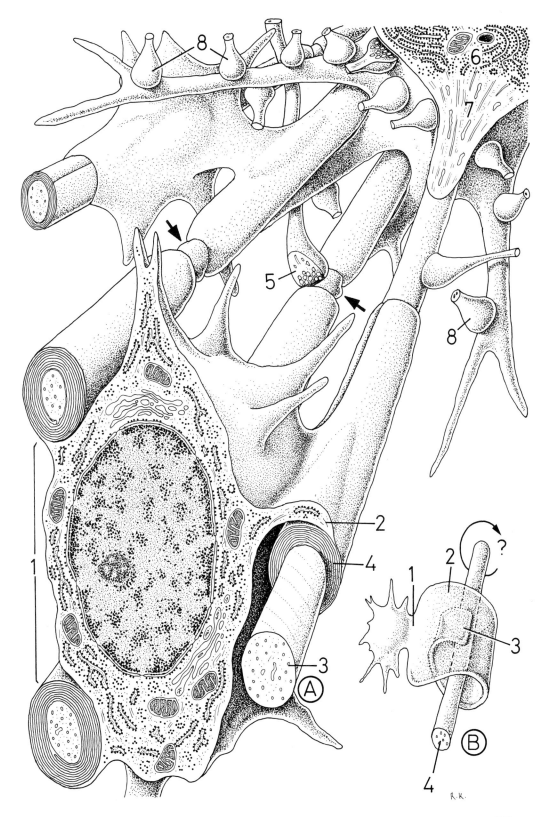

Plate 157. Form of Nerve Cells Under the Light Microscope

Excitability, conduction, and processing of information are the basic properties of nerve cells (neurocytes, ganglion cells). Their morphology can be studied following Golgi staining.

As was evident in the previous plates, both simple and complex ganglion cells exist, however they are always constructed according to the same principle. The structure of a ganglion cell will be discussed with the example of a simple bipolar nerve cell (Fig. A 1) from Corti's ganglion.

The dendrite (Fig. A 2) conducts the stimulus in the direction of the arrows toward the cell body, i.e., it is the receptive part of the cell. In the perikaryon, the information is processed, altered, and then conducted to the axon (Fig. A 3), also in the direction of the arrow. An axon thus has an effector function. The framework of neurofibrils (Fig. A 4), which occurs in all nerve cells, can also be seen.

Nerve cells with all their processes represent genetic, morphological, biological, and regenerative units of the nervous tissue which are termed neurons. Together with elements of the neuroglia they form the nervous system.

The behavior of the processes and the form of the perikaryon of a nerve cell can vary considerably according to function. Thus, e.g., multipolar ganglion cells from the spinal cord are distinguished by an irregular perikaryon (Fig. D 1); they have many dendrites (Fig. D 2) extending in various directions and only one, occasionally 1-m-long, axon (Fig. D 3) with lateral ramifications, the collaterals or paraxons (Fig. D 4). It is apparent that the axon, in contrast to the dendrites, always has the same diameter.

Pyramidal cells from the gray matter of the cerebrum can be observed in section to have a triangular perikaryon (Fig. E 1) and numerous horizontally branching dendrites (Fig. E 2), bearing small dendritic spines. At the base of the triangle, an axon and its paraxons (Fig. E 3) leave the perikaryon.

The body of Purkinje cells (Fig. G 1) from the cerebellum is in the form of a pear and the branching dendrites (Fig. G 2), with great numbers of spinous processes (Fig. G, arrows), give the cell a treelike appearance. There is one long axon (Fig. G 3). These cells belong to the Golgi type I neurons, i.e., nerve cells that are characterized by many dendrites and long axons that run into the white matter.

Small neurons, e.g., basket cells (Fig. F) and granule cells (Fig. H) of the cerebellum, are rich in dendrites (Figs. F 1, H 1) but only bear short axons (Figs. F 2, H 2), which never leave the gray matter. They are termed Golgi type II neurons and serve to conduct impulses to several neurons within the gray matter.

This plate also shows two exceptions to the rule according to which every nerve cell has one axon and several dendrites. Spinal ganglion cells (Fig. B) are considered biaxonal and adendritic, and amacrine cells (Fig. C) from the retina are nerve cells that apparently only possess dendrites.

Magnifications: Figs. A–H, ×600

REFERENCES

Bray D, Gilbert D (1981) Cytoskeletal elements in neurons. Ann Rev Neurosci 4:505–523

Peters A, Palay SL, Webster H de F (1970) The fine structure of the nervous system: The cells and their processes. Harper and Row, New York

Pfenninger KH (1978) Organization of neuronal membranes. Ann Rev Neurosci 1:445–471

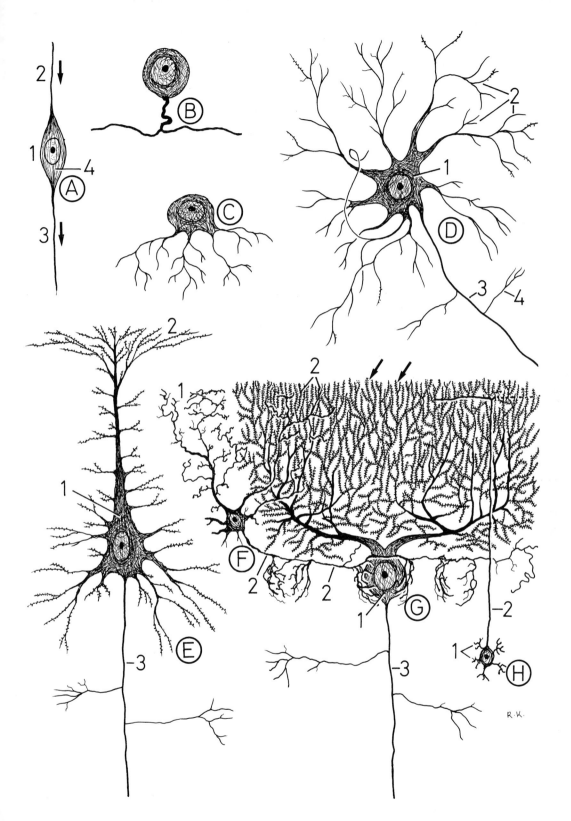

NERVOUS TISSUE

Plate 158. Various Types of Nerve Cell

According to the criteria of shape of the cell body and the number of processes, the following types of neuron can be distinguished.

A. Apolar nerve cells are neurocytes without dendrites and axons which only occur at the beginning of histogenesis (Fig. **A1**). Subsequently, they develop into the various nerve cells (Fig. **A2**) listed below. Broadly speaking, the hair cells (Fig. **A3**) of the vestibuloacoustic organ and taste buds can also be classed as apolar cells.

B. Unipolar nerve cells only appear in the form of rod (Fig. **B1**) and cone cells (Fig. **B2**) of the retina. Both have axonlike processes (arrows) stemming from a single cell pole.

C. Pseudounipolar nerve cells are initially bipolar neurons (Fig. **C1**) whose processes gradually join together close to their points of origin (Fig. **C2**, in the direction of the arrows) as a result of asymmetrical cell growth. The T-shaped process (Fig. **C3**) then develops that is characteristic of the ganglion cells of the spinal ganglia and ganglia of the cerebral nerves, except for the ganglion of the vestibuloacoustic nerve.

D. Bipolar nerve cells have one dendrite (Fig. **D1**) and one axon (Fig. **D2**), which are very difficult to differentiate morphologically. Bipolar nerve cells are found in the retina (as the second neuron of the visual pathway) and as the first neurons of the vestibuloacoustic organ.

E. Multipolar nerve cells are the most frequent kind of neuron. As stated in the previous plate, there are two types of multipolar neuron: Golgi type I, with long axons, and Golgi type II, with short axons.

REFERENCES

Akert K (1971) Struktur und Ultrastruktur von Nervenzellen und Synapsen. Klin Wochenschr 49:509–519

Hyden H (ed) (1967) The neuron. Elsevier, Amsterdam

Picard D, Michel-Bechtet M, Athouel AM, Rua S (1972) Granules neurosécrétoires, lysosomes et complexe GRL dans le noyau supra-optique du rat. Bipolarité des complexes golgiens. Exp Brain Res 14:331–353

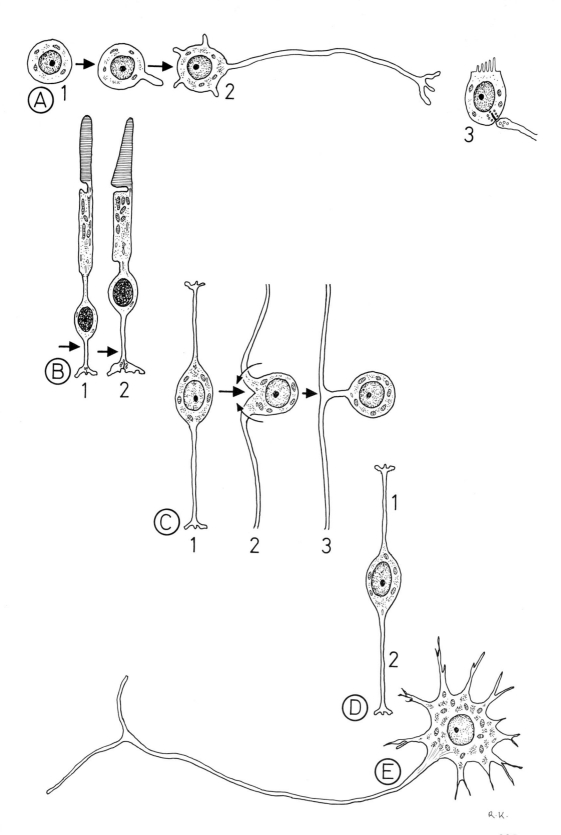

Plate 159. Arrangement of Nerve Cells

In the gray matter of the cerebrum and cerebellum, the neurons form more or less well-defined layers. In other parts of the central nervous system, they occur in groups or nuclei in which nerve cells do not have a particular order within the neuropil. Purkinje cells (1) of the cerebellum, however, are an exception to this rule.

In plan view, these neurons appear as already depicted in Plate 157. Purkinje cells have been shown to be completely flattened elements, tightly packed and arranged in a consecutive fashion. Their highly branching dendrites (2), with dendritic spines (3), run in a plane perpendicular to the axis of the folia of the cerebellum.

Transverse sections show the Purkinje cells (1a) to be spindle- or bottle-shaped, narrow, and with poorly developed dendrites.

A small granule cell (4) makes synaptic contact by means of its T-shaped branched axon (5) with large numbers of apposed Purkinje cells.

Magnification: × 1,500

REFERENCES

Fox CA, Barnard JW (1957) A quantitative study of the Purkinje cell dendritic branchlets and their relations to afferent fibers. J Anat 91:299–313

Hanna RB, Hirano A, Pappas GD (1976) Membrane specializations of dentritic spines and glia in the weaver mouse cerebellum: a freeze-fracture study. J Cell Biol 68:403–410

Herndon RM (1963) The fine structure of the Purkinje cell. J Cell Biol 18:167–180

Scherini E, Bolchi F, Biggiogera M, Bernocchi G (1981) Further evidence of different morphofunctional aspects in the Purkinje cell population of adult rat cerebellum. An ultrastructural study. J Submicrosc Cytol 13:17–29

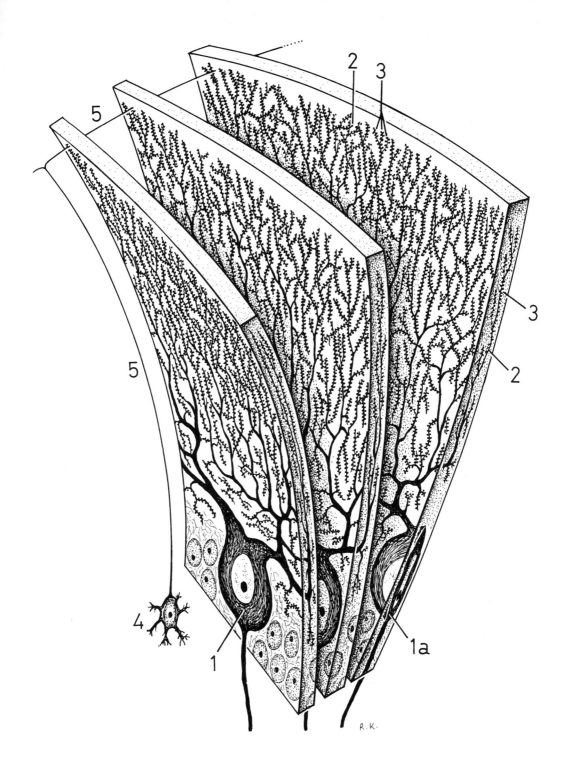

NERVOUS TISSUE

Plate 160. Nerve Cell and Neuropil

Several dendrites (1) and a myelinated axon (2) leave the voluminous perikaryon of a pyramidal cell. The cytoplasm of the nerve cells, here termed neuroplasm, contains a nucleus (3), with a distinct nucleolus (4), and a well-developed, frequently multiple Golgi apparatus (5). Cisternae of rough endoplasmic reticulum are also developed to a high degree and largely occur as groups of parallel double lamellae which form the characteristic Nissl bodies (6). They are abundant throughout the whole of the neuroplasm and dendrites. Nissl bodies are only absent from the axon and from the small zone where the axon leaves the cell body, the axon hillock (7). Many mitochondria, lysosomes, and occasionally lipofuscin granules can be seen between the lamellae of rough endoplasmic reticulum.

The 20-nm-thick neurotubules and approximately 7-nm-thick neurofilaments, which are very numerous in the neuroplasm, together form the light-microscopically visible neurofibrils.

Since nerve cells have completely lost the ability to divide, centrioles are not found in the neuroplasm.

Large numbers of boutons terminaux (8) and several astrocyte processes (9) are in contact with the plasmalemma of nerve cells. One of these processes (arrow) simultaneously contacts the neurocyte and a capillary (10).

The pyramidal cell in this plate has been drawn so that it projects from the tightly packed neuropil (11), which has here been cut as a cube. In this way, it is easier to follow the path of the glial and nerve cell processes and obtain a clearer impression of the closed cellular union which is the nervous tissue of the CNS. (See Plates 44, 45, 119–121 in KRSTIĆ 1979.)

Magnification: × 3,500

REFERENCES

Broadwell RD, Cataldo AM (1983) The neuronal endoplasmic reticulum: Its cytochemistry and contribution to the endomembrane system: I. Cell bodies and dendrites. J Histochem Cytochem 31:1077–1088

Brunk U, Ericsson JLE (1972) Electron microscopical studies on rat brain neurons. Localization of acid phosphatase and mode of formation of lipofuscin bodies. J Ultrastruct Res 38:1–15

Chan-Palay V (1973) The cytology of neurons and their dendrites in the simple mammalian nucleus lateralis. An electron microscope study. Z Anat Entwickl Gesch 141:289–317

Conradi S, Kellerth JO, Berthold CH (1979) Electron microscopic studies of serially sectioned cat spinal α-motoneurons: II. A method for the description of architecture and synaptology of the cell body and proximal dendritic segments. J Comp Neurol 184:741–754

Palay SL, Chan-Palay V (1973) High voltage electron microscopy of the central nervous system in Golgi preparations. J Microsc 97:41–47

Pfenninger KH (1973) Synaptic morphology and cytochemistry. Progr Histochem Cytochem 5:1–86

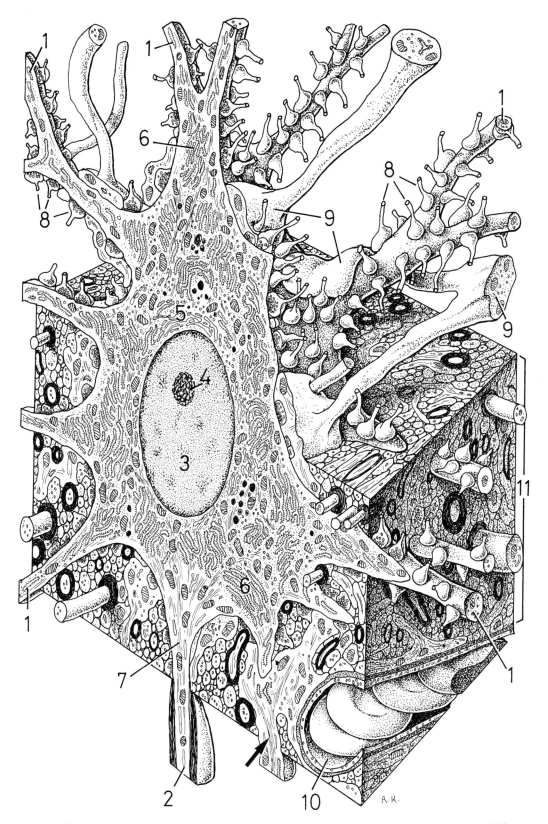

Plate 161. Nerve Cell. Neurofibrils

The light-microscopic appearance of neurofibrils has already been depicted in Plate 157. The present, somewhat schematized, reconstruction was prepared from a study of 2- to 3-µm-thick sections in the high-voltage electron microscope.

The nucleus (**1**) is located in the center of the nerve cell body and is surrounded by flattened, perforated lamellae of the multiple Golgi apparatus (**2**). These lamellae, recognizable in sections as Golgi fields, are connected by means of numerous thin tubules (**3**).

Neurofibrils (**4**), which are made up of neurofilaments, run through all parts of the perikaryon and extend into the dendrites (**5**) and axon (**6**). In this manner, a fibrillar framework develops within every nerve cell, which not only imparts mechanical strength, but is possibly also responsible for conducting vesicular structures and impulses through the cell body.

Mitochondria (**7**) are contained within the interstices of the neurofibrillar network. (See Plates 31–33, 66 in KRSTIĆ 1979.)

Magnification: ×4,000

REFERENCES

Bargmann W (1966) Neurosecretion. Int Rev Cytol 19:183–201

Bargmann W (1971) Die funktionelle Morphologie des endokrinen Regulationssystems. In: Bargmann W, Kühnau J, Siebenmann RE, Steiner H, Uehlinger E (eds) Endokrine Regulations- und Korrelationsstörungen. Handbuch der allgemeinen Pathologie, vol 8/1, pp 1–106. Springer, Berlin Heidelberg New York, pp 1–106

Kilarski W, Koprowski H (1976) Observation of whole, cultured human brain cells using 100 kilovolts electron microscopy. M Microsc Biol Cell 25:73–80

Leterrier JF, Liem RKH, Shelanski ML (1982) Interactions between neurofilaments and microtubule-associated proteins: a possible mechanism for intraorganellar bridging. J Cell Biol 95:982–986

Normann TC (1976) Neurosecretion by exocytosis. Int Rev Cytol 46:1–78

Rash JE (1983) The rapid-freeze technique in neurobiology. Trends Neurosci 6:208–212

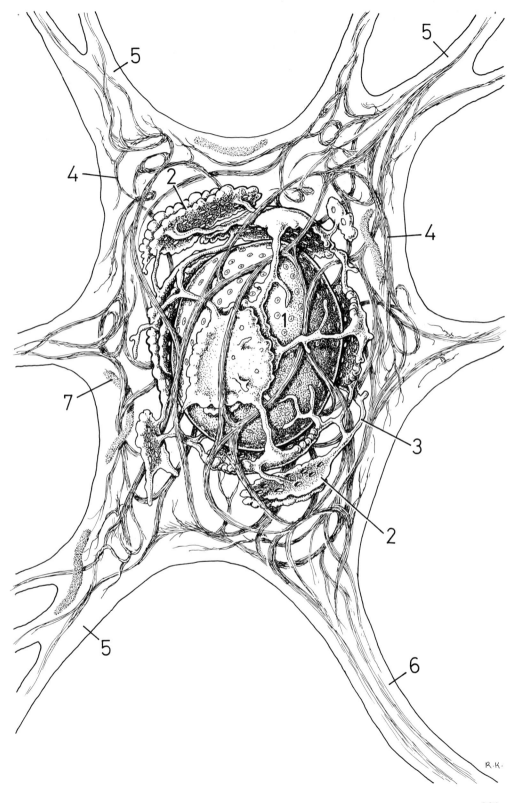

Plate 162. Neurosecretory or Neuroendocrine Nerve Cells.
Example: Cells of the Nucleus Supraopticus and Paraventricularis

Certain nerve cells respond to stimuli not by producing bioelectric impulses but by secretory activity. These so-called neuroendocrine cells occur in groups, nuclei, predominantly in the hypothalamus (Fig. A, inset). A three-dimensional, schematic section of the hypothalamus is presented in Fig. B.

Several boutons terminaux (Fig. B 1) of other nerve cells make synaptic contacts with the stellate neurons of the nucleus supraopticus (NSO) and nucleus paraventricularis (NPV). A few dendritic processes (Fig. B 2) of neurons situated in the vicinity of the ventricular wall project through the ependyma (Fig. B 3) and reach the cerebrospinal fluid; the cells concerned are neurosecretory cerebrospinal fluid-contacting neurons (see Plate 147). Axons (Fig. B 4) of NSO and NPV neurons extend to the neurohypophysis (Fig. B 5). Some of these axons come into contact with capillary loops (Fig. B 6) of the hypophyseal stalk, whereas the expanded terminations of other axons rest on capillaries (Fig. B 7) of the neurohypophysis. A series of numerous thickenings, Herring bodies (Fig. B 8), can be seen along the axons. All these axons together comprise the neurosecretory hypothalamohypophyseal tract (Fig. B 9).

Special staining techniques reveal neurosecretory cells (Fig. C) to be spindle-shaped or stellate elements, usually close to capillaries. Axons (Fig. C 1) rich in secretory product run between the nerve cells.

The inset in Fig. C corresponds to Fig. D. Like all nerve cells, neurosecretory cells contain Nissl bodies (Fig. D 1), mitochondria, and lysosomes. Osmiophilic, 100- to 200-nm-wide, secretory granules (Fig. D 2), containing neurohormones and surrounded by a unit membrane, detach themselves from the Golgi apparatus and move into the axon (Fig. D, E, F 3). The nerve cell is separated from the capillary (Fig. D 4) by an extremely thin investment of astrocytes (Fig. D 5) and two basal laminae (Fig. D 6). Boutons terminaux (Fig. D 7) are located on the cell body and its processes.

Figure E is a continuation of Fig. D and shows a small Herring body, which contains numerous neurosecretory granules (Fig. E 2) as well as mitochondria and cisternae of smooth endoplasmic reticulum (Fig. E 8). Neurohormones in the neurosecretory granules are bound to a carrier, neurophysin.

Lower down, the neurosecretory axon (Fig. F 3) terminates in the immediate proximity of a capillary (Fig. F 9) with fenestrated endothelium. The nerve ending contains mitochondria, cisternae of smooth endoplasmic reticulum, neurotubules, many neurosecretory granules, and synaptoid vesicles (Fig. F 10). Synapses (Fig. F 11) with other axons also exist.

Neurosecretory granules (Fig. F 2) when required are expelled from the nerve endings into the pericapillary spaces (Fig. F 12) by exocytosis. From here, the neurohormones, now freed from the neurophysin, enter the blood circulation and are transported to the effector cells, where they carry out their specific function.

The neurosecretory cells of the NSO synthesize antidiuretic hormone, and those of the NPV oxytocin. (See Plates 32, 33, 64 in KRSTIĆ 1979.)

Magnifications: Fig. C, ×400;
Figs. D–F, ×17,000

REFERENCES

Boudier JA, Boudier JL, Massacrier A, Cau P, Picard D (1979) Structural and functional aspects of lysosomes in the neurosecretory neurons. Biol Cell 36:185–192

Farner DS, Lederis K (eds) (1981) Neurosecretion. Plenum, New York

Krieger D, Hughes JC (1980) Neuroendocrinology. Blackwell, Oxford

Nordmann JJ, Labouesse J (1981) Neurosecretory granules: Evidence for an aging process within the neurohypophysis. Science 211:595–597

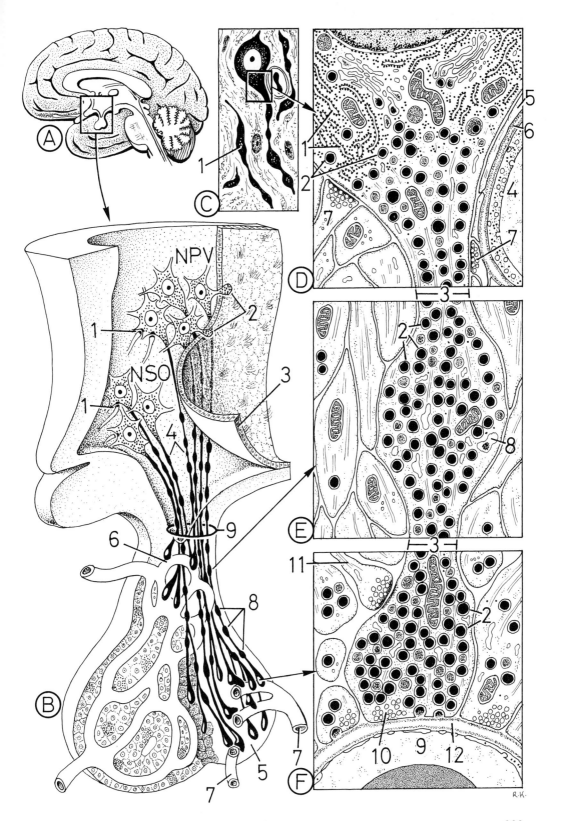

Plate 163. Neurosecretory Nerve Cell. Simplified General Scheme of the Mode of Action of Polypeptide Hormones

Advances in neuroendocrinology have shown that neurosecretory cells are at the beginning of a chain of hormonal events. Recently, the mode of action of hormones has been investigated more closely at the molecular level. This simplified scheme of the mechanisms of action of polypeptide hormones was prepared on the basis of these new findings.

Synaptic contacts (**1**), as mentioned in Plate 162, and other possible sources of excitation cause the neuroendocrine cell (**2**) to be stimulated in the direction of the unshaded arrows. The cell synthesizes its hormone and secretes it (symbolized as black spherules, **3**) into the blood circulation. Since the hormone carries information determining the behavior of certain cells, it is termed a "first messenger."

Of particular importance in the further action of the hormone is a specific union between the hormone and a genetically determined receptor sites in the plasmalemma of the target cell. Several such receptor sites can be seen as hemispherical indentations (**4**) at the level of the outer layer of the cell membrane (**5**). Thus, "recognition" of the hormone is dependent on the form of the receptor on the target cell membrane: It is believed that the hormone fits into the receptor sites like a key into a lock.

The hormone-receptor complex (**6**) activates an enzyme, adenyl cyclase (**7**), located on the inner side of the cell membrane, which forms cyclic adenosine monophosphate (cAMP) from adenosine triphosphate (ATP) in the presence of Mg^{2+} ions. In this way, the hormonal information becomes amplified 10^5- to 10^9-fold and is transferred to the cytoplasm without the hormone itself entering the cell. Since cAMP occurs in practically all cells it is referred to as the "second messenger." It activates the specific enzymes and/or genes of target cells, in addition to affecting their permeability. As a response, the target cells secrete specific products (**8**) which can act on other cells (**9**) or inhibit the activity of the hormone-producing cells in the form of a negative feedback (**10**). After stimulation of adenyl cyclase by the hormone-receptor complex, the hormone becomes inactivated (**11**).

The high specificity of the hormone is also indicated in this plate. It is clear that the morphological (and chemical) receptors of the other two cells (**12, 13**) are incapable of binding the structurally different hormone.

This scheme applies to the action of epinephrine and all polypeptide hormones. For all practical purposes, it can be assumed that the mode of action of neurohormones is based on the same mechanism. A discussion of the mechanism of action of steroid hormones is beyond the scope of this book. (See Plates 153, 154 in KRSTIĆ 1979 and biochemistry and physiology texts for further information.)

REFERENCE
Litwack G (ed) (1982) Biochemical actions of hormones. Academic, New York

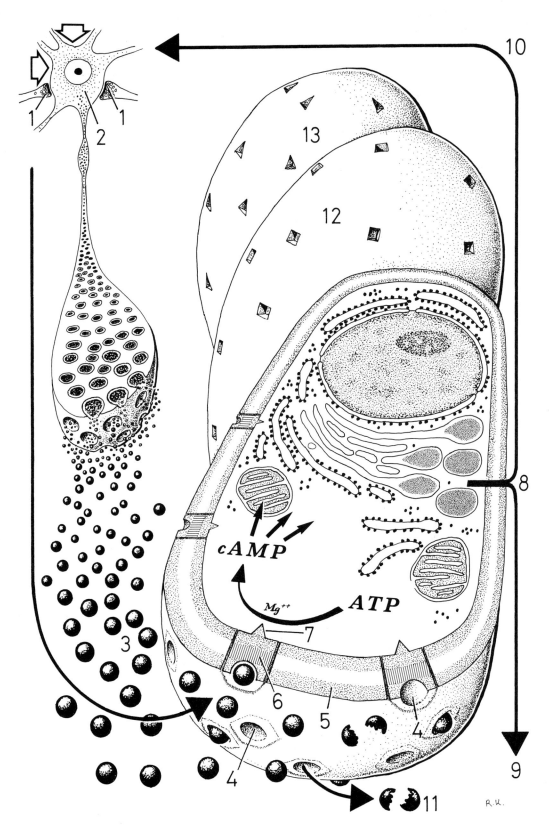

Plate 164. Nerve Fibers

A nerve fiber signifies an axon and its envelopes. Two types of nerve fiber exist – myelinated and unmyelinated.

Myelinated, rapidly conducting nerve fibers have a diameter of 1–20 µm and are up to 1 m long. In a transverse section stained to show the myelin sheath, the nerve fiber (Fig. A 1) appears as a dark circle with a pale centrum, representing the axon (Figs. A 2, B 2). The Schwann's cells (Figs. A 3, B 3) surround the whole structure in the form of a signet ring. In routine preparations, the lipid-containing myelin sheath is dissolved, leaving only a spongelike neurokeratin material (Fig. B 4), corresponding to the proteinaceous components of myelin.

In longitudinal sections, the axon (Fig. C 1), the myelin sheath (Fig. C 2), and the Schwann's cell (Fig. C 3) can be distinguished. Nodes of Ranvier (Fig. C 4) are present in the peripheral (PNS) and central (CNS) nervous systems at points where the myelin sheath is interrupted. The internode (Fig. C 5) is an approximately 50 µm–1-mm-long segment between two nodes of Ranvier and corresponds to one Schwann's cell (or one oligodendrocyte process in the CNS, see Plate 146). The pale, oblique lines within the myelin sheath, the Schmidt-Lanterman incisures (Fig. C 6), are almost exclusively restricted to nerve fibers of the PNS.

In the electron microscope, transversely sectioned nerve fibers are seen to comprise, proceeding from the center to the exterior, the axon with neurofilaments (Fig. D 1), neurotubules (Fig. D 2), cisternae of smooth endoplasmic reticulum (Fig. D 3), and mitochondria (Fig. D 4). The axon is delimited by the axolemma (Fig. D 5), separating it from the myelin sheath (Fig. D 6). The myelin sheath is composed of numerous myelin lamellae, which originate from the plasmalemma of the Schwann's cell. The exact manner in which these lamellae wrap around the axon is not precisely known. It is clear that the myelin sheath is a component of the Schwann's cell.

Schwann's cells (Fig. D 7), which represent the inner envelope of the nerve fibers, sometimes called the neurolemma, belong to the peripheral glia and have an ellipsoidal nucleus, poorly developed organelles, but a large number of free ribosomes. Two closely apposed cell membranes of the Schwann's cell on the left form the external mesaxon (Fig. D 8), which runs through the Schwann's cell cytoplasm.

The outer nerve fiber sheath or endoneurial sheath, which only occurs in the PNS, is made up of the basal lamina (Fig. D 9) of the Schwann's cell and a network of reticular and collagen microfibrils (Fig. D 10). Part of the basal lamina has been lifted back so that the fingerlike processes (Fig. D 11) of Schwann's cells are exposed in the region of the node of Ranvier (Fig. D 12).

Whereas in myelinated nerve fibers only one axon is contained within one Schwann's cell, in unmyelinated nerve fibers this cell embraces several axons (Fig. D 13). The axons usually possess a mesaxon (Fig. D 14), though many occur as so-called naked axons (Fig. D 15). Unmyelinated fibers are not segmented (see Plate 146), but in the PNS they have an endoneurial sheath consisting of a basal lamina (Fig. D 9) and interlaced collagen and reticular microfibrils (Fig. D 10).

There are no morphological criteria that allow a differentiation to be made between afferent and efferent nerve fibers. (See plates 86–88 in KRSTIĆ 1979.)

Magnifications: Figs. **A–C**, × 2,000; Fig. **D**, × 14,000

REFERENCES
Carey JD, Eldridge CF, Cornbrooks CJ, Timpl R, Bunge RP (1983) Biosynthesis of type IV collagen by cultured rat Schwann cells. J Cell Biol 97:473–479
Carlsen F, Behse F (1980) Three dimensional analysis of Schwann cells associated with unmyelinated nerve fibres in human sural nerve. J Anat 130:545–557
Bunge MB, Williams AK, Wood PM, Uitto J, Jeffrey JJ (1980) Comparison of nerve cell and nerve cell plus Schwann cell cultures, with particular emphasis on basal lamina and collagen formation. J Cell Biol 84:184–202

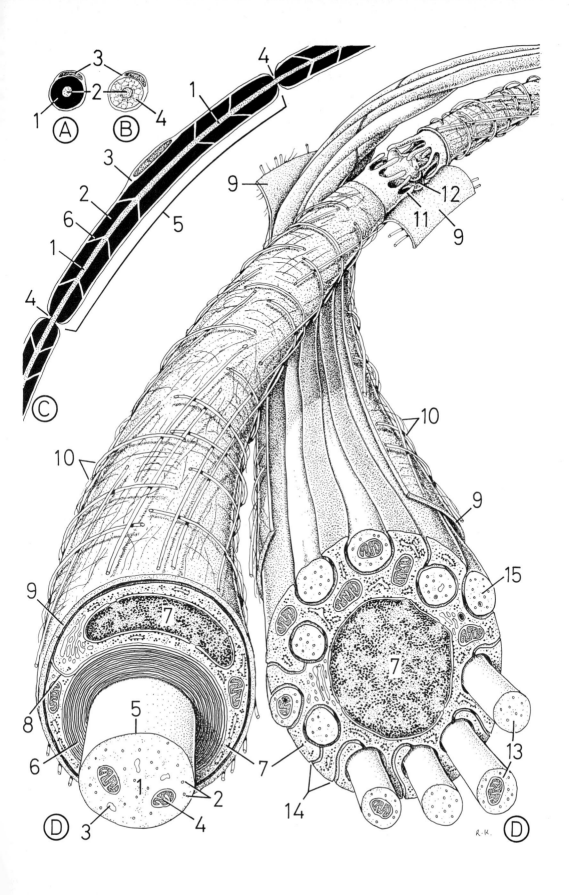

Plate 165. Nerve Fiber. Node of Ranvier

The small drawings at the bottom right show the light-microscopic appearance of the node of Ranvier following azan staining (Fig. A), osmium fixation (Fig. B), and silver staining (Fig. C). In the latter method, diffusion of the silver nitrate produces a cross (Ranvier's cross, Fig. C1) in the region of the node.

Between the myelin-free paranodal end portions of two Schwann's cells (Fig. D1), the axon (Fig. D2) becomes slightly thickened. At this point, the axolemma (Fig. D3) comes into contact with the interdigitating processes (Fig. D4) of Schwann's cells and of the basal lamina (Fig. D5) of nerve fibers. Thus, in this region the axon directly contacts the endoneurial sheath. The interior of the axon contains neurofilaments (Fig. D6), neurotubules (Fig. D7), and cisternae of smooth endoplasmic reticulum (Fig. D8). (Mitochondria are not depicted.) Cytoplasmic processes (Fig. D9) from the Schwann's cells appose the external surface of the axon. These processes are separated from one another by mesaxons (Fig. D10) and connected with the axolemma by means of dense bars (Fig. D11). On the external surface of the axon, the cytoplasmic processes are fluted in a spiral fashion (Fig. D12). In the upper Schwann's cell, a few mesaxons and processes have been partially removed. Here it is apparent that the innermost myelin lamellae are the shortest, and thus it may be assumed that the rolling of the myelin lamellae proceeds roughly according to the mechanism depicted in Plate 155. In both Schwann's cells, the external mesaxons (Fig. D13) and the fibrous elements of the endoneurial sheath (Fig. D14) outside the basal lamina (Fig. D5) are evident.

The velocity of propagation of stimuli varies between 0.5 and 120 m/s in myelinated nerve fibers. It is dependent on the structure of the particular nerve fiber.

In myelinated nerve fibers, all excitatory and conducting processes of the action potentials occur at the nodes of Ranvier. The action potential jumps from one node to the next, a phenomenon termed saltatory conduction. The myelin sheaths of the internodal segments can thus be regarded as a form of insulation. (See Plate 88 in Krstić 1979 and physiology texts for further information.)

Magnifications: Figs. A–C, ×900; Fig. D, ×55,000

REFERENCES

Caley DW, Butler AB (1974) Formation of central and peripheral myelin sheaths in the rat. An electron microscopic study. Am J Anat 140:339–349

Ghabriel MN, Allt G (1982) The node of Ranvier. In: Harrison RJ, Nava-Ratam V (eds) Progress in anatomy, vol 2. Cambridge University Press, Cambridge

Glees P, Meller K (1968) Morphology of neuroglia. In: Bourne GH (ed) The structure and function of nervous system: I. Structure. Academic, New York, pp. 301–325

Raine CS, Finch H, Masone A (1983) Axoplasmic asymmetry at the node of Ranvier. J Neurocytol 12:533–536

Waxman SG (1974) Ultrastructural differentiation of the axon membrane at synaptic and nonsynaptic central nodes of Ranvier. Brain Res 65:338–342

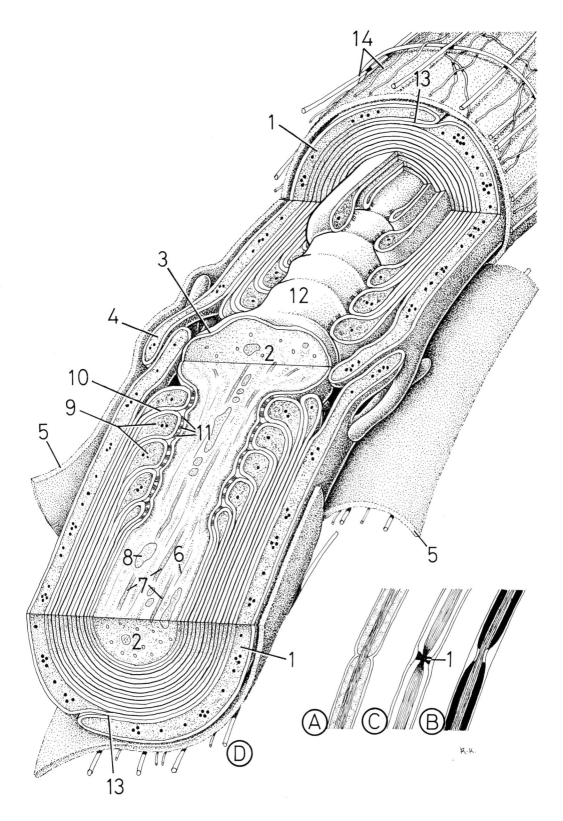

Plate 166. Nerve Fiber. Schmidt-Lanterman Incisures

As already seen in Plate 164, Schmidt-Lanterman incisures are found within the myelin sheath of peripheral nerve fibers. The morphology of these incisures is dependent on the preparation and staining techniques employed. Thus, for example, Fig. **A** shows the incisures following OsO_4 fixation and Fig. **B** after iron hematoxylin staining. In Fig. **B**, the incisures correspond to the so-called Golgi funnels, which delimit the Schmidt-Lanterman segments (Fig. **B**).

In the electron microscope, the Schmidt-Lanterman incisures appear as zones in which the myelin lamellae (Fig. **C1**) run separately. Cytoplasm of the Schwann's cell occurs between the lamellae. An external mesaxon (Fig. **C2**), the basal lamina (Fig. **C3**), and the endoneurial sheath (Fig. **C4**) are also represented. The function of the Schmidt-Lanterman incisures is still uncertain. It remains to be clarified whether they impart plasticity to nerve fibers during bending, stretching, etc. The fact that these structures almost never appear in the CNS, i.e., where nerve fibers are not subject to mechanical force, would support this hypothesis.

Magnifications: Figs. **A**, **B**, × 2,000; Fig. **C**, × 55,000

REFERENCES

Celio MR (1976) Die Schmidt-Lanterman'schen Einkerbungen der Myelinscheide des Mauthner-Axons: Orte longitudinalen Myelinwachstums? Brain Res 108:221–235

Friede RL, Samorajski T (1969) The clefts of Schmidt-Lanterman: a quantitative electron microscopic study of their structure in developing and adult sciatic nerves of the rat. Anat Rec 165:89–102

Metuzals J (1965) Ultrastructure of the nodes of Ranvier and their surrounding structures in the central nervous system. Z Zellforsch 65:719–759

Mira J-C (1976) Études quantitatives sur la régénération des fibres nerveuses myelinisées. Arch Anat Microsc Morphol Exp 65:225–284

Shanklin WM, Azzam NA (1964) Histological and histochemical studies on the incisures of Schmidt-Lanterman. J Comp Neurol 123:5–10

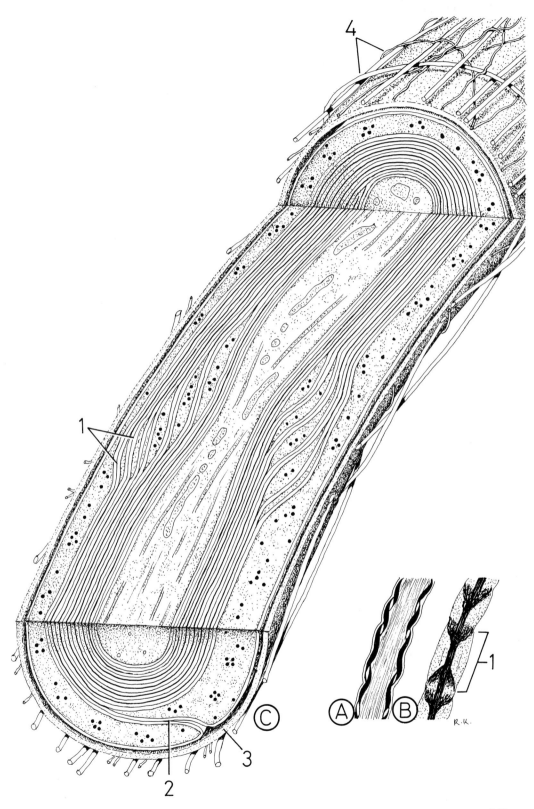

Plate 167. Regeneration of Nerve Fibers
(Modified from WILLIS and WILLIS 1972)

Severing of nerve fibers immediately leads to loss of the ability to conduct impulses and within a few weeks the fibers degenerate. Nonetheless, nerve fibers are capable of regeneration under certain circumstances. The process is depicted in this plate.

Following separation of an axon (Figs. **A 1–E 1**) from the body of the nerve cell (Figs. **A 2–E 2**), the myelin sheath becomes fragmented during the first 3 days and after about 2–3 weeks transforms into plaquelike fatty droplets (Figs. **A 3, B 3**). Simultaneously, the severed axon disintegrates. This secondary or Wallerian degeneration extends as far as the sensory and motor nerve endings (e.g., motor end plates, Figs. **A–E 4**). Schwann's cells (Figs. **A 5–E 5**), on the other hand, do not degenerate. Cross sections through the disintegrating nerve fibers appear in small drawings to the left of the main figures.

Severance of the axon can affect the perikaryon of the nerve cell, causing it to swell. The Nissl bodies disappear (chromatolysis or tigrolysis), and the nucleus is displaced toward the cell membrane ("fish-eye cells"). Provided the damage does not occur in the immediate vicinity of the perikaryon, the nerve cells recover rapidly from the trauma. The ascending degeneration of the nerve fibers extends to the next node of Ranvier (**R**).

From the 2nd week to the 2nd month after injury, the myelin sheath lipids gradually become broken down, phagocytized by microglia in the CNS or by macrophages (Fig. **B 6**) in the PNS, and transported away. The Schwann's cells at the ends of the proximal and distal stumps begin dividing and approach one another as early as the 1st week (arrows). This gives rise to the bands of Büngner (Fig. **B 7**), which – as tubular glial strands – bridge the gaps. The axon (Fig. **C 1**) growing from the proximal stump utilizes these glial strands to locate the target organ. As the axon advances – about 1–2 mm/day – through the glial strands, it becomes invested by a myelin sheath. During this regeneration phase, which can last up to 3 months, muscle fibers (Fig. **C 8**) as well as other effectors atrophy owing to inactivity.

Contact with the target organ is initially reestablished by a thin, weak, myelinated axon (Fig. **D 1**). In the course of the maturation phase, which can last a few months, the diameter and performance of the regenerating nerve fiber increase.

In the case of amputation or large-scale destruction of nerve fibers without surgical intervention, connective tissue inserts itself between the stumps, such that the glial strands cannot form a bridge over the gap. The proximal Schwann's cells then proliferate and, with the outgrowing axon and adjacent connective tissue, form a swollen structure, the so-called amputation neuroma (Fig. **E 9**).

In humans, nerve fibers of the CNS are incapable of regeneration.

REFERENCES

Adrian EK, Elliot MJ (1972) Cell proliferation in the spinal cord following dorsal root transection. Anat Rec 172:259–260

Bray GM, Aguayo AJ (1974) Regeneration of peripheral unmyelinated nerves. Fate of the axonal sprouts which develop after injury. J Anat 117:515–529

Willis RA, Willis AT (1972) The principle of pathology and bacteriology. Butterworths, London

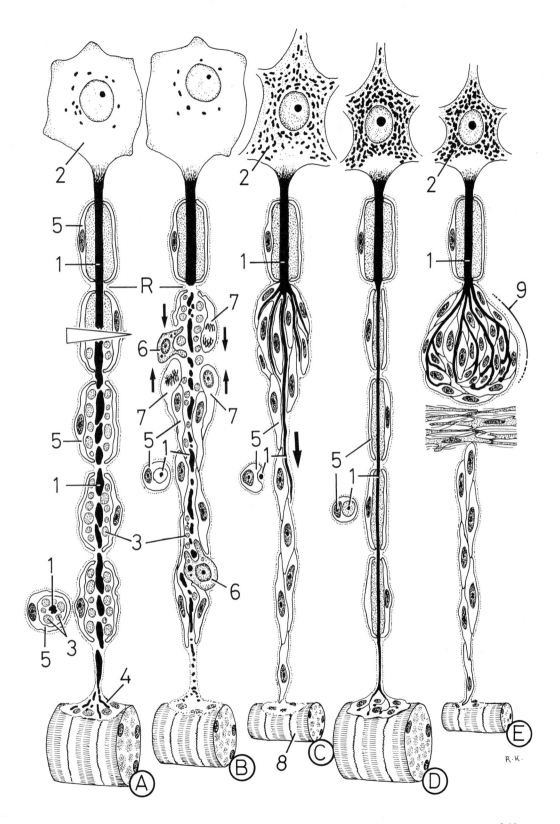

Plate 168. Synapses. Classification

The intercellular contact sites where impulses are transmitted from one neuron to another, or from a neuron to the target organ, are termed synapses. They are classed according to their mode of function into electrical (Fig. A) and chemical (Figs. B–F) synapses. The latter type is the more widely distributed.

Electrical or electrotonic synapses (Fig. A), the morphology of which corresponds to that of a nexus, are particularly frequent in muscular tissue (see Plates 119, 120, 136, 137), though relatively uncommon in nervous tissue. Where these synapses occur, the electrical resistance of the plasmalemma is very low, and this favors impulse conduction.

The relatively rare chemical synapses, synaptic bars (Fig. B), are found, as already indicated in Plate 158, in the receptor cells of the vestibuloacoustic organ and in pinealocytes. They are made up of an osmiophilic, 180- to 320-nm-long and about 35-nm-wide bar (Fig. B2), surrounded by synaptic vesicles (Fig. B1). It is not known what transmitter substance these approximately 40- to 80-nm-wide vesicles contain. According to some sources, it could be gamma-aminobutyric acid (GABA).

Synaptic ribbons (Fig. C1) are also a rare type of chemical synapse found in the axon endings of the rod and cone cells of the retina (see Plate 158). The nature of the transmitter in the surrounding synaptic vesicles has not been defined with certainty.

The most widely distributed chemical synapses are those with boutons terminaux. The synapse here consists of a broad presynaptic portion of axon, the bouton terminal itself (Fig. D1), a synaptic cleft (Fig. D2), and a postsynaptic element (Fig. D3). The widened terminal portion of the axon contains mitochondria, cisternae of smooth endoplasmic reticulum, neurofilaments (Fig. D4), neurotubules (Fig. D5), and two types of synaptic vesicle. One kind of vesicle is small and apparently empty (Fig. D6); the other type is significantly larger (Fig. D7), but much less frequent and filled with a fine-granular, highly osmiophilic substance. The neurotransmitter acetylcholine has been identified in the former, and 5-hydroxytryptamine (serotonin) or possibly dopamine in the latter. A presynaptic density (Fig. D8) borders the synaptic cleft.

The synaptic cleft is approximately 20 nm wide and filled with a fine-granular, weakly osmiophilic material. The adjacent postsynaptic membrane (Fig. D9) features a feltlike dense region, the postsynaptic density (Fig. D10). Since the vesicles in these synapses are spherical, they are referred to as S-type or type I synapses (after GRAY).

In the F-type of synapse with a narrower and almost apparently empty synaptic cleft (Fig. E1), the synaptic vesicles are predominantly flattened (hence F-type synapse), slightly biconcave, and with an unknown transmitter. This type of synapse is also referred to as type II (after GRAY).

In the synapses thus far described, impulses are only transmitted in one direction, i.e., from the pre- to the postsynaptic portion. In reciprocal synapses (Fig. F), however, the impulses can travel in both directions. This latter type of synapse is rare; it has been described in the olfactory bulb of dogs.

The mode of action of chemical synapses may be summarized as follows. When an impulse arrives, the synaptic vesicles release their neurotransmitter into the synaptic cleft. The postsynaptic membrane thus becomes depolarized and the information from one cell is transferred to the other. The morphological appearance of the vesicles does not allow a definitive conclusion to be made regarding the function of the particular synapse. It is now known that the majority of acetylcholine-containing type I synapses have excitatory effects. Conversely, the tempting hypothesis of an inhibitory action of type II synapses requires further investigation. (See Plates 119–128 in KRSTIĆ 1979 and physiology texts for further information.)

Magnifications: Fig. A, × 100,000; Fig. B, × 40,000; Fig. C, × 30,000; Figs. D–F, × 33,000

REFERENCES

Bodian D (1970) An electron microscopic characterization of classes of synaptic vesicles by means of controlled aldehyde fixation. J Cell Biol 44:115–124

Gershon MD, Schwartz JH, Kandel ER (1981) Morphology of chemical synapses and patterns of interconnection. In: Kandel ER, Schwartz JH (eds) Principles of neural science. Elsevier, New York

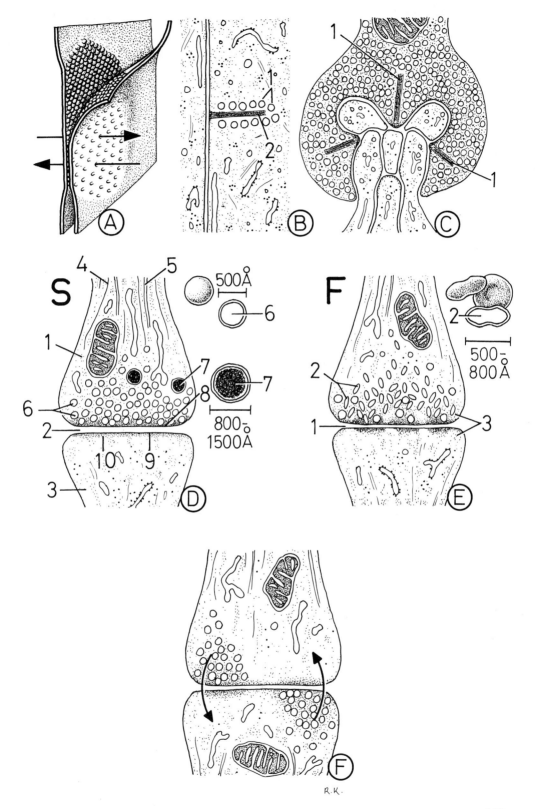

Plate 169. Simplified Classification of Interneuronal Synapses (Modified from ANDRES 1975)

The boutons terminaux are basically capable of ending on the body or soma (**S**), dendrites (**D**), and axons (**A**) of a nerve cell, and accordingly it is possible to distinguish axosomatic, axondendritic, and axoaxonic synapses. These, in turn, can be subdivided into various combinations depending on the morphology of the point of contact and the relationship with the boutons terminaux.

Thus, the following can be found on the soma:
1. Simple axosomatic synapses
2. Invaginated axosomatic synapses
3. Axosomatic spinous synapses;

On the dendrites:
4. Simple axodendritic synapses
5. Axodendritic spinous synapses
6. Crest synapses
7. Branched spinous synapses
8. "En passant" synapses
9. Axodendritic reciprocal synapses
10. Polysynaptic endings
11. Interdigitated spinous synapses;

On the axon hillock:
12. Axoaxonic synapses
13. Axoaxonic inhibitory synapses (morphologically difficult to distinguish);

Between axons:
14. Axoaxonic synapses
15. "En passant" synapses.

It should here be stated that this classification of synapses is not exhaustive. (See Plates 119–125 in KRSTIĆ 1979.)

REFERENCES

Andres KH (1975) Morphological criteria for the differentiation of synapses in vertebrates. J Neural Transmiss Suppl 12:1–37

Benshalom G (1979) Ultrastructure of an excitatory synapse. Cell Tissue Res 200:291–298

Bodian D (1972) Neuron junctions: A revolutionary decade. Anat Rec 174:73–82

Galbraith S (1980) Electron-microscopic study of the synapses of the cerebral cortex in man. Acta Anat 107:46–51

Gray EG, Guillery RW (1966) Synaptic morphology in the normal and degenerating nervous system. Int Rev Cytol 19:111–182

Heuser JE, Reese TS (1977) Structure of the synapse. In: Kandel ER (ed). The nervous system, vol 1. Americam Physiological Society, Bethesda (Handbook of physiology)

Jones DG (1978) Some current concepts of synaptic organization. Adv Anat Embryol Cell Biol 55/4:1–69

Kasa P, Jancso G, Karcsu S, Toth L (1974) Analysis of synaptic vesicles in F-type axon terminals. Acta Histochem 49:46–50

Manina AA (1979) The synapses of the nervous system. Int Rev Cytol 57:345–383

Robertis E De (1967) Ultrastructure and cytochemistry of the synaptic region. Science 156:907–914

Loos H Van der (1963) Fine structure of synapses in the cerebral cortex. Z Zellforsch 60:815–825

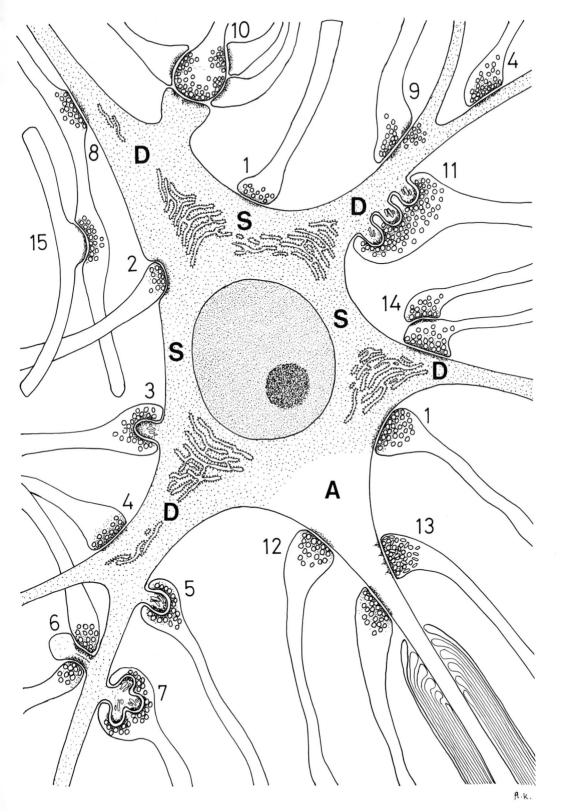

Plate 170. Structural Elements of the Peripheral Nervous System

From a histological point of view, the following parts of the PNS are important:
1. Spinal ganglia
2. Spinal nerves
3. Peripheral nerve endings
3a. Efferent (motor end plates, autonomic nerve endings)
3b. Afferent (neuromuscular spindles, Golgi tendon organ, terminal corpuscle)
4. Sympathetic trunk with its ganglia (**4a**) and prevertebral sympathetic ganglia (**4b**).

The following is a brief description of the neuronal connections between components of the PNS. Arrows indicate the direction of impulse conduction.

Axons (**5**) of the large, multipolar motor neurons (**6**) in the anterior gray column of the spinal cord make up the major portion of the peripheral nerves (**2**) and innervate the skeletal musculature as efferent nerve fibers.

From the sympathetic nerve cells (**7**) in the intermediolateral nucleus of the lateral gray column of the spinal cord, axons run through the white rami communicantes (**9**) to the ganglia of the sympathetic trunk (**4a**) or continue further to the prevertebral ganglia (**4b**). Axons of the sympathetic neurons (**7**) of the spinal cord synapse with dendrites of multipolar nerve cells (**10**) in the ganglia of the sympathetic trunk (**4a**). From here, the postganglionic unmyelinated axons of the multipolar nerve cells enter the spinal nerve via the gray rami communicantes (**11**). These axons (interrupted line) assure the autonomic innervation of glands, muscles, etc. (**3a**).

Postganglionic nerve fibers run from the prevertebral ganglia (**4b**), e.g., celiac ganglion, superior mesenteric ganglion, etc. to the inner organs (e.g., intestinal tract), where they synapse with intramural ganglion cells (**12**).

From the periphery, i.e., via sensory nerve endings (**3b**) of various types (free and encapsulated nerve endings, neuromuscular spindle, Golgi tendon organ), afferent impulses are transmitted along nerve fibers to pseudounipolar ganglion cells (**13**), located in the spinal ganglia (**1**). From here, the stimuli are transmitted via the cellulifugal branch (**14**) of the T-process to the spinal cord, where a few collaterals (**15**) join the motor neurons (**6**). In this way, a simple reflex arc is formed. (See anatomy texts for further information.)

REFERENCES

Guerin J, Bioulac B, Henry P, Loiseau P (1979) Système nerveux végétatif. Sandoz-Edition, Basle

Oehmichen M, Torvik A (1976) The origin of reactive cells in retrograde and Wallerian degeneration. Cell Tissue Res 173:343–348

Poritsky R (1969) Two and three dimensional ultrastructure of boutons and glial cells on the motoneuronal surface in the cat spinal cord. J Comp Neurol 135:423–452

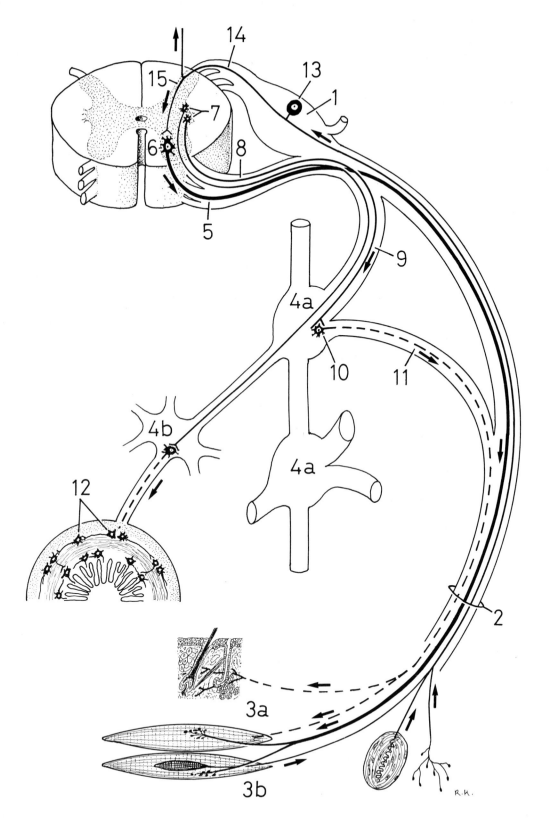

Plate 171. Spinal Ganglion. Light-Microscopic Appearance

A spinal ganglion (Fig. A 1) is included within the posterior root (Fig. A 2) of the spinal nerve and possesses a bilaminar capsule. The inner layer is a continuation of the arachnoid membrane (see Plate 174), which as the perineurial epithelium (Fig. A 3) extends over the spinal ganglion and nerve. The outer capsule layer (Fig. A 4), composed of dense connective tissue, stems from the dura mater (see Plate 174) and is gradually continuous with the perineurial connective tissue and epineurium (Fig. A 5) of spinal nerves. The anterior (Fig. A 6) and posterior roots together form a spinal nerve (Fig. A 7).

The perineurial epithelium (Fig. A 3) has been partially removed so as to expose its lamination and the flattened cells (Fig. A 8) of the surface facing the ganglion. The space (Fig. A 9) between the capsule and tissue of the ganglion has been drawn wider for the sake of clarity.

The spinal ganglion cells (Fig. A 10) occur in clusters in the ganglion interior. Many nerve fibers (Fig. A 11) run between the groups of cells.

After routine staining, ganglion cells (Fig. B 1) appear round with a diameter of up to 100 μm. They are characterized by a spherical, clear nucleus (Fig. B 2) with a very well-developed nucleous (Fig. B 3), and the cytoplasm contains Nissl bodies and lipofuscin granules (Fig. B 4). Sections occasionally reveal a pale axon hillock (Fig. B 5).

Satellite cells or amphicytes (Fig. B 6) encapsulate the ganglion cells. Since the contact between the two is not particularly firm, ganglion cells sometimes become detached from the amphicytes as a result of the shrinkage (Fig. B 7) induced by fixation.

Loose connective tissue of the endoneurium (Fig. B 8), containing unmyelinated and myelinated nerve fibers (Fig. B 9), is found outside the satellite cells.

Following silver staining, many neurofibrils (Fig. C 1) and, in particular, their forked T-processes (Fig. C 2) can be seen in the perikarya of the spinal ganglion cells. Numerous nerve fibers (Fig. C 3) appear as black lines in the endoneurium.

Magnifications: Fig. **A**, × 20; Figs. **B**, **C**, × 350

REFERENCES

Andres KH (1961) Untersuchungen über den Feinbau von Spinalganglien. Z Zellforsch 55:1–48

Nishimura T, Kon I, Awataguchi S, Ishida M, Yamamoto N (1965) Submicroscopic studies on the spiral ganglion in guinea pigs: I. The fine structure of the normal spiral ganglion. Hirosaki Med J 17:1–19

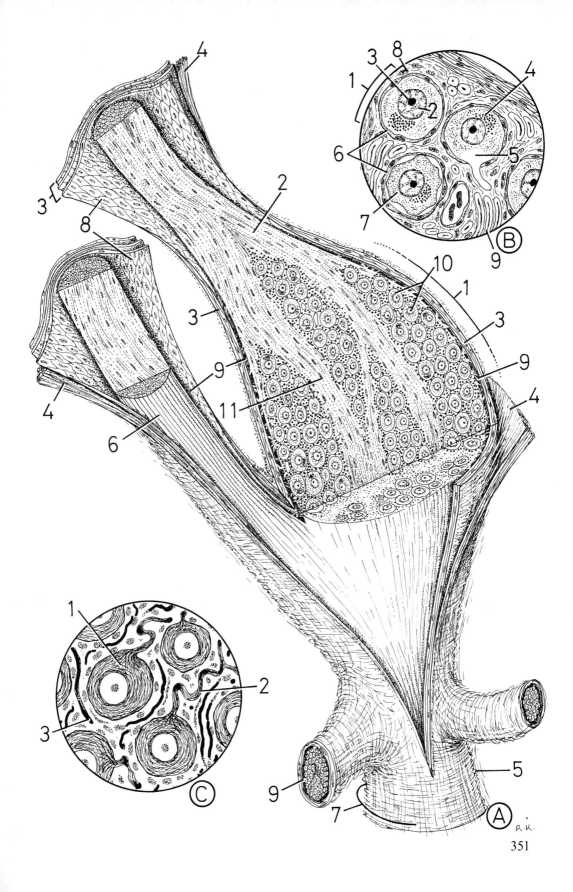
351

Plate 172. Spinal Ganglion. Pseudounipolar Spinal Ganglion Cell

The body of a nerve cell contains a large nucleus with a nucleolus, scattered Nissl bodies (1), very extensive Golgi fields (2), and numerous mitochondria and lysosomes. Neurofilaments and neurotubules constitute cytoskeletal components.

Satellite cells (3) closely surround the perikaryon of nerve cells and there is no intervening basal lamina. In profile, satellite cells appear flattened and in plan view are seen to be stellate, highly interdigitated cells with a heterochromatin-rich nucleus, osmiophilic cytoplasm, and sparse organelles. They are separated from the endoneurium by a basal lamina (4), to which the fibrous structures (5) of the loose endoneurial connective tissue adhere.

In the direct vicinity of the nerve cells are scattered fibrocytes (6), capillaries (7), and several myelinated (8) and unmyelinated (9) nerve fibers. Each of these structures of course has its own basal lamina.

As shown in the light-microscopic image of the previous plate, pseudounipolar nerve cells have a T-shaped process (10), partially covered by satellite cells. This process has been drawn considerably shorter here owing to lack of space.

It can be observed that the two horizontal T-branches are enveloped by a Schwann's cell (11), which supplies them with a myelin sheath (12). A pseudounipolar ganglion cell is thus morphologically a biaxonal element. The impulse-conducting (afferent) process deriving from the periphery can only be electrophysiologically defined as a dendrite (13), and that running into the spinal cord as an axon (14); this distinction cannot be made morphologically.

Magnification: ×5,500

REFERENCES

Pannese E (1969) Electron microscopical study on the development of the satellite cell sheath in spinal ganglia. J Comp Neurol 135:381–422

Pannese E (1981) The satellite cells of the sensory ganglia. Adv Anat Embryol Cell Biol 65:1–111

Rambourg A, Marraud A, Chretien M (1973) Tri-dimensional structure of the forming face of the Golgi apparatus as seen in the high voltage electron microscope after osmium impregnation of the small nerve cells in the semilunar ganglion of the trigeminal nerve. J Microsc 97:49–57

Wuerker RB, Kirkpatrick JB (1972) Neuronal microtubules, neurofilaments, and microtubules. Int Rev Cytol 33:45–75

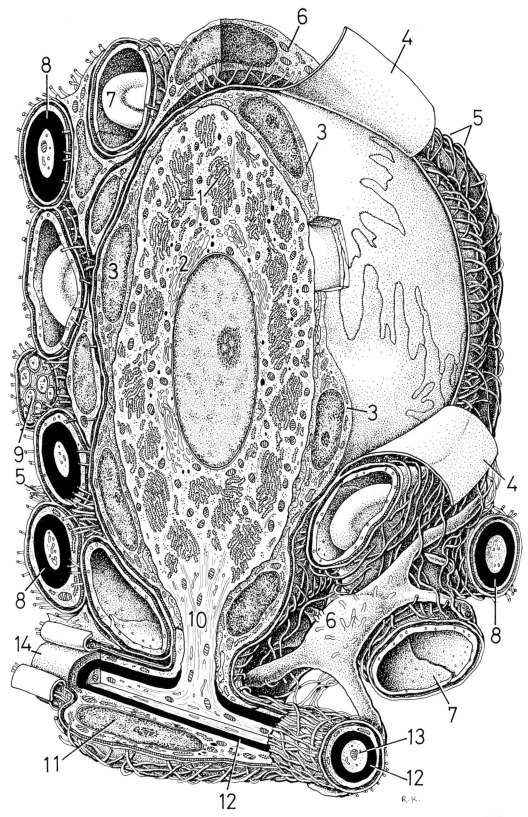

Plate 173. Bipolar Nerve Cell from Corti's Ganglion

Unlike other ganglion cells of the organism, the bipolar nerve cells of Corti's or the spiral ganglion are invested by myelin lamellae. These bipolar ganglion cells are well protected by the bony pillar of the cochlea, the modiolus (Fig. A 1). Their dendrites extend to the hair cells, and the ascending axons contact the nerve cells of the acoustic nuclei in the rhomboid fossa (see anatomy texts for further information).

Staining of the myelin sheaths (Fig. B 1) reveals that not all cells of the spiral ganglion are myelinated. In the surrounding endoneurium (since although these nerve cells are deeply embedded in skull bone they belong to the PNS), myelinated nerve fibers (Fig. B 2), loose fibrous connective tissue, and blood capillaries are found.

Ganglion cells are spindle-shaped and have only two processes – a dendrite (Fig. C 1) and an axon (Fig. C 2). The cell body contains a spherical, frequently indented nucleus with dispersed chromatin and a large nucleolus. As in every nerve cell, Nissl bodies are numerous, spreading into the dendrite but not into the axon hillock (Fig. C 3).

Large numbers of other organelles occur in the bipolar ganglion cells. A few lysosomes and residual bodies are found between the ergastoplasmic cisternae.

The outer surface of bipolar cells is covered by myelin lamellae (Fig. C 4) furnished by Schwann's cells (Fig. C 5). This envelope continues over the dendrite and axon and thus makes it morphologically difficult to differentiate the two. As stated previously, Schwann's cells are separated from the surrounding loose endoneurial connective tissue by a basal lamina (Fig. C 6).

The purpose of this extremely good insulation of the bipolar ganglion cells is not fully apparent. It is possible that it aids the tonal "resolving power" (acuity) of the ear. (See Plate 107 in KRSTIĆ 1979.)

Magnifications: Fig. **A**, × 20;
Fig. **B**, × 600; Fig. **C**; × 7,000

REFERENCES

Awataguchi S, Nishimura T, Kon I, Ishida M, Yananoto N (1965) Submicroscopic studies on the spiral ganglion in guinea pigs: II. The transfigurations of the fine structures of the spiral ganglion exposed to the acoustic stimulation. Hirosaki Med J 17:20–32

Keithley EM, Feldman ML (1979) Spiral ganglion cell counts in an age-graded series of rat cochleas. J Comp Neurol 188:429–442

Peach R (1972) Fine structural features of light and dark cells in the trigeminal ganglion of the rat. J Neurocytol 1:151–160

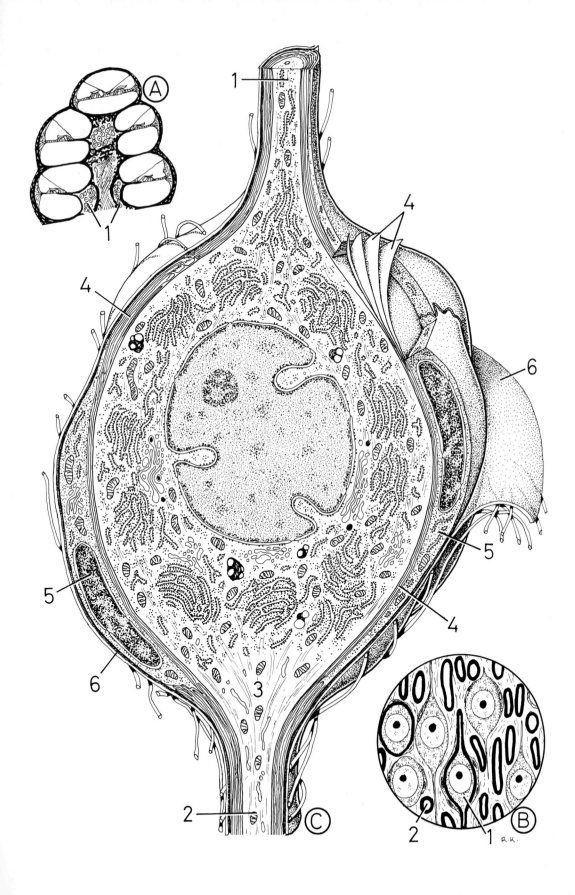

Plate 174. Peripheral or Spinal Nerve. General View

As stated in Plate 170, union of the anterior (**1**) and posterior (**2**) roots gives rise to a peripheral or spinal nerve (**3**). The perspective in this drawing has been exaggerated for greater clarity. This plate also shows the transition from the envelopes of the spinal cord to the sheaths of a spinal nerve.

The spinal cord is, like all parts of the CNS, enveloped by the pia mater (**4**), which has been cut at the point of exit of the motor nerve roots and drawn back slightly on either side.

The second of the leptomeninges, the arachnoid membrane (**5**), continues as the perineurial epithelium (**6**) into the region of the spinal nerves, where several nerve fibers (**7**) unite to form a nerve fascicle (**8**).

The perineurial sheath is made up of 3–15 concentric layers of very flattened cells of perineurial epithelium. Perineurial epithelial cells (**9**) appear clearly on the surface facing the spinal cord and nerve fibers. It is evident from the drawing that the subarachnoid space (**10**) communicates with the interior of every nerve fascicle [see arrow from the spinal ganglion (**11**) to the exposed bundle of nerve fibers].

The external surface of the arachnoid membrane lies close to the dura mater (**12**), which is continuous with the perineurial connective tissue and epineurium (**13**). The strong, slightly undulating, longitudinal collagen fibers (**14**) of the epineurium bind all the nerve fascicles together and form a nerve (**3**).

Sympathetic trunk ganglia (**15**) can also be seen in this plate. Myelinated white rami communicantes (**16**) run to the sympathetic trunk ganglia, and unmyelinated gray rami communicantes (**17**) leave the ganglia and join the spinal nerve (see Plate 170). Hence, it is the presence or absence of the myelin sheath that gives the rami communicantes a white or gray appearance upon macroscopic inspection.

Since all spinal nerves contain myelinated motor and sensory nerve fibers in addition to autonomic (sympathetic and parasympathetic) fibers, they are referred to as mixed peripheral nerves.

REFERENCES

Himango WA, Low FN (1971) The fine structure of a lateral recess of the subarachnoid space in rat. Anat Rec 171:1–20

Kahn R (1976) Membrane receptors for hormones and neurotransmitters. J Cell Biol 70:261–286

McCabe JS, Low FN (1969) The subarachnoid angle: an area of transition in peripheral nerve. Anat Rec 164:15–34

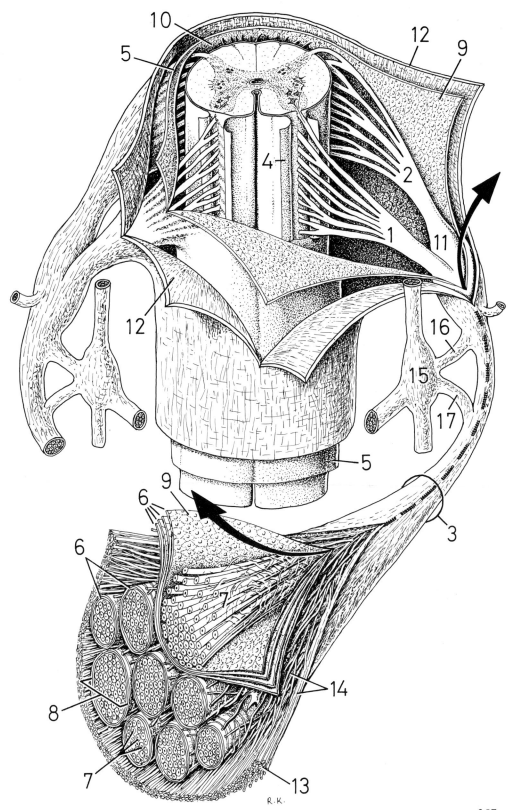

Plate 175. Structure of a Spinal Nerve

Every peripheral nerve is made up of a variable number of rounded nerve fascicles (Fig. A 1) and every fascicle comprises a variable number of nerve fibers (Fig. A 2). The individual fascicles are ensheathed by concentric lamellae of the perineurium (Fig. A 3). The nerve fibers are surrounded by a delicate fibrous connective tissue, the endoneurium (Fig. A 4).

A surface nerve fascicle has been cut and its perineurium pulled back to expose the perineurial epithelium (Fig. A 5) and the undulating course of the nerve fibers.

The strong collagen fibers of the epineurium (Fig. A 6) are longitudinally directed and also undulate, which gives the nerve a certain degree of elasticity. The epineurium holds the whole fascicle together, but also extends between the bundles of nerve fibers with its blood vessels (Fig. A 7), lymphatics, and adipose tissue (Fig. A 8).

A further layer made up of loose connective tissue, the paraneurium (Fig. A 9), surrounds the nerve. The structures in the rectangular inset in Fig. A correspond to Fig. B, and those in the square to Fig. C.

As already seen in Plate 164, axons (Fig. B 2), myelin sheaths (Fig. B 3), and nodes of Ranvier (Fig. B 4) are evident in longitudinal sections of myelinated nerve fibers (Fig. B 1). Blood capillaries (Fig. B 5) occur in the endoneurium.

Routine staining of a transverse section reveals the concentric layers of the perineurium (Fig. C 1), the epineurium (Fig. C 2), and, between the nerve fibers (Fig. C 3), the loose endoneurium (Fig. C 4) with blood capillaries (Fig.gC 5). Note the stellate neurokeratin network (Fig. C 6) and axons (Fig. C 7).

Magnifications: Fig. A, × 35;
Figs. B, C, × 700

REFERENCES

Bishcoff A, Moor H (1967) Ultrastructural differences between the myelin sheaths of peripheral nerve fibers and CNS white matter. Z Zellforsch 81:303–310

Davison AN, Peters A (1970) Myelination. Thomas, Springfield

Landon DN (ed) (1967) The peripheral nerve. Chapman and Hall, London

Spencer PS, Lieberman AR (1971) Scanning electron microscopy of isolated peripheral nerve fibers. Z Zellforsch 119:534–551

Steer JM (1971) Some observations on the fine structure of rat dorsal spinal nerve roots. J Anat 109:467–485

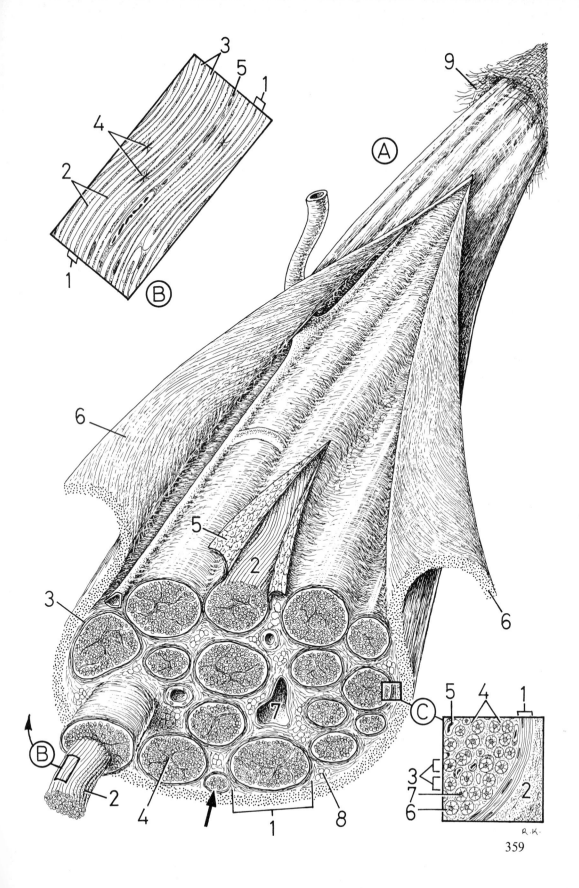

Plate 176. Structure of the Perineurium. Transverse Section

The perineurium (1), as already stated in Plates 174 und 175, is made up of 3–15 concentric layers of very flattened epithelial cells (2), termed perineurial epithelium or endothelium, and perineurial connective tissue.

The epithelial cells, which are only about 0.1–0.3 µm thick, are joined to one another in the overlapping areas by four to five zonulae occludentes (3) and desmosomes (4). Each epithelial cell lamella is covered on both sides by a basal lamina (5). Longitudinally oriented collagen microfibrils (6) are largely found between the concentric perineurial layers. The epithelial cells themselves have very flattened nuclei with condensed chromatin, inconspicuous organelles, and an exceptionally large number of micropinocytotic vesicles (7) in the thin cytoplasmic extension.

The outermost perineurial epithelial lamella is in contact with the relatively thin perineurial connective tissue sheath (8). Further toward the exterior are attached the longitudinal collagen fibers of the epineurium (9). These two connective tissue sheaths, which merge into one another without a sharp border, can, for all practical purposes, be regarded as a continuation of the dura mater. Within the nerve fascicle are found a myelinated (10) and an unmyelinated (11) nerve fiber and fibrous and cellular components of the endoneurium – collagen microfibrils (12), fibrocytes (13), etc. The endoneurial sheath (14) of both nerve fibers can also be seen.

Magnification: × 17,500

REFERENCES

Akert K, Sandri C, Weibel ER, Peper K, Moor H (1976) The fine structure of the perineural endothelium. Cell Tissue Res 165:281–295

Burkel WE (1967) The histological fine structure of perineurium. Anat Rec 158:177–190

Shinowara NL, Michel ME, Rapoport SI (1982) Morphological correlates of permeability in the frog perineurium: Vesicles and "transcellular channels." Cell Tissue Res 227:11–22

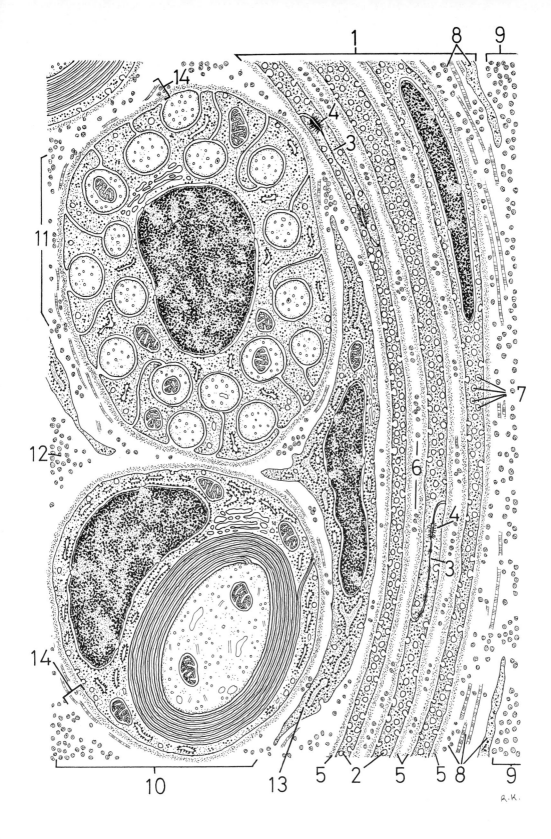

Plate 177. Nerve Fascicle with Perineurium. Three-Dimensional View

The perineurium (1) of a nerve fascicle has been partially removed so as to provide a better impression of its layered structure. The epithelial lamellae (2) are made up of very flattened, elongated cells. One cell has been lifted back at the point where it overlaps another cell, revealing the characteristic structure of a zonula occludens (3). Numerous openings (4) are visible on the external and internal surfaces of the epithelial cells, corresponding to the micropinocytotic vesicles. Basal laminae (5) cover the epithelial cells, between which run many longitudinal collagen microfibrils (6). Myelinated (7) and unmyelinated (8) nerve fibers, which, like the blood capillaries (9), possess their own basal laminae, are located in the fascicle interior. The basal lamina (11) of a myelinated nerve fiber has been cut and partly pulled away (lower left) in order to expose the insertion of the outer mesaxon (10). Numerous longitudinally oriented collagen microfibrils (12) and fibrocytes (13) also occur in the endoneurium.

The outer covering of the nerve fascicle is composed of perineurial connective tissue (14). The epineurium has been omitted from this plate.

The multilamellar perineurial epithelium represents a highly selective diffusion barrier between the nerve fibers and the surrounding epineurial connective tissue. Morphological evidence of this is the presence of innumerable micropinocytotic vesicles in the firmly joined epithelial cells. Experiments have shown that, e.g., ferritin, horseradish peroxidase, and dyes are incapable of penetrating the perineurial layers. For this reason, the term "blood-nerve barrier" (BNB), comparable with the blood-brain barrier, is employed.

Magnification: × 3,000

REFERENCES

Aker D (1972) A study of hematic barriers in peripheral nerves of albino rabbits. Anat Rec 174:21–38

Livingston RB, Pfenninger K, Moor H, Akert K (1973) Specialized paranodal and interparanodal glial-axonal junctions in the peripheral and central nervous system: a freeze-etching study. Brain Res 58:1–24

Reale E, Luciano L, Spitznas M (1976) Freeze-fracture aspects of the perineurium of spinal ganglia. J Neurocytol 5:385–394

Revel J-P, Hamilton DW (1969) The double nature of the intermediate dense line in peripheral nerve myelin. Anat Rec 163:7–16

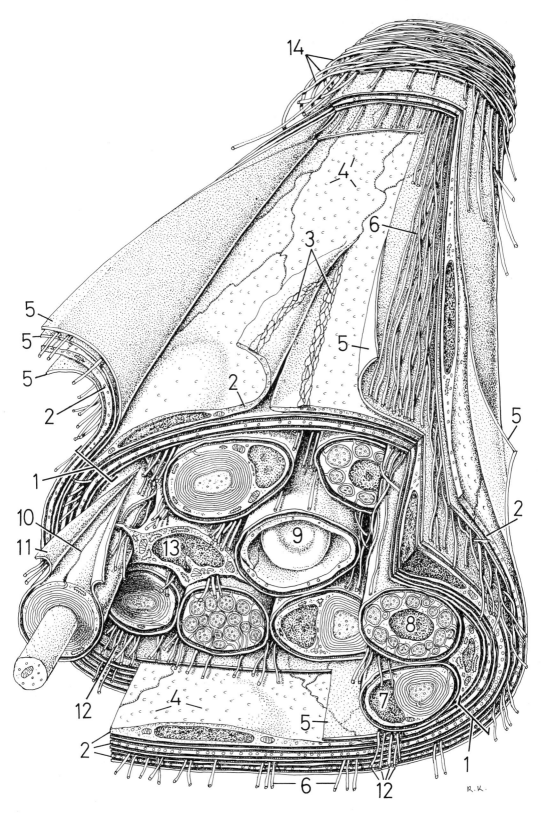

Plate 178. Endings of Efferent Nerve Fibers. Motor End Plate

Motor end plates (Fig. A 1) are barely visible in routine preparations but become distinct following gold chloride staining. In the vicinity of the motor end plates, the efferent nerve fibers branch in a grapelike manner ("en grappe") over the striated skeletal muscle fiber (Fig. A 2). Nuclei (Fig. A 3) belonging to the teloglial cells, i.e., the specially differentiated Schwann's cells overlying the muscle fiber, appear between the ramifications of the nerve fibers. Other nuclei belong largely to the muscle fibers. An area corresponding to Fig. A is three-dimensionally presented in Fig. B.

Three myelinated efferent nerve fibers (Fig. B 1) with nodes of Ranvier (Fig. B 2) contact the skeletal muscle fibers (Fig. B 3). Flattened teloglia cells (Fig. B 4) mark the points of contact. The middle nerve fiber has been cut so as to display its axon (Fig. B 5) projecting into a depression of the muscle fiber.

A nerve fiber can innervate a variable number of muscle fibers, depending on the region of the body. In muscles with very precise functions, e.g., in the extrinsic musculature of the eyes and the muscles of the larynx, one nerve fiber may innervate a single muscle fiber – as illustrated in this plate. In muscles where contraction demands less precision, one nerve fiber supplies a whole bundle of muscle fibers. A motor unit is therefore defined as the area of muscle innervated by a motor neuron. Satellite cells (Fig. B 6) and the endomysium (Fig. B 7), which has been largely removed, are also evident in the drawing. Perpendicular to the muscle fiber run the precapillaries (Fig. B 8), from which arise the longitudinally oriented capillaries (Fig. B 9). The latter anastomose with one another by means of transverse branches (Fig. B 10). (See Plate 129 in KRSTIĆ 1979.)

Magnifications: Fig. A, ×750; Fig. B, ×2,000

REFERENCES

Coers C (1967) Structure and organization of the myoneural junction. Int Rev Cytol 22:239–268

Couteaux R (1973) Motor end plate structure. In: Bourne GH (ed) The structure and function of muscle. Academic, New York

Dreyer F, Peper K, Akert K, Sandri C, Moor H (1973) Ultrastructure of the "active zone" in the frog neuromuscular junction. Brain Res 62:373–380

Gauthier GF (1976) The motor end plate. In: Landon DN (ed) The peripheral nerve. Chapman and Hall, London

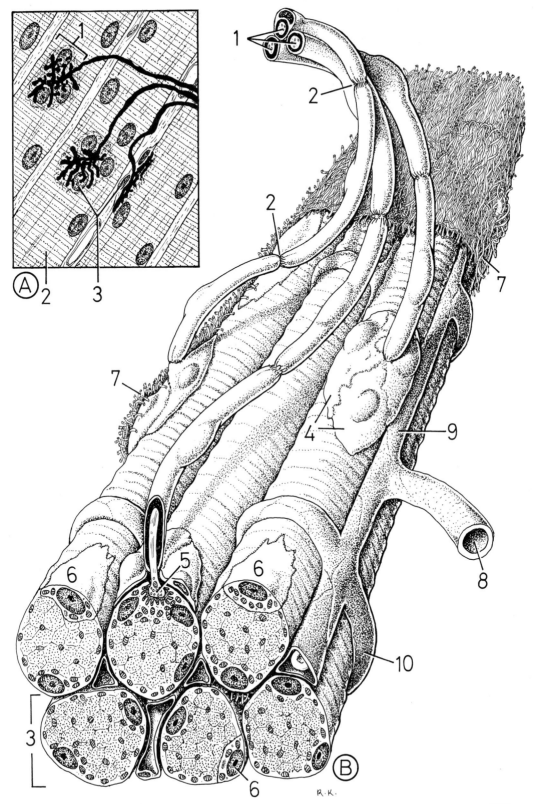

Plate 179. Motor End Plate. Three-Dimensional View

Every axon (1) loses its myelin sheath (2) in the vicinity of its terminal branching, and the Schwann's cells (3) flatten and transform into teloglia cells (4). The basal lamina (5) of the Schwann's cells simultaneously covers the teloglia cells and is continuous with the basal lamina (6) of the muscle fiber (7). In the slightly protruding sole plate of the muscle fiber, the axon ramifies into several short end branches (8), which become thicker toward the end and fit into the synaptic troughs.

The axolemma does not, however, come into direct contact with the plasmalemma of the muscle fiber, since the fissure that separates them contains a glycoprotein-rich amorphous substance (arrowheads), which outside the synapse is continuous with the basal lamina (6) of the muscle fiber. The terminal portion of an axon branch (9) has been lifted back so that the synaptic clefts of the subneural apparatus (10) can be observed from above. On the underside of the end branch, the parallel arrangement of the membrane-bound particles of the so-called active zones (11) can be distinguished.

The transversely sectioned axon ending at the base of the plate reveals numerous synaptic vesicles (12), which although they appear empty do in fact contain acetylcholine. The microfolds of the subneural apparatus (13), which are filled with a fine-granular amorphous substance, are evident below the end branch.

In the vicinity of the end plate, the sarcoplasm contains many mitochondria but is devoid of myofibrils (14).

The function of a motor end plate can be summarized as follows. When an impulse reaches the terminal ramifications of a motor axon, the depolarizing acetylcholine is released into the synaptic cleft (15) between the axolemma and sarcolemma, thereby bringing about muscular contraction. Prior to each new transmission of impulses, the acetylcholine is destroyed by the enzyme cholinesterase.

The openings of the T-tubules (16) and the fibrous feltwork of the endomysium (17) can be observed. (See Plate 130 in KRSTIĆ 1979 and physiology texts for further information.)

Magnification: × 20,000

REFERENCES

Ellisman MH, Rash JE, Staehelin LA, Porter KR (1976) Studies of excitable membranes: II. A comparison of specializations at neuromuscular junctions and nonjunctional sarcolemmas of mammalian fast and slow twitch muscle fibers. J Cell Biol 68:752–774

Kelly AM, Zacks SI (1969) The fine structure of motor endplate morphogenesis. J Cell Biol 42:154–169

Padykula HA, Gauthier GF (1970) The ultrastructure of the neuromuscular junctions of mammalian red, white and intermediate skeletal muscle fibers. J Cell Biol 46:27–41

Uehara Y, Campbell GR, Burnstock G (1976) Muscle and its innervation. Arnold, London

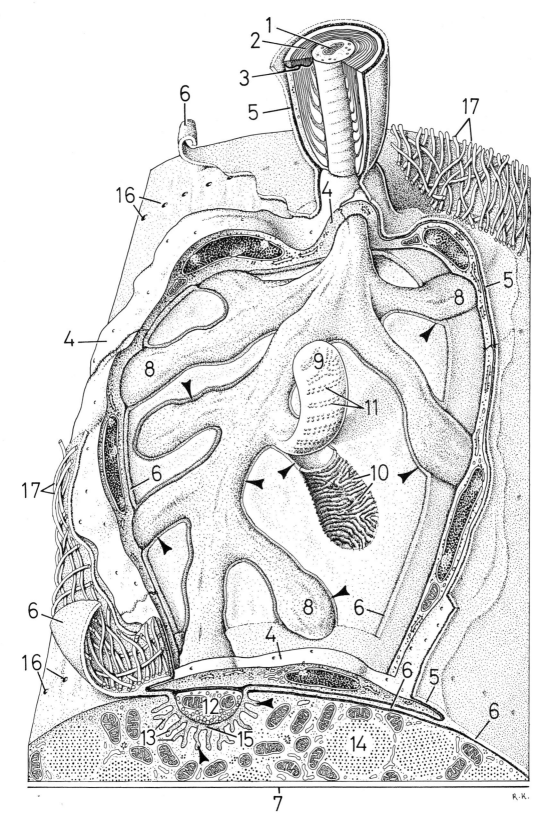

Plate 180. Endings of Autonomic Nerves in the Smooth Musculature

Following silver staining (Fig. A), the adrenergic autonomic nerve fibers and their endings in the smooth musculature appear to have a netlike arrangement. Fluorescent-microscopic analysis (Fig. B) reveals a higher concentration of catecholamines in the autonomic terminal branchings, especially in the small varicosities (Fig. B 1).

In Fig. C, the ramification of a muscle cell bundle (Fig. C 1) is depicted. The spindle-shaped smooth muscle cells (Fig. C 2) are contained within a network of more or less transversely or longitudinally oriented collagen and reticular fibrils (Fig. C 3).

The autonomic nerve endings (Fig. C 4) have lost their Schwann's sheath and are "naked" between the muscle cells. Axoaxonal "en passant" synapses (Fig. C 5) are frequent in the network of autonomic nerve fibers. Junctions between sympathetic and parasympathetic axons have also been detected. No differentiations comparable with motor end plates occur between muscle cells and autonomic fibers. It is only occasionally that the swollen axonal terminations contact indirectly (through the basal lamina) the sarcolemma. Even less frequently, the axonal terminations enter specially prepared depressions of the sarcolemma (see Plate 120).

A varicosity (Fig. C 6) is shown at higher magnification in Fig. D. The presence of 50- to 80-nm-wide synaptic vesicles with osmiophilic cores (Fig. D 1), which are distinct in the transmission electron microscope, characterizes both the varicose dilations and the axonal end branches (see Plate 120). Other vesicles (Fig. D 2), fewer in number but up to 150 nm across, are found among the synaptic vesicles. Mitochondria (Fig. D 3) and cisternae of smooth endoplasmic reticulum (Fig. D 4) also occur in the axoplasm.

The small synaptic vesicles with the osmiophilic cores contain norepinephrine, and dopamine is found in the larger vesicles. Both transmitters are released from the endings as required and exert their effects on the whole sarcolemmal surface of the muscle cells. The excitatory waves produced are propagated between the muscle cells through nexus zones.

Cholinergic nerve endings are distinguished by a large number of apparently empty synaptic vesicles. It is, however, very difficult by means of morphological criteria to determine the nature of autonomic end branches. (See Plate 131 in KRSTIĆ 1979.)

Magnifications: Figs. A, B, ×200; Fig. C, ×2,000; Fig. D, ×30,000

REFERENCES

Avery JK, Cox CF, Chieco DJ Jr (1980) Presence and location of arenergic nerve endings in the dental pulps of mouse molars. Anat Rec 198:59–71

Chiba T, Yamauchi A (1970) On the fine structure of the nerve terminals in the human myocardium. Z Zellforsch 108:324–338

Hand AR (1972) Adrenergic and cholinergic nerve terminals in the rat parotid gland. Electron microscopic observations on permanganate-fixed glands. Anat Rec 173:131–140

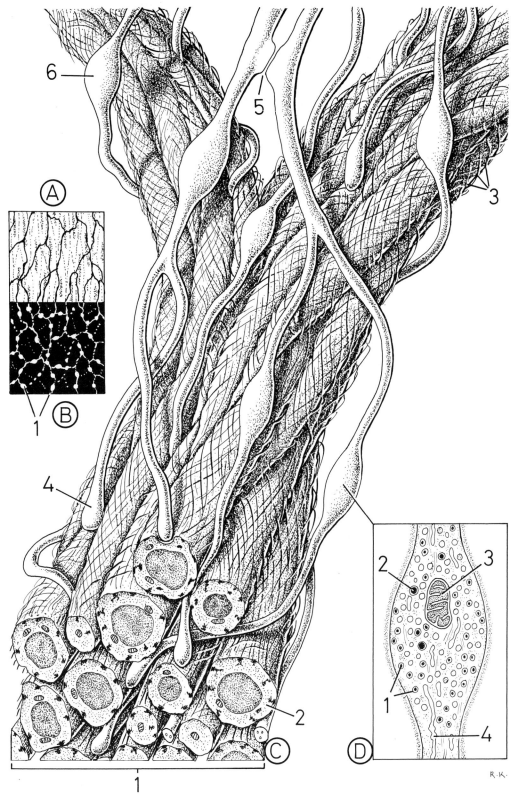

Plate 181. Endings of Afferent Nerve Fibers. Neuromuscular Spindle

Neuromuscular spindles are small organs, 2–10 mm long and 0.5 mm thick, embedded in the skeletal musculature. They are involved in proprioception. Figure A shows a transverse section through a neuromuscular spindle from the extraocular muscles, and Fig. B a neuromuscular spindle from the musculi lumbricales. Both structures are surrounded by a fairly thick capsule (Figs. A 1, B 1), in which the intrafusal muscle fibers (Figs. A 2, B 2) and nerve fibers (Figs. A 3, B 3) can be recognized under the light microscope. Skeletal muscle fibers (Figs. A 4, B 4) surround the neuromuscular spindles.

In a light-microscopic magnification of a longitudinal section (Fig. C), the capsule (Fig. C 1) of the neuromuscular spindle is less easy to distinguish, though two types of intrafusal muscle fiber are now evident. The nuclei of the first type are stacked on top of one another and the fibers are thus termed nuclear chain fibers (Fig. C 2). In the other type, the nuclei are clustered in a central saclike distension, and these fibers are called nuclear bag fibers (Fig. C 3).

Following silver staining, the ramification of nerve fibers in the neuromuscular spindle can be traced. The annulospiral sensory nerve endings (Fig. D 1) are conspicuous and entwine around every intrafusal muscle fiber over an approximately 300-μm-long section. Other nerve endings extend to the intrafusal muscle fibers and form motor end plates (Fig. D 2). Outside the neuromuscular spindle, the extrafusal nerve fibers that innervate skeletal musculature are seen to have considerably larger motor end plates (Fig. D 3). In the electron-microscopic image of a transversely sectioned neuromuscular spindle, the outer capsule (Fig. E 1), which is regarded as an extension of the perineurium, can be clearly seen. The capsule (see also Plate 176) has a lamellar structure, and capillaries (arrows) run between the layers. The outer and inner capsules (Fig. E 2) delimit the periaxial space (Fig. E 3) with its unmyelinated (Fig. E 4) and myelinated (Fig. E 5) nerve fibers.

Nuclear chain fibers (Fig. E 6) are narrow, whereas nuclear bag fibers (Fig. E 7) are distinguished by several nuclei in the broader middle sections. Fibrocytes (Fig. E 8) and myelinated and unmyelinated nerve fibers can be observed between intrafusal muscle fibers. The neuromuscular spindle also of course contains capillaries.

Magnifications: Figs. A–D, ×250;
Fig. E, ×2,000

REFERENCES

Adal MN, Chew Cheng SB (1980) Capsules of duck muscle spindles. Cell Tissue Res 211:465–474

Banks RW (1981) A histological study of the motor innervation of the cat's muscle spindle. J Anat 133:571–591

Barker D (1974) The motor innervation of muscle spindles. In: Beallairs R, Gray EG (eds) Essays on the nervous system. Clarendon, Oxford

Rumpelt H-J, Schmalbruch H (1969) Zur Morphologie der Bauelemente von Muskelspindeln bei Mensch und Ratte. Z Zellforsch 102:601–630

Scalzi HA, Price HM (1971) The arrangement and sensory innervation of the intrafusal fibers in the feline muscle spindle. J Ultrastruct Res 36:375–390

Taxi J (ed) (1980) Ontogenesis and functional mechanisms of peripheral synapses. Elsevier, Amsterdam

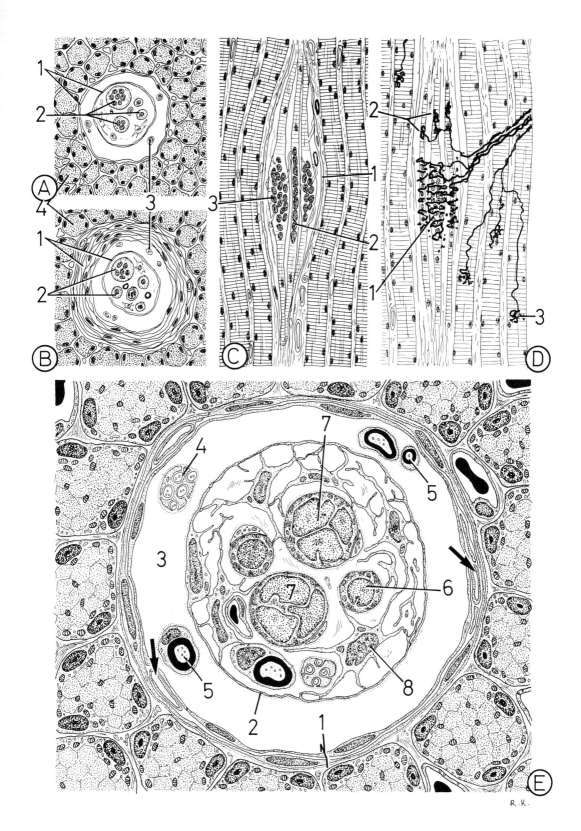

Plate 182. Endings of Afferent Nerve Fibers. Neuromuscular Spindle. Three-Dimensional View

In a neuromuscular spindle that has been opened along its longitudinal axis, an outer (**1**) and inner (**2**) capsule can be distinguished. The outer capsule represents a continuation of the perineurium (**3**). Thin, transversely striated intrafusal nuclear chain fibers (**4**) and nuclear bag fibers (**5**) are bound at their ends to the inner surface of the outer capsule by means of reticular and collagen microfibrils (**6**), originating from the connective envelope of these muscle fibers.

Annulospiral endings of the thick, sensory nerve fibers (Ia or Aff. I) wind around the middle of each muscle fiber. The thin, likewise sensory, nerve fibers (II or Aff. II), the so-called flower-spray endings, are located almost exclusively on the terminal sections of nuclear chain fibers.

In addition to sensory innervation, intrafusal and extrafusal muscle fibers receive motor innervation in the form of γ-nerve fibers, which terminate on intrafusal muscle fibers as motor end plates (**7**, Eff. II) and grapelike endings (**8**, Eff. III). It should be mentioned here that in differentiating the various types of nerve fiber, metallic staining methods are superior to electron microscopy. Fibrocytes (**9**) also occur in the interior of the neuromuscular spindle. Outside the spindle capsule, α-nerve fibers terminate on thick skeletal muscle fibers (**10**).

The endomysium (**11**) surrounds the neuromuscular spindle and connects it to the extrafusal muscle fibers. This fact is of great significance in the function of the neuromuscular spindle.

Neuromuscular spindles are mechanoreceptors which provide information about the length of the muscle both to the cerebellum and, via a collateral, directly to the multipolar motor nerve cells of the gray matter of the spinal cord. Thus, for example, contraction of the intrafusal nerve fibers caused by γ-nerve fibers is transmitted to the annulospiral endings, which then excite the motor neurons by their collaterals. As a response, the impulses are conducted from the motor neurons via α-fibers to the extrafusal muscle fibers, which bring about a contraction in accordance with the strain on the muscle involved. In the same way, sudden stretching of muscle leads to immediate muscular contraction (e.g., patellar reflex).

Magnification: $\times 1{,}500$

REFERENCES

Banks RW, Harker DW, Stacey MJ (1977) A study of mammalian intrafusal muscle fibers using a combined histochemical and ultrastructural technique. J Anat 123:783–796

Bridgman CF, Shumpert EE, Eldred E (1969) Insertions of intrafusal fibers in muscle spindles of the cat and other mammals. Anat Rec 164:391–402

Dow PR, Shinn SL, Ovalle WK Jr (1980) Ultrastructural study of a blood-muscle spindle barrier after systemic administration of horseradish peroxidase. Am J Anat 157:375–388

Kucera J (1981) Histochemical profiles of cat intrafusal muscle fibers and their motor innervation. Histochemistry 73:397–418

Mayr R (1970) Zwei elektronenmikroskopisch unterscheidbare Formen sekundärer sensorischer Endigungen in einer Muskelspindel der Ratte. Z Zellforsch 110:97–107

Soukup T (1976) Intrafusal fiber types in rat limb muscle spindles. Histochemistry 47:43–57

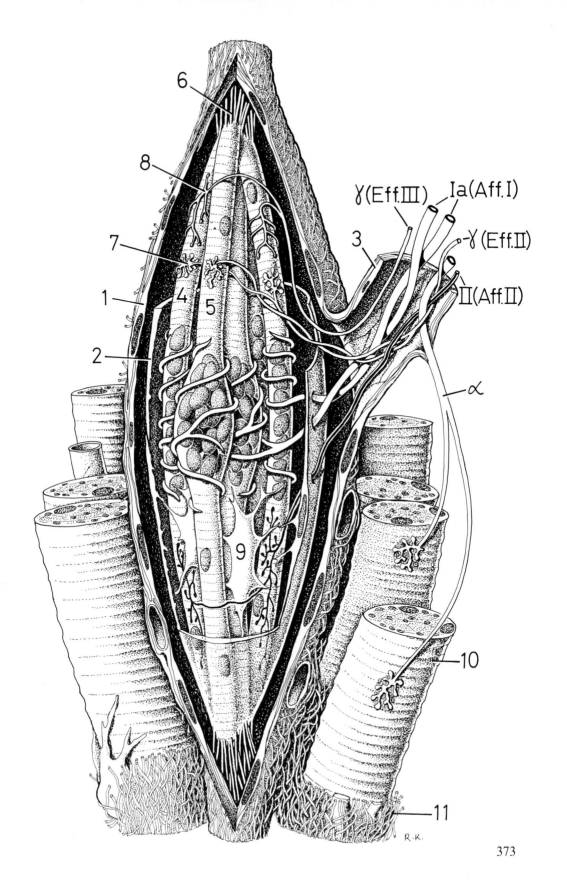

Plate 183. Endings of Afferent Nerve Fibers. Intrafusal Muscle Fibers

This plate shows a nuclear chain (Fig. A 1) and a nuclear bag fiber (Fig. A 2) at higher magnification and drawn to reveal the internal structure. In both types of muscle fiber, the myofibrils (Fig. A 3) – like those of the myotubes (see Plate 124) – are displaced toward the plasmalemma and absent in central sections of the fibers. These regions are thus incapable of contraction. Leptomeric myofibrils (Fig. A, arrows; Fig. B), which frequently occur in intrafusal muscle fibers, contain structures that are reminiscent of Z-lines (Fig. B 1). These lines run for a distance of 100 nm, mainly perpendicular to the orientation of the myofibrils. Since the leptomeric bands change their direction, a zebra pattern develops, which is visible in longitudinal sections. Leptomeric myofibrils are rarely found in skeletal and cardiac musculature, but are more common in muscle fibers that still bear some embryonic characteristics (Purkinje cells of the impulse-conducting system, intrafusal muscle fibers). The function of leptomeric fibrils remains to be established. It is believed that they are precursors of normal striated myofibrils, according to which the leptomeric bands would correspond to Z-lines.

Among the nervous elements, it is possible to distinguish the annulospiral endings (Fig. A 4), flower-spray fibers (Fig. A 5), and the two types of motor γ-fiber (Figs. A 6, A 7), described in Plate 182. The area contained within the circle in the upper right is enlarged in Fig. C.

An annulospiral unmyelinated ending (Fig. C 1) is separated from the plasmalemma of the muscle fiber by an approximately 20-nm-wide cleft. Zonulae adherentes (Fig. C 2) are occasionally seen in this region. In the interior of the ending, numerous mitochondria, cisternae of smooth endoplasmic reticulum, and vesicles are found.

The basal lamina (Fig. C 3) covering the muscle fiber also coats the annulospiral ending. Unlike a motor end plate, therefore, the annulospiral ending is hypolemmal, i.e., it is located beneath the basal lamina of the muscle fiber.

The endings of the flower-spray fibers have a similar structure to annulospiral fibers. All γ-motor end plates have exactly the same morphology as that described in Plate 179.

The collagen microfibrils (Fig. A 8) of the lower section of the intrafusal muscle fibers continue into a fine tendon (Fig. 9) (see Plate 182).

Magnifications: Fig. A, $\times 4,000$; Fig. B, $\times 10,000$; Fig. C, $\times 25,000$

REFERENCES

Adal MN (1969) The fine structure of the sensory region of cat muscle spindles. J Ultrastruct Res 26:332–354
Kucera J (1982) A study of sensory innervation to long nuclear chain intrafusal fibers in the cat muscle spindle. Histochemistry 75:113–121
Kucera J (1982) The topography of long nuclear chain intrafusal fibers in the cat muscle spindle. Histochemistry 74:183–197
Kucera J, Hughes R (1983) Histological study of motor innervation to long nuclear chain intrafusal fibers in the muscle spindle of the cat. Cell Tissue Res 228:553–547
Ovalle WK Jr (1972) Motor nerve terminals on rat intrafusal muscle fibers, a correlated light and electron microscopic study. J Anat 111:239–252
Payne CM (1982) A quantitative analysis of leptomeric fibrils in an adriamycin/carnitine chronic mouse model. J Submicrosc Cytol 14:337–345

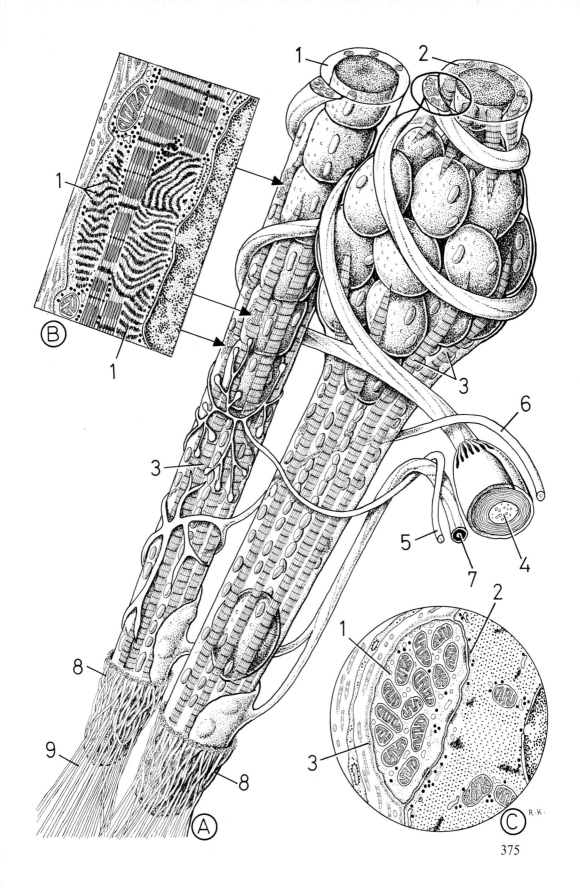

Plate 184. Endings of Afferent Nerve Fibers.
Golgi Tendon Organ or Neurotendinal Spindle

At the point of transition between a muscle and a tendon, sporadic Golgi tendon organs are found. Following silver staining, a rich ramification of sensory nerve fibers (Fig. **A 1**) becomes evident. In the upper part of the drawing, striated skeletal muscle fibers (Fig. **A 2**) penetrate the tendon organ. Below, the capsule, which is difficult to differentiate under the light microscope, with its inner collagen fibers is continuous with the tendon (Fig. **A 3**). In Fig. **B**, the capsule has been opened to expose the interior of the tendon organ.

As in the neuromuscular spindle, the lamellar structure of the capsule (Fig. **B 1**) of the Golgi tendon organ is a direct continuation of the perineurium (Fig. **B 2**). The extremely flattened cells of the perineurial epithelium (Fig. **B 3**) can be recognized both on the inner and outer surfaces of the capsule.

The entire Golgi tendon organ is invested by a feltwork made up of reticular and collagen microfibrils (Fig. **B 4**). All fibrous structures are continuous with the tendon fibers (Fig. **B 5**). Other tendon fibers (Fig. **B 6**) surround the whole tendon organ.

Several anastomosing and loosely arranged collagen fibers (Fig. **B 7**) run through the interior of the Golgi tendon organ. These collagen fibers are continuations of the connective envelope of muscle fibers (Fig. **B 8**) that enter the tendon organ.

The myelinated nerve fibers (Fig. **B 9**) that penetrate the Golgi tendon organ soon lose the myelin sheath but not the Schwann's sheath. The richly branched axons wrap around the collagen fibers and terminate in nodular thickenings (Fig. **B 10**).

The relationship between the collagen fibers and nerve endings is shown three-dimensionally in Fig. **C** (modified after SCHOULTZ and SWETT 1972). The collagen fibers (Fig. **C 1**) press deep into the terminal thickenings of the axon (Fig. **C 2**). It appears as if the neuronal elements occupy all the space between the collagen fibers. The axons are accompanied by Schwann's cells (Fig. **C 3**).

The mode of function of the Golgi tendon organ can be summarized as follows. Muscular contraction causes the loosely arranged intrafusal collagen fibers to become taut. In this way, the fibers come closer to one another, in the direction of the arrows, and compress the axon thickenings and this induces stimulation of the nerve endings. The impulses are then conducted to the spinal cord.

Magnifications: Fig. **A**, × 150; Fig. **B**, × 500; Fig. **C**, × 22,000

REFERENCES

Bridgman CF (1968) The structure of tendon organs in the cat: a proposed mechanism for responding to muscle tension. Anat Rec 162:209–220

Bridgman CF (1970) Comparisons in structure of tendon organs in the rat, cat and man. J Comp Neurol 138:369–372

Schoultz TW, Swett JE (1972) The fine structure of the Golgi tendon organ. J Neurocytol 1:1–26

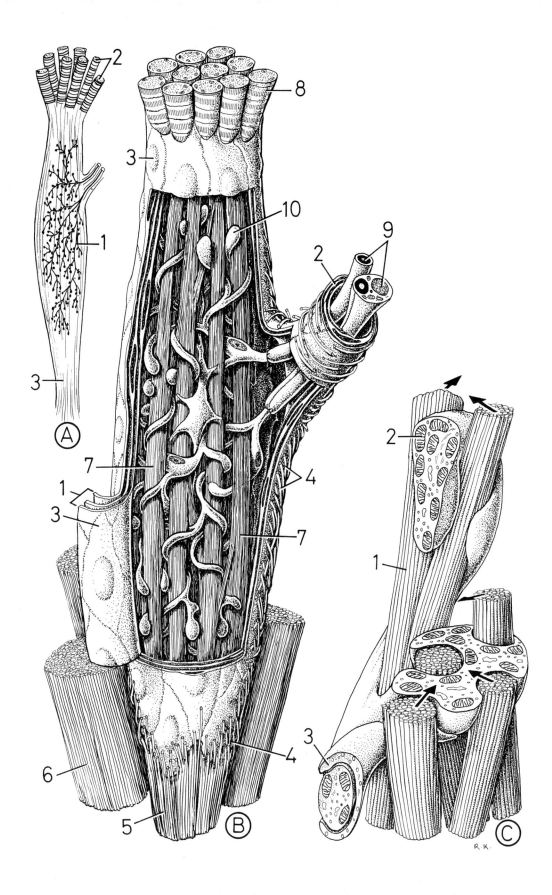

Plate 185. Endings of Afferent Nerve Fibers in Epithelial and Connective Tissues

Figure A shows a schematized section of skin, in which a detailed view of all layers other than the epidermis (Fig. A 1) has been omitted. Endoepidermal nerve endings (Fig. A 2) terminate both freely and on certain cells (Fig. A 3). Other free nerve fibers are found associated with the hair follicle sheath (Fig. A 4).

In the connective tissue beneath the epidermis, encapsulated terminal corpuscles occur, of which only Meissner's (Fig. A 5) and Vater-Pacini (Fig. A 6) corpuscles are dealt with in this book.

In addition to the nerve fibers, a hair shaft (Fig. A 7), sebaceous (Fig. A 8) and sweat (Fig. A 9) glands, and the arrector muscle (Fig. A 10) are presented in this drawing. The inset corresponds to Fig. B (modified after HALATA 1975).

Nerve endings lose their myelin sheath as they enter the epidermis, and upon penetrating the epithelial basal lamina (Fig. B 1) lose their Schwann's sheath (Fig. B 2). The basal lamina of the Schwann's cells (Fig. B 2) is continuous with that of the epidermis. Axons are located in the intercellular spaces and in their vertical path through the germinative layer almost extend as far as the stratum granulosum (Fig. B 3). Similar endoepithelial nervous ramifications also occur in the cornea and various mucous membranes.

Perception of pain and temperature is very probably achieved through stimulation of the free nerve endings.

Another form of endoepithelial ending found in glabrous and hairy skin is shown on the right in Fig. B. Two axons lose their myelin sheath (Fig. B 4) and for a short distance are surrounded only by the Schwann's cells (Fig. B 2). The axons penetrate the epidermis as unmyelinated fibers and run to the specially differentiated Merkel's cells (Fig. B 5). Here, each nerve ending becomes broader, creating a Merkel's disc (Fig. B 6), which forms a synapselike junction (Fig. B 7) with the adjacent Merkel's cell. Merkel's cells form several digitiform processes and attach to adjacent cells by small desmosomes; their cytoplasm contains electron-dense, approximately 100-nm-wide, membrane-bound granules, the significance of which is unknown.

Merkel's discs respond to pressure and vibration stimuli, and so are classed among the mechanoreceptors.

Magnifications: Fig. A, ×70; Fig. B, ×1,200

REFERENCES

Cauna N, Ross LL (1960) The fine structure of Meissner's touch corpuscles of human fingers. J Biophys Biochem Cytol 8:467–482

Halata Z (1975) The mechanoreceptors of the mammalian skin. Adv Anat Embryol Cell Biol 50 (5):1–77

Nurse CA, Mearow KM, Holmes M, Visheau B, Diamond J (1983) Merkel cell distribution in the epidermis by quinacrine fluorescence. Cell Tissue Res 228:511–524

Pellegrini MS (1972) Richerche ultrastrutturali sulle giunzioni mioneurali nella tunica muscolare de alcuni organi del ratto. Arch Ital Anat Embriol 77:69–81

Saxod R (1980) Development of Merkel corpuscles in the chicken beak. The problem of their origin and identity. Biol Cell 37:61–66

Straile WE (1969) Encapsulated nerve end-organs in the rabbit, mouse, sheep and man. J Comp Neurol 136:317–336

Turner DF (1983) The morphology and distribution of Merkel cells in primate gingival mucosa. Anat Rec 205:197–205

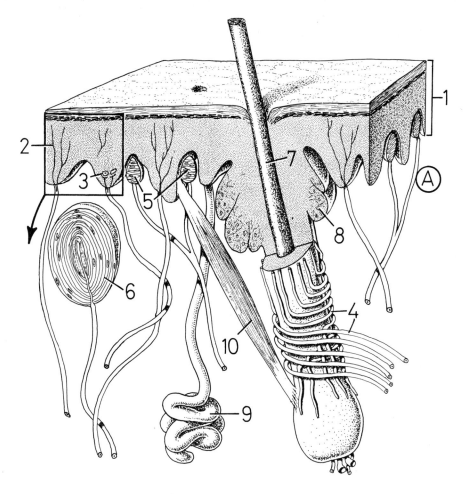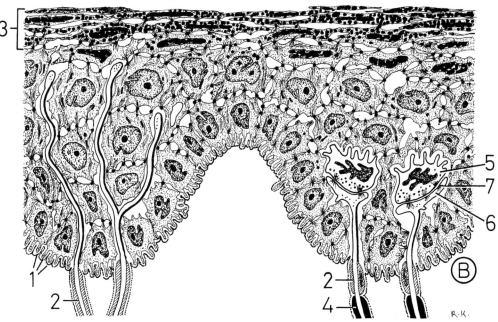

Plate 186. Endings of Afferent Nerve Fibers Around the Hair Follicle (Modified from ANDRES and DÜHRING 1973; HALATA 1975; MUNGER 1971)

Every hair follicle with its sheaths is surrounded by several longitudinally arranged afferent nerve endings. A transverse section through a hair follicle reveals first of all the hair shaft (Fig. **A 1**) and then the inner (Fig. **A 2**) and outer (Fig. **A 3**) epithelial root sheaths. Between the outer epithelial and connective tissue sheaths (Fig. **A 4**) ist found a basal lamina (Fig. **A 5**).

The connective tissue sheath encloses several sensory nerve fibers (Fig. **A 6**), running parallel to the axis of the hair. The basal lamina (Fig. **A 7**) of the nerve fibers is continuous with that of the outer epithelial root sheath (see also Fig. **B 6**).

The nerve fibers terminate in a cylindrical portion (Fig. **A 8**), where a flattened nerve ending (Fig. **A 9**) is sandwiched between two Schwann's cells (Fig. **A 10**). A narrow belt of the compressed axon protrudes between the Schwann's cells. A few collagen microfibrils (Fig. **A 11**) and elastic fibers (Fig. **A 12**) of the connective tissue root sheath can be seen between the nerve endings.

Figure **B** depicts the sandwiching of an axon in cross section. In the upper part of the drawing, peripheral cell regions (Fig. **B 1**) belonging to the outer epithelial root sheath can be seen. The basal lamina (Fig. **B 2**) is attached to the cells by means of numerous hemidesmosomes (Fig. **B 3**). Schwann's cells (Fig. **B 4**) flanking the axons are characterized by a pale cytoplasm and several micropinocytotic vesicles. The flattened nerve ending (Fig. **B 5**) contains a remarkably large number of mitochondria and is located only 100–200 nm from the hair sheath epithelium. A three-dimensional reconstruction of the sensory nerve endings is given in Figs. **C** and **D**.

In the lower part of Fig. **C**, the penultimate normally structured Schwann's cell (Figs. **C 1**, **D 1**) ends in series of cytoplasmic processes (Fig. **C 2**) described in Plate 165. The cell is of course enveloped by a basal lamina (Figs. **C 3**, **D 3**). The line of origin of the outer mesaxon (Fig. **D 4**) is evident.

After it leaves the penultimate Schwann's cell, the axon (Figs. **C 5**, **D 5**) becomes shaped by the last two glial cells into a spearhead form. In Fig. **D**, where the nerve ending (Fig. **D 5**) has been partially removed, micropinocytotic vesicles (Fig. **D 6**) on the surface facing the axon and the nucleus (Fig. **D 7**) of a Schwann's cell can be observed.

Every movement of the hair stimulates the nerve endings and the impulses are conducted to the CNS.

Magnifications: Fig. **A**, ×2,500; Fig. **B**, ×18,000; Figs. **C**, **D**, ×14,000

REFERENCES

Andres KH, Düring M v (1973) Morphology of cutaneous receptors. In: Autrum H, Jung R, Lowenstein WR, MacKay DM, Teuber HL (eds) Handbook of sensory physiology, vol 2. Springer, Berlin Heidelberg New York

Halata Z (1975) The mechanoreceptors of the mammalian skin. Adv Anat Embryol Cell Biol 50(5):1–77

Halata Z, Munger BL (1980) The sensory innervation of primate eyelid. Anat Rec 198:657–670

Munger BL (1971) Patterns of organisation of peripherical sensory receptors. In: Autrum H, Jung R, Loewenstein, WR, MacKay DM, Teuber HL (eds). Handbook of sensory physiology, vol 1. Springer, Berlin Heidelberg New York

Tay SSW, Wong WC, Ling EA (1983) An ultrastructural study of small granule-containing cells in the heart of the monkey (Macaca fascicularis). J Anat 136:35–45

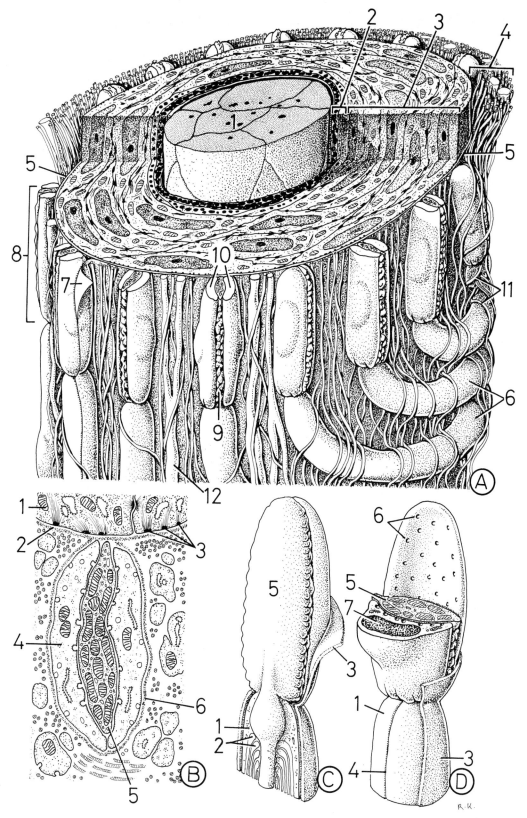

Plate 187. Endings of Afferent Nerve Fibers. Meissner's Corpuscle (Modified from ANDRES and DÜHRING 1973)

Meissner's corpuscles belong to the so-called encapsulated end organs. Following metallic staining, the corpuscles (Fig. A; see Plate 185), which occur at the tip of a connective tissue papilla, appear as ovoid, 50- to 150-µm-long and 60-µm-broad structures. They are made up of stacked, partially flattened, pear-shaped tactile cells (Fig. A1), whose nuclei are predominantly located at the periphery of the corpuscle. Between the tactile cells, there is a network of unmyelinated nerve fibers (Fig. A2), which become myelinated upon leaving the corpuscle (Fig. A3).

Meissner's corpuscles are surrounded by a capsule (Fig. A4). Collagen fibrils (Fig. A5) irradiate from the capsule to the basal cells of the epidermis, and blood capillaries (Fig. A6) are always found in the vicinity of the capsule.

In Fig. B, prepared from electron-microscopic observations, the Meissner's corpuscle appears like an egg in an eggcup. A large number of flattened cytoplasmic layers (Fig. B1) of the tactile cells (Fig. B2) form the ovoid structure. Tactile cells are sometimes considered specially differentiated Schwann's cells. Closer study reveals that many cytoplasmic lamellae leave the peripheral nuclear regions and that these lamellae interdigitate in a fairly complex manner with similar processes of the opposite tactile cells.

Myelinated nerve fibers (Fig. B3) enter the corpuscles from below, lose their myelin sheath, and thread between the cytoplasmic lamellae in tortuous and repeatedly branching spirals (Fig. B4). Within the corpuscle, the axons are enveloped only by a Schwann's sheath. Every axon terminates in a flattened, clublike swelling (Fig. B5), which can synapse with the tactile cells. The transversely cut nerve fiber indicated by an arrow is three-dimensionally presented in Fig. C.

A mitochondria-rich axon (Fig. C1) is enclosed within flattened Schwann's cells (Fig. C2), which in turn are invested by a basal lamina (Fig. C3).

Tactile cells are contained within an incomplete capsule (Fig. B6) belonging to the perineurium. Like all hitherto described perineurial envelopes, it is composed of several layers of perineurial epithelial cells. The space between the tactile cells and the capsule has been drawn significantly broader for greater clarity.

Many collagen microfibrils and fibers (Fig. B7) join the tactile and capsule cells with the branched basal processes of the basal epithelial cells. The Meissner's corpuscle thus appears to hang from the epidermis. These connections between the terminal corpuscle and epidermis are important in the function of the former. Every deformation of the epidermis induced by pressure is transferred to Meissner's corpuscles. Since these corpuscles are particularly numerous in the tips of the fingers and toes, it is supposed that they are responsible for touch and pressure perception.

Magnifications: Fig. A, ×800; Fig. B, ×3,500; Fig. C; ×15,000

REFERENCE

Castano P, Ventura RG (1979) The Meissner's corpuscle of the green monkey (Cercopithecus aethiops L).: II. The connective tissue component. Some considerations from a functional standpoint. J Submicrosc Cytol 11:185–191

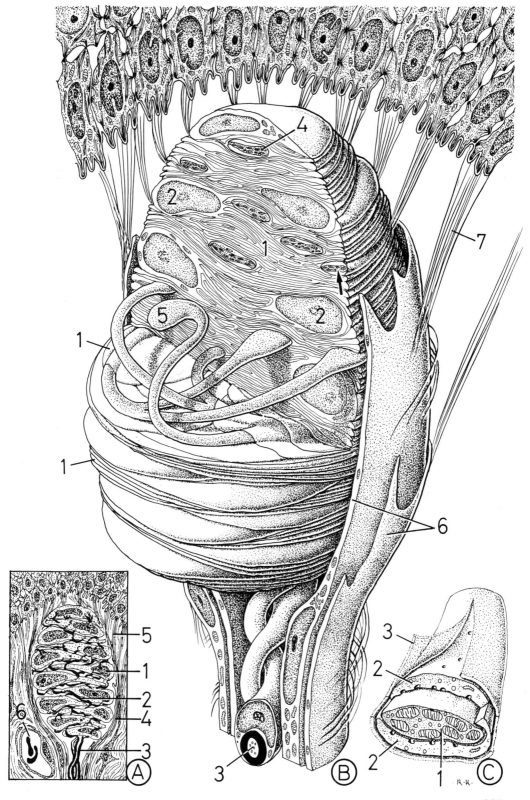

Plate 188. Endings of Afferent Nerve Fibers. Vater-Pacini Corpuscle

Vater-Pacini corpuscles are the largest of all encapsulated nerve endings; they can attain a length of 3–4 mm and a width of 2 mm. They are found in various inner organs and in the subcutis. Vater-Pacini corpuscles are particularly frequent in the skin of the finger pads and palm. A large number of these bodies have also been identified along arteriovenous anastomoses.

Under the light microscope, a transverse section through a terminal corpuscle reveals first a capsule (Fig. A 1) and then an onionlike lamellar outer core (Fig. A 2). In the center is an inner core (Fig. A 3) with the axon.

In Fig. B, which is based on the results of ultrastructural analysis, the lamellar structure of the capsule (Fig. B 1), surrounded by blood capillaries (Fig. B 2), and connective tissue fibers can be observed. The outer core is made up of 10–60 thin, concentric, impervious cell lamellae (Fig. B 3), formed of flattened fibrocytes. The liquid-filled spaces between the lamellae contain collagen microfibrils (Fig. B 4) and occasional blood capillaries (Fig. B 5).

The capsule and lamellae are a direct continuation of the perineurium (Fig. B 6), however, unlike the perineurium of a nerve fascicle the spaces between the epithelial cells here are exceptionally narrow (see Plate 176).

On the right of the picture, a nerve fiber (Fig. B 7) and two blood capillaries (Fig. B 8) can be seen entering the corpuscle. The axon (Fig. B 9) loses its myelin sheath (Fig. B 10) in this region, but continues to be accompanied by Schwann's cells, which form the inner core (Fig. B 11). Only the portion of axon located within the inner core has receptor properties. The axon terminates in a nodular thickening (Fig. B 12). A simplified cross section is shown in Fig. C to clarify the complex structure of the inner core.

The inner core is made up of two symmetrical systems of tightly interdigitating half lamellae (Fig. C 1) of Schwann's cells. The lamellae are partially separated by a radial cleft (Fig. C 2). In the center of the inner core is the sensitive portion of the axon (Fig. C 3).

Every pressure-induced deformation of the Vater-Pacini corpuscle is transmitted by the liquid-filled spaces between the lamellae of the outer core, and this stimulates the unmyelinated axon segments of the inner core.

The ability of Vater-Pacini corpuscles to react to pressure explains why they occur not only in skin but also in the vicinity of arteriovenous anastomoses, where they are thought to regulate blood pressure. Vater-Pacini corpuscles are stimulated by vibrations as well as by pressure.

Magnifications: Fig. A, $\times 70$; Fig. B, $\times 1,200$; Fig. C, $\times 1,700$

REFERENCE

Ide C, Saito T (1980) Electron microscopic histochemistry of cholinesterase activity of Vater-Pacini corpuscle. Acta Histochem Cytochem 13:298–305

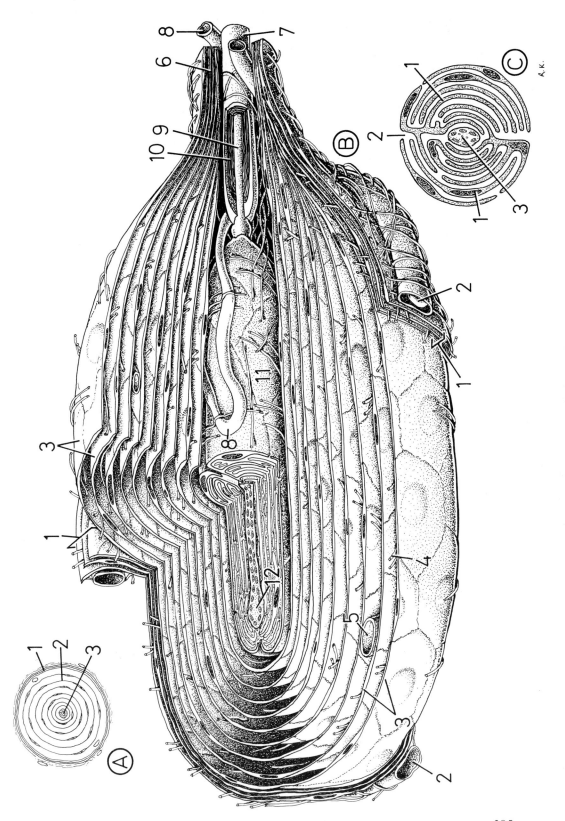

Plate 189. Sympathetic Trunk. Light-Microscopic Appearance of Autonomic Ganglion and Its Nerve Cells

A schematic representation of the connections between the spinal cord and the sympathetic ganglia has already been given in Plate 170.

Branches project from the sympathetic trunk ganglia (Fig. **A1**) to the prevertebral ganglia (Fig. **A2**). A horizontal section through a prevertebral ganglion (arrow) is shown in Fig. **B**.

Like all parts of the peripheral nervous system, the sympathetic ganglia and their branches are surrounded by perineurium (Fig. **B1**), which is supported on its outer surface by a fibrous network predominantly made up of collagen fibers. The interior of the ganglion is dominated by a well-developed endoneurium (Fig. **B2**), which groups the nerve cells in small clusters (Fig. **B3**). A few rami communicantes (Fig. **B4**) can be seen at the periphery of the ganglion. The two insets in Fig. **B** are enlarged in Figs. **C** and **D**.

Routinely stained sections (Fig. **C**) reveal the capsule (Figs. **C**, **D1**) and, in the interior of the ganglion, the vascular endoneurium (Figs. **C2**, **D2**) containing many myelinated and unmyelinated nerve fibers. The spindle-shaped or polygonal sympathetic cells (Figs. **C3**, **D3**) are smaller than spinal ganglion cells. The nuclei of sympathicus cells are spherical and clear but have one or two relatively large nucleoli.

Nissl bodies and lipofuscin granules occur more frequently in sympathetic ganglion cells than in spinal ganglion cells. A somewhat irregular layer of satellite cells (Fig. **C4**) separates the nerve cells from the endoneurium.

Following silver staining (Fig. **D**), the precise form of the sympathetic ganglion cells can be discerned, and it is evident that they are multipolar nerve cells. The cell interior is seen to contain many neurofibrils (Fig. **D4**). Dendrites and axons of the multipolar nerve cells are difficult to distinguish. Some observations indicate that sympathetic nerve cells may even have several axons. The cell processes near the perikaryon form a fairly dense convoluted network. Many nerve fibers (Fig. **D5**) in synaptic contact with the ganglion cells can also be seen in the endoneurium. Amphicytes do not appear so clearly after silver staining.

Magnifications: Fig. **B**, × 20; Figs. **C**, **D**, × 350

REFERENCES

Brooks CMC (1983) Newer concepts of the autonomic systems's role derived from reductionist and behavioral studies of various animal species. J Auton Nerv Syst 7:199–212

Grillo MA (1966) Electron microscopy of sympathetic tissues. Pharmacol Rev 18:387–399

Sulkin DF, Sulkin NM, Rothrock ML (1968) Fine structure of autonomic ganglia in recovery following experimental scurvy. Lab Invest 19:55–66

Zypen E Van der (1967) Elektronenmikroskopische Befunde an der Endausbreitung des vegetativen Nervensystems und ihre Deutung. Acta Anat (Basel) 67:481–515

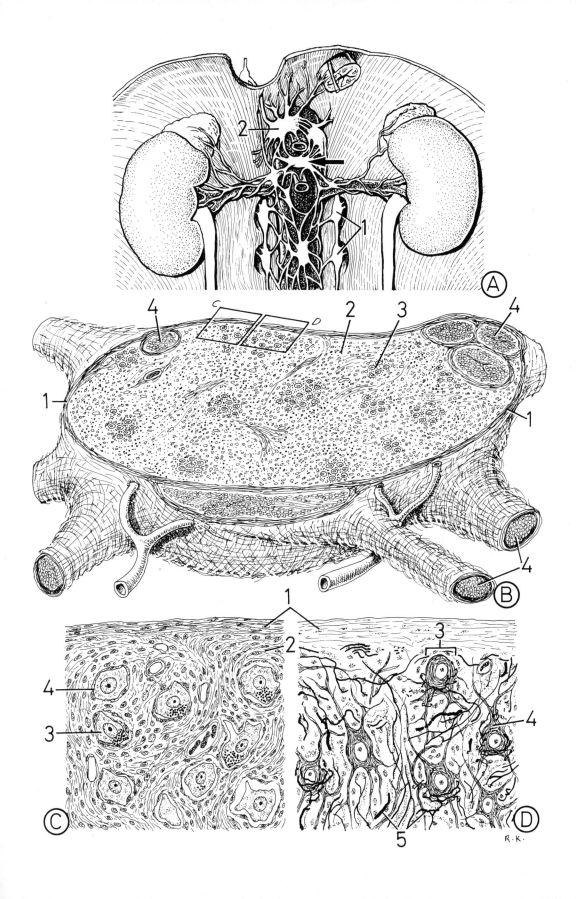

Plate 190. Sympathetic Trunk. Multipolar Autonomic Nerve Cell

The stellate body of a sympathetic autonomic neuron contains a large nucleus with a well-developed nucleolus. Polynuclear autonomic nerve cells are also occasionally found.

Nissl bodies (1), the most conspicuous element of the neuroplasm, are very well developed in the vegetative cells and extend into the dendrites (2). They are absent only in the axon hillock (3).

The Golgi apparatus is variously developed and around it occur granules and vesicles. Vegetative nerve cells also contain numerous mitochondria. Lipofuscin granules are seen in older individuals.

Some data, though not fully corroborated, suggest that vegetative nerve cells may even be capable of division after birth. However, centrioles have as yet to be detected in vegetative ganglion cells. Neurotubules and neurofilaments run in various directions through the cytoplasm.

Several cell extensions leave the perikaryon. Boutons terminaux (4) of other autonomic axons, either naked (5) or accompanied by a Schwann's sheath (6), contact the dendrites. Axoaxonal synapses (7) are of course also present.

The outer surface of the perikaryon is surrounded by several satellite cells (8). The investment of an autonomic nerve cell by satellite cells becomes less complete the greater the distance from the sympathetic trunk. In the sympathetic trunk, the satellite cell envelope appears practically uninterrupted, as in the cell presented here, whereas intramural ganglion cells are only surrounded by an endoneurial sheath. There is no basal lamina between the satellite cell and nerve cell. Occasionally, however, satellite cells cover axosomatic synapses (9). The basal lamina (10) of satellite cells is continuous with that of Schwann's cells (11). Numerous autonomic nerve fibers (12) run through the endoneurium in the vicinity of the autonomic neurons.

It is not possible morphologically to differentiate between autonomic nerve cells of the sympathetic and parasympathetic nervous systems. (See Plates 119, 120 in KRSTIĆ 1979.)

Magnification: × 10,000

REFERENCES

Chang PL, Taylor JJ, Wozniak W, Young PA (1976) An ultrastructural study of sympathetic ganglion satellite cells in the rat: II. Effect of preganglionic sympathectomy. J Neural Transm 38:43–47

Lemos C De, Pick J (1966) The fine structure of thoracic sympathetic neurons in the adult rat. Z Zellforsch 71:189–206

Eränkö O, Soinila S, Päivärinta H (eds) (1980) Histochemistry and cell biology of autonomic neurons, SIF cells, and paraneurons. Advances in biochemistry and psychopharmacology, vol 25. Raven, New York

Forssmann WG (1964) Studien über den Feinbau des Ganglion cervicale superius der Ratte. Acta Anat (Basel) 59:106–140

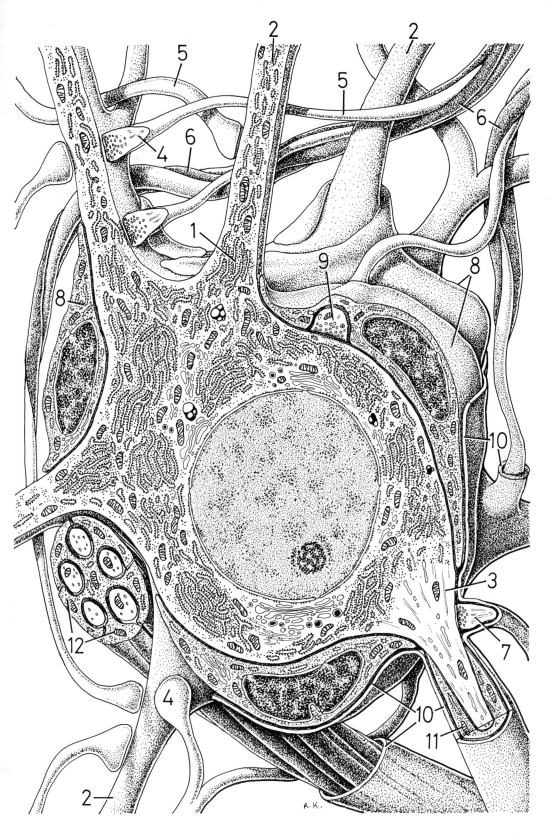

General Reading

Alberts B, Bray D, Lewis J, Raff M, Roberts K, Watson JD (1983) Molecular biology of the cell. Garland, New York

Allen JM (ed) Molecular organization and biological function. Harper and Row, New York (1967)

Babel J, Bischoff A, Spoendlin H (1970) Ultrastructure of the peripheral nervous system and sense organs. Thieme, Stuttgart

Bargmann W (1977) Histologie und mikroskopische Anatomie des Menschen, 7th edn. Thieme, Stuttgart

Bielka H (ed) (1969) Molekulare Biologie der Zelle. VEB Fischer, Jena

Bloom W, Fawcett DW (1975) A textbook of histology, 10th edn. Saunders, Philadelphia

Breatnach AS (1971) An atlas of the ultrastructure of human skin. Churchill, London

Bucher O (1980) Cytologie, Histologie und mikroskopische Anatomie des Menschen, 10th edn. Huber, Bern

Carr KE, Toner PG (1982) Cell structure. An introduction to biomedical electron microscopy. 3rd edn. Churchill Livingstone, Edinburgh

Constantinides P (1974) Functional electronic histology. Elsevier, Amsterdam

DeRobertis EDP, Nowinski WW, Saez FA (1975) Cell biology, 5th edn. Saunders, Philadelphia

Ebe T, Kobayashi S (1972) Fine structure of human cells and tissues. Igaku Shoin, Tokyo

Fawcett DW (1981) The cell, 2nd edn. Saunders, Philadelphia

Freeman WH, Bracegirdle B (1969) An atlas of histology. Heinemann, London

Fujita T, Tanaka K, Tokunaga J (1981) SEM atlas of cells and tissues. Igaku-Shoin, Tokyo

Ham AW, Cormack DH (1979) Histology, 8th edn. Lippincott, Philadelphia

Hamilton WJ, Boyd JD, Mossman HW (1972) Human embryology, 4th edn. Hefer, Cambridge; Williams and Wilkins, Baltimore

Hirsch GC, Ruska H, Sitte P (eds) (1973) Grundlagen der Cytologie. VEB Fischer, Jena

Ganong WF (1981) Review of medical physiology, 10th edn. Lange Medical, Los Altos

Junqueira LC, Carneiro J (1980) Basic histology, 3rd edn. Lange Medical, Los Altos

Kessel RG, Kardon RH (1979) Tissues and organs: a text-atlas of scanning electron microscopy. Freeman, San Francisco

Krstić RV (1979) Ultrastructure of the mammalian cell. An atlas. Springer, New York

Krstić RV (1984) Illustrated encyclopedia of human histology. Springer, Berlin

Laguens RP, Gómez Dumm CLA (1969) Atlas of human electron microscopy. Mosby, St. Louis

Lentz TL (1971) Cell fine structure. An atlas of drawings of whole-cell structure. Saunders, Philadelphia

Leonhardt H (1985) Histologie und Cytologie des Menschen, 7th edn. Thieme, Stuttgart

Maillet M (1975) Abrégé de cytologie. Masson, Paris

Maillet M (1976) Les épithéliums de revêtement. Vigot Frères, Paris

Maillet M (1976) Le tissue musculaire. Vigot Frères, Paris

Maillet M (1977) Les épithéliums glandulaires. Vigot Frères, Paris

Maillet M (1977) Le tissue nerveux. Vigot Frères, Paris

Matthews JL, Martin JH (1971) Atlas of human histology and ultrastructure. Lea and Febiger, Philadelphia

Metzner H (ed) Die Zelle. Struktur und Funktion. Wissenschaftliche Verlagsgesellschaft, Stuttgart

Motta P (1984) Anatomia microscopica. Atlante di microscopia ottica ed elettronica, 3rd edn. Dr. Francesco Vallardi, Milan

Motta P, Andrews PM, Porter KR (1977) Microanatomy of cell and tissue surfaces: An atlas of scanning electron microscopy. Lea and Febiger, Philadelphia

Orci L, Perrelet A (1975) Freeze-etch histology. A comparison between thin sections and freeze-etch replicas. Springer, Berlin

Reale E (1973) Elektronenmikroskopie der Zellen und Gewebe. Fischer, Stuttgart

Rhodin JAG (1974) Histology. A text and atlas. Oxford University Press, London

Sandborn EB (1970) Cells and tissues by light and electron microscopy. Vols I and II. Academic, New York

Schade JP (1971) Die Funktion des Nervensystems. Fischer, Stuttgart

Weiss L (ed) (1983) Cell and tissue biology, 5th edn. Macmillan, New York

Wheater PR, Burkitt HG, Daniels VG (1979) Functional histology. Churchill Livingstone, Edinburgh

Index

All figures correspond to the *plate numbers* and not to page numbers. **Boldface** figures indicate the main treatment of the subject.

A-band **125–130**
Absorption cavities **105**
Absorptive cells **14**, 31
Acetylcholine 168, 179
Acinus 33, **38**, 42, 43
ACTH 73
Actin myofilaments 119–122, 127, 130, 131, 136
– –, molecular structure **131**
Actin myosin interaction **131**
Actinin 122
Active zones, *see* Zones, active
Adenosine monophosphate, cyclic, *see* cAMP
Adenyl cyclase **163**
Adipocyte
–, brown **59–61**
–, immature 48
–, mature 48
–, serous 58
–, white 49, **57, 58**, 59, 62, 77
Adipose cell, *see* Adipocyte
Adipose tissue, brown 49, **59, 60**
– –, –, occurrence 59
– –, –, organization 59
– –, –, vascularization 59, 60
– –, immature 92
– –, multilocular, *see* Adipose tissue, brown
– –, white 5, 49, **56, 57**
– –, –, classification 56
– –, –, innervation 57
– –, –, occurrence 56
– –, –, organization 56, 57
– –, –, structural 56
– –, –, vascularization 57
Adrenal gland 68
– medulla 144
Adrenocorticotropic hormone, *see* ACTH
Alpha nerve fibers 182
Amacrine cells 157
Ameboidism 66, 67, 111
Ameloblasts 28
Amniotic cavity 1, 25
– epithelium **25**, 52
– fluid 25
– vacuoles 25
Amphicytes, *see* Satellite cells
Amputation neuroma 167
Anisotropy, skeletal musculature 130
Annulospiral nerve ending, *see* Nerve endings, annulospiral
Antidiuretic hormone 162

Antilymphocyte serum 115
Aponeurosis 49, **81**
APUD system 144
Arachnoid membrane 153, 171, 174
Area postrema 154
Areolar tissue, *see* Connective tissue, loose
Arteriole 45
–, afferent 15
–, efferent 15
Artery
–, coronary 139
–, interlobular 15
Articular capsule 114
– cartilage, *see* Cartilage, articular
– surface 101, **103**
– –, artificial **6**
Asbestos cartilage 85
Astrocyte
–, fibrous 144, 146, 147, **148, 151**
–, protoplasmic 144, 146, 147, **148–150**, 153
–, velate **148, 152**
ATP 131, 150
Atrial granules 137
Atrioventricular node 140
Atrophy **4**
Auricle 86
Autografting, *see* Transplantation, autoplastic
Autonomic ganglia 145, 146, **189**
– ganglion cells, *see* Nerve cells, autonomic
Axolemma 164, 179
Axon 61, 146, 149, 152, **155, 157**, 160–162, 164–167, 170, 173, 175, 178, 179, 185–188
–, hillock 155, 171, 190
–, naked **164**, 180, 190
–, neurosecretory **162**
–, postganglionic 170
–, preganglionic 170
–, supraependymal **147**
–, thickenings 184, 187, 188
Azure II 69

Bands, of Büngner 167
Bars, synaptic, *see* Synaptic bars
Basal
– cells **17–20, 22–24, 26**, 187
– labyrinth 11, 12, 25
– lamina 7, 9–11, 13–15, 17–21, 23–26, 31, 38, 39, 41, 58, 61, 119, 120, 123, 124, 127, 129, 132, 133, 136–138, 141, 149, 150, 152, 154, 156, 164, 166, 172, 173, 176, 179, 183, 185, 186, 190

393

Basal
- plate (placenta) 52
- processes 24
Basalis (mucous coat of uterus) 47
Basket cells (cerebellum) 157
Bile pigments (RHS) 68
Bilirubin 55
Bladder, urinary 117
Blastema 50
Blastocoele 1
Blastocyst 1, 47
Blastomere 1, 2
Blood
- corpuscles, red, see Erythrocytes
- transfusion 6
- vessels, artificial 6
Blood-brain barrier 148, 153, **154**
- -, areas of brain without **154**
Blood-nerve barrier **177**
Bone, see Bony tissue
-, cancellous **106**
-, cartilage, see Bone, endochondral
-, compact **106**
-, endochondral 91, 95, 99, 101, **102**
- formation, chondral, see Bone formation, indirect
- -, direct **91–94**
- -, indirect **95–104**
- -, intramembranous, see Bone formation, direct
- -, perichondrial **95**
- -, secondary 95, **105**
-, ground substance 94
-, immature 91, 95, 99, 105
-, mature 95, **108**
-, matrix 91, 92, 94
-, membranous 92, 95
-, woven, see Bone, immature
Bony tissue **91–112**
Boutons terminaux 149, 155, 160, 162, **168, 169**, 190
Bowman's membrane 22, 81
Bridges, interfibrillar 80
Bronchi 117
Bronchioles 117
Brown adipose cells **59–61**
Bundle, of His 140
Bursa, of Fabricius 70

C cells 45, 144
Ca^{2+} ions 119, 129, 131, 138
Calcification, limit of 99
- zone **99**
Calcitonin 45, 94
Callus, bony **112**
-, fibrocartilaginous **112**
Cambrium 97, 104, **112**
cAMP 57, 58, 61, 105, 138, **163**
Canaliculi
-, intercellular 39
-, intracellular 37
- of bone **107, 109, 110**

Canalis nasopharyngicus 27
Capillaries 12, 14, 26, 37–39, 43–45, 50, 56–60, 62, 63, 74, 77, 85, 87, 91–93, 100, 102, 105, 113, 125, 128, 134, 135, 146–156, 160, 162, 172, 173, 175, 177, 178, 181, 187, 188
Capsular matrix **84**, 85, **89**
Capsule (lymph node) 53, 54
Cardiac muscle cells 134, **135, 137**
- - - fibers **134–136**, 138
- - -, longitudinal section 134, 135, **137, 138**
- - -, transverse section **136**
- musculature 116, **134–142**
- -, necrosis **139**
- -, transplantation **139**
- -, vascularization **139**
- valves, artificial 6
Cardiodilatin 137
Carotenoids 58
Cartilage, see also under Cartilaginous tissue
-, articular **103**
-, bone, see Bone, endochondral
-, capsule, see Capsular matrix
-, cells, see Chondrocytes
-, ground substance **83, 84**, 89
-, matrix **83**, 87, 88, 89, 96, 97
Cartilaginous tissue **83–89**
-, elastic 48–49, **86, 87**
-, -, chondrocytes **86–88**
-, -, elastic fibers **86**
-, -, isogenic groups **86**
-, -, perichondrium **86, 87**
-, fibrous 48, 49, **89**, 95
-, -, chondrocytes **89**
-, -, collagen fibers **89**
-, -, lacunae **89**
-, -, territorial matrix **89**
-, hyaline 5, 48–49, **84, 85**, 90, 96, 98, 112
-, -, appositional growth **83, 85**, 95
-, -, -, interstitial **83, 85**, 95, 101
-, -, calcified **96, 99–102**, 103
-, -, chondrocytes **83–85, 88**
-, -, fibrils, systems of **84**, 85
-, -, histogenesis **83**
-, -, intercellular substance **83–85**
-, -, transplantation 114
-, mesenchymal 83
Cartilaginous debris, mineralized **100**
Casein granules 40
Catecholamines 180
Cells, basal 17, 18, 19
-, bipolar 26
-, centroacinar 38
-, chromaffin 144
-, ciliated **16, 18**
-, columnar 17
-, cuneate 18
-, endocrine 18, 37
-, facet 19
-, fixed (loose connective tissue) **62–65**
-, perichondral 85
-, secretory (oviduct) 16

–, serous   39
–, slender   **35**
–, superficial, *see* Cells, facet
–, wandering (loose connective tissue)   **62**, 66, **67**, 69–73
–, –, role in graft rejection   **115**
Cell groups, isogenic   84, 85
– interdigitations   16
– protrusions (thyroid gland)   46
– union, closed   **2**, 27, 29, 160
– –, wide-meshed   **2**, 27–28
Cement line   **107**, 108
Cementocytes   **113**
Cementum   113
Central canal   143
– nervous system   2, 145
– – –, relationship between neurons and glia cells   **146**
Cerebellar glomeruli   148, 152
Cerebellum   145, 148
Cerebrospinal fluid   12
– – contacting neurons   **147**, 162
Cerebrum   145
Chief cells (gastric glands proper)   37
Cholinesterase   179
Chondral bone   **102**
Chondroblasts   83
Chondroclasts   48, **97–100, 111**
Chondrocytes   48, **83–89, 96**, 98, 99, 103
–, degenerated   96, **98**
–, hypertrophied   **96, 99, 100**
Chondroitin 4- and 6-sulfates   64, 84, 85, 88, 91
Chondrones   **84**, 103
Chordal cells   **90**
– reticulum   90
– tissue   **90**
Chorionic plate (placental)   52
– villi   25
Choroid   74
– plexus   **12**, 144, 154
Chromaffin cells, *see* Cells, chromaffin
Chromatolysis   **167**
Chylomicrons   **31**
Cilia   7, 8, 16, 18
Ciliated cells *see* Cells, ciliated
– epithelium   **18**
Circular myofibrils, hypolemmal   140, **142**
Cleft, synaptic, *see* Synaptic cleft
Cochlea   26
Cohnheim's fields   **125**, 134
Collagen fibers, *see* Fibers, collagen
– microfibrils, *see* Microfibrils, collagen
– secretory granules   65
Collaterals, *see* Paraxons
Collecting tubule (kidney)   **15**
Colloid   32, 45–46
Cone cells   13
Connecting portion (nephron)   15
Connective tissue   1, 2, **3, 48–82**
– –, "cellular"   49, **76**
– –, classification   49

– –, dense irregular   49, 53, **78**
– –, dense regular   49, **79**
– –, elastic   **82**
– –, embryonic, *see* Mesenchyme, Connective tissue, gelatinous
– –, gelatinous   49, **52**
– –, loose   14, 47, 49, **62–73**, 79, 92, 118
– –, –, special forms   49, **74–77**
– –, origin   **48**
– –, perineurial   **171, 174, 176**
– –, reticular   5, 49, **53**, 54, 56
– –, retiform   49, **77**
– –, transplantation   **114**
Connexons   136, 137
Cornea
–, endothelium   10, 81
–, epithelium   22, 81
–, transplantation   114
Corneal stroma   22, **81**
Corpus cavernosum penis   117
Cortex (lymph nodes)   53
Corti, organ of   26
Cortisol   73, 88, 115
Cranial bone   92
Crescents, serous, *see* Demilunes, serous
Crusta (transitional epithelium)   19
Crypts, intestinal   47, 118
Cuticular capsule   30
Cytostatics   115

Decidual cells   76
Degeneration   **4**
Demilunes, serous   34, **39**
Dendrites   146, 149, 152, **157, 159–161**, 173, 190
Dendritic spines   150, 157, 159
Dense bodies   119–121, **122**, 123
– connective tissue, *see* Connective tissue, dense
– core vesicles   120, 180
Density, postsynaptic   **168**
–, presynaptic   **168**
Dental lamina   27, 28
– cementum   49, **113**
– papilla   27, 28
– pulp   113
– sac   28
Dentin   28, 49, **113**
Dentinal matrix vesicles   113
– tubules   **113**
Dermatan sulfate   65
Descemet's membrane, *see* Membrane, Descemet's
Desmin microfilaments, *see* Intermediate microfilaments
Desmosomes   22, 23, 136, 141
Diads   **138**
Diametric growth of long bones   **104**
Diaphysis   95–98, 99, 101–103
Differentiation   **2**
Disc, articular   89
–, intercalated   **134–138, 141**
–, intervertebral   89, 90
–, membranous   13

395

Discoid vesicles, *see* Vesicles, discoid
Distal tubule (nephron)   15
Dopa reaction   24
Dopamine   180
Duct, cochlear   26
Ductus deferens   117
– epididymidis   17, 117
Duodenum   43
Dura mater   92, 153, 174
– –, transplantation   114
Dyad   138

Ectoderm   1, 48, 143
Elastic fibers, *see* Fibers, elastic
Elastic connective tissue, *see* Connective tissue, elastic
Elastic ligaments, *see* Connective tissue, elastic
Elastin   65
Embryoblast   1, 25
Enamel   28, 113
Enamel epithelium, inner   **28**
– –, outer   **28**
– organ   **27, 28**
– pulp   27, 28
End corpuscles, encapsulated, *see* Meissner's corpuscles, Vater-Pacini corpuscles
End plates, motor   **178, 179**, 181–183
Endings, *see* Nerve endings
Endocardium   117
Endochondral bone, *see* Bone, endochondral
Endochondral ossification zone   **99**
Endoderm   1, 48
Endolymph   26
Endometrium   47
Endomysium   **125, 126**, 134, 135, 137, 178, 179, 182
Endoneurial sheath   **164, 165**, 166, **176**
Endoneurium   171, **172**, 173, 175, **176, 177**, 189
Endoplasm   **119**, 134, **137**
Endotendineum   **79**
Endothelial cells   **9**, 48, 68
– – of cornea   **10**
Enterocytes, *see* Absorptive cells
Ependyma   162
Ependymal cells   143, 144, **147**
– fibers   147
Ependymoblasts   143, 144
Ependymocytes, *see* Ependymal cells
EPF (exophthalmus-producing factor)   57
Epidermis   8, **23, 24**, 185, 187
Epimysium   125
Epinephrine   58
Epineurium   171, **175, 176**
Epiphyseal centers   **99, 101**
Epiphyseal disc, *see* Epiphyseal plate
Epiphyseal plate   **101, 103**
Epiphyses (bony tissue)   **95–97**, 98, 99, **101–103**
Epiploic appendices   56
Epitendineum   **79**
Epithelial migration   47
– tissue   1, **3, 7–47**

Epithelium, atypical   **27–29**
–, excretion   **31**
–, gas exchange   **31**
–, gliding   **31, 57**
–, keratinized stratified squamous   5, 7, 8, **23, 31**, 47
–, localization   **8**
–, morphological changes   **30**
–, nonkeratinized stratified squamous   5, 7, 8, **22**
–, protection   **31**
–, pseudostratified columnar   5, 7, 8, **17, 18**, 36, 47
–, regeneration   47
–, resorption   **11, 14, 15, 31**
–, secretion   11, 12, 16, **25, 26, 31**, *see also under* Glands
–, simple columnar   7, 8, **14, 15**, 30, 31, 143
–, simple cuboidal   7, 8, **11–13**, 25, 30, 76, 123
–, simple squamous   7, 8, **9, 10**, 31, 57
–, stratified columnar   7, 8, **21**
–, transitional   5, 7, 8, **19, 20**
–, transplantation   47
–, vascular   **26**
– with cilia   7, 8, **16**
Ergastoplasm   72
Erosion zone   **99, 100**
Erythroblast   55, 68
Erythrocytes   67
–, nucleated   97
–, old   68
Esophagus   62, 117
Estradiol   88
Excretory duct (of glands)   32, 34, 38, **42**, 43, 47
Exocytosis   9, 162
Explantation   6
Eye color   74

F-actin   131
Facet cells, *see* Cells, facet
Facial bones   92
Fascia   **125**
– adherens   136, 141, 142
– lata   114
Fat absorption   **31**
– cell, *see* Adipocyte
– droplets, *see* Lipid droplets
Fatty acids   58
Fertilization   1
Fiber bundles, collagen   78–80
Fibers
–, collagen   62, 63, 74, 76, 77, 89, 184, 187
–, elastic   62–64, **65**, 74, 77, 78, 81, **82**, 86, **87, 88**, 119, 120, 186
–, masked   **84, 85**
–, reticular   53, 57, 58, 62, 76, 77, 104
Fibrae tendinae, *see* Tendon fibers
Fibrils, collagen   **107**
Fibroblast   48, **62, 64**
Fibrocartilage, *see* Cartilaginous tissue, fibrous
Fibrocyte   48, **62–64**, 76–78, 82, 85, 172, 176, 177, 182
Fibronectin   65

Fibrous vagina   79
Filopodia   55, 67
"First messenger"   **163**
"Fish-eye cells"   167
Flower-spray endings   **182, 183**
Follicles, see Thyroid follicles
Follicular phase of menstrual cycle   16, 47
Foramen nutricium   98, **106**
Fracture   112
–, setting   112
Fusiform vesicles see Vesicles, discoid

G-actin   131
GABA   168
Galea aponeurotica   92
Gallbladder   117
Gamma nerve fibers   **182, 183**
Gamma globulins   70, 72
Gamma-aminobutyric acid, see GABA
Ganglia, autonomic, see Autonomic ganglia
–, prevertebral, see Ganglia, sympathetic
–, spinal, see Spinal ganglia
–, sympathetic   144, 145, **146**, 170, **189**
Ganglion cells, see Nerve cells
Gap junction, see Nexus
Gastric gland proper   37
– pits   37
Gelatinous tissue   **52**
Germ layer, bilaminar   1
– –, trilaminar   1
Germinal epithelium   76
– layer   **22, 23**, 41, 47, 185
Gianuzzi's crescents, see Demilunes, serous
Glands   **32–46**
–, alveolar, unistratified   **33, 34, 44**
–, –, pluristratified   **33, 34, 41**
–, apocrine   **40**
–, compound   **34**
–, classification   **33**
–, ductless, see Glands, endocrine
–, development   **32**
–, endocrine   **43–46**
–, endoepithelial   32, **36**
–, exocrine   **33, 42**
–, –, classification   **33**
–, –, development   **32**
–, –, form   **34**
–, –, intercalated ducts   **38**, 39, 42
–, –, striated ducts   **42**
–, exoepithelial   **32**
–, gastric proper   37
–, heterocrine   **33, 37**
–, holocrine   **41**
–, homocrine   **33**
–, lobes   **42**
–, lobules   **42**
–, mucous   **33**
–, sebaceous   **34, 41**, 47, 185
–, seromucous   33, **39**
–, serous   **33**
–, terminal secretory portions   **33, 34**

–, tubular, simple   **33, 34, 37**
–, –, branched   **34**
–, tubuloacinar   **33, 34, 38**
–, –, compound   **42**
–, tubuloalveolar   **33, 34, 39**
–, unicellular   **35**
–, uterine   47
Glandulae sebaceae, see Glands, sebaceous
– uterinae, see Glands, uterine
Glia   146, 147, **148–156**
– of CNS   146, **147–156**
– of peripheral nervous system   146, **147**
Glia cells of CNS   146, **147–156**
– – of peripheral nervous system   146, **147**, 164, **172**
Glial fibers   147
– fibrillary acidic protein   150
Glioblasts   143, **144**
Gliofibrils   150, 151
Gliofilaments   151
Gliosis   151
Glucagon   44
Glycogen   21, 136–138, 141, 151
Glycoproteins   65
Glycosaminoglycans   65
Goblet cells   14, **18**, 32, **35**
Golgi funnel   166
– tendon organ   **184**
– – –, capsule   **184**
– type I/II neurons   **157**, 158
Graft rejection   **115**
Grafting, see Transplantation
Granulation tissue   47
Granule cell (cerebellum)   152, 159
Granules, eosinophilic   **73**
–, metachromatic   **69**
Granulocyte, eosinophilic   62, 63, **73**
Grapelike endings, motor   182
Gray rami communicantes, see Rami communicantes, gray
Ground substance, connective tissue   52, 62, 63, **64**
– –, cartilage   84, **85**, 87–89, 96
Groups, isogenic   84, 85, **99**, 103
Growth, appositional, see Cartilaginous tissue, hyaline
–, definition   **4**
–, interstitial, see Cartilaginous tissue, hyaline
Growth hormone   84, 94

H-band   **127, 130**
Hair   92
– follicle   186
– – sheaths   41, 47, 185, 186
– – –, connective tissue   186
– – –, inner   186
– – –, outer   186
Hassall's corpuscles   **27, 29**
Haversian canals   **105–109**, 112
– lamellae, see Osteons
– spaces   **105**

Heart 134, *see also under* Cardiac
Helper cells 115
Hemidesmosomes 22, 186
Hemocytoblast 48
Hemoglobin metabolism 68
Heparin 65, 69
Herring bodies **162**
Heterotransplantation, *see* Transplantation, heteroplastic
"Hibernation gland" 59
Hillock, axon 155, 160, 171
Hilus (lymph node) 53
His bundle, *see* Bundle of His
Histamine 69, 73
Histiocyte 48, 54, **62, 63, 66, 68**, 77, 115
Histocompatibility **6**
Histology **2**
Histophysiology **2**
Homografting, *see* Transplantation, homoplastic
Hormone-receptor complex 163
Hormones 32
Hortega cells, *see* Microglia
Howship's lacuna 93, **111**
5-HT, *see* Serotonin
Hyaluronic acid 63, 65
Hyaluronidase 52, 63
Hyaluronidase-resistant substance 16
Hydroxyapatite 91, 96, 100, 111
Hydroxyproline 65
Hyperplasia **4**
Hypertrophic zone 96, **98–100**
Hypertrophy **4**
Hypophysis 68, 154, 162
Hypothalamohypophyseal tract 162
Hypothalamus 162

I-band **125–130**
Immature bone, *see* Bone immature
Immunity, cellular **115**
–, humoral **115**
Immunoblasts **115**
Immunoglobulins, *see* Gamma globulins
Implantation **6**
Impulse-conducting system **140–142**
Infarction 139
Inner circumferential lamellae **106**
– synovial sheet 79
Insulin 44
Intercalated disc, *see* Disc, intercalated
– ducts **39**
Intercellular matrix substance 16
– space 2
– substance 2, 50, 63, **65**, 90
– –, calcified **91, 94, 96, 99, 100**
Intermediate cells (Stria vascularis) 26
– microfilaments 119–121, 136
– microglia cells 156
Internode 155, **164**
Intestinal villi, *see* Villi, intestinal
Intracellular canaliculi, *see* Canaliculi, intracellular
Involution **4**

Iris, color of 74
Iron 55
– metabolism (RHS) 68
Islet cells 44
– of Langerhans, *see* Pancreatic islets
Isogenic groups, *see* Groups, isogenic
Isografting, *see* Transplantation, isoplastic
Isoprismatic epithelium 11

Jejunum 14
Junction, myotendinal, *see* Myotendinal junction

Keratan sulfate 65, 81
Keratin 23
– scales 23, 31
Keratinization 23
Keratocytes 10, **81**
Keratohyalin granules 23, 29
Keratoplasty 81, 114
Killer cells **115**
Kinetosomes 18
Kinocilia, *see* Cilia
Knee cap 80
Kupffer's cells 68

Labyrinth, basal, *see* Basal labyrinth
Lacunae (cartilaginous tissue) **84, 85**, 96, 98, 99, **100**
Lamellae, concentric **105–110**
–, interstitial **105, 106**, 107
– of bone **105–110**
Lamellar bone **105–108**
Lamina muscularis mucosae 62, 118
– propria 62
Laminin 65
Langerhans, islets of, *see* Pancreatic islets
Larynx 43, 86
Lens 30
– development 30
– ectoderm 30
– fibers **30**
– nucleus 30
– vesicles 30
Leptomeninges 144, 153
Leptomeric myofibrils, *see* Myofibrils, leptomeric
Leukotrienes 69
Lieberkühn, crypts of, *see* Crypts, intestinal
Ligament, spiral 26
Ligamenta flava 82
Ligaments, elastic 49, **82**
–, yellow, *see* Ligaments, elastic
Ligamentum nuchae 82
Link glycoprotein 65
Lipase 31, 61
Lipid droplets 31, 40, 41, **56–60**, 61
– metabolism (RHS) 68
Lipoblast 48
Lipochrome 58
Lipocyte, *see* Adipocyte
Lipofuscin granules 137, 171
Lipolysis 58

Lipotropins 57
Liquor amnii, *see* Amniotic fluid
Littoral cells 53, 54, 68
Longitudinal growth of long bones **99, 101, 103**
L-system 124, 128, **129, 138**
Lumen (glands) 33, 38
Luteal phase of menstrual cycle 16
Lymph capillaries 31, 53, 62, 115
– nodes 53–54, 68, 115
Lymphatic tissue 53
Lymphocytes 53–54, **62, 63**, 70, 77, 115
–, B **70, 115**
–, T **70, 115**
Lymphokines 115
Lymphoreticular tissue, *see* Lymphatic tissue
Lymphotoxin 115
Lysosomes 55, 66

Macroglia 147
Macrophages 53, **66, 67, 71**, 109, 115, 133
Mammary gland 114
– –, lactating 40
Mantle zone (neural tube) 143
– –, derivative cells **144**
Marginal folds 9
– zone (neural tube) 143
Marrow cavity, *see* Medullary cavity
Mast cell 29, 48, **62, 63, 69**, 77
Mastocyte, *see* Mast cell
Matrix, interterritorial 84, **87**
–, territorial, *see* Capsular matrix
– vesicles 88, 94
– – dentinal, *see* Dentinal matrix vesicles
Median eminence 154
Medulla (lymph node) 53
Medullary cavity, definitive 106
– –, primitive **98**
– plate, *see* Neural plate
– tube, *see* Neural tube
Medulloblasts 144
Meissner's corpuscles 185, **187**
Melanin granules 13, 23, 24, 31, 74
Melanoblasts 144
Melanocytes 23, **24, 31, 74, 75**, 144
Melanosomes 24, 75
Membrana limitans gliae perivascularis **148, 149, 153, 154**, 156
– – – superficialis 148, **153, 154**
Membrane, Descemet's 10
–, elastic 87
–, undulating 55, 67
Memory cells 115
Meniscus, articular 89
Menstrual cycle 47
Merkel's cells **185**
– disc 185
Meromyosin, heavy **131**
–, light **131**
Mesaxon, external 164–166, 177, 186
Mesaxons 165
Mesectoderm 144

Mesenchymal cells 48, **50, 51**, 91, 93, 97, 98
Mesenchyme 27–28, 48, 49, **50**, 91
Mesoderm 1, 48
Mesoglia, *see* Microglia
Mesotendineum 79
Mesothelial cells 9, 31, 50, 56, 57, 77
Mesothelium, *see* Mesothelial cells
Metachromasia 63, **69**, 84
Metaplasia **5**, 56
Microfibrillar proteins of elastic fibers and elastin **65**
Microfibrils, collagen 23, 24, 52, 60, 61, 63, 64, **65**, 80–83, 85, 87–90, 94, 110, 119, 120, 123, 126, 128, 132, 136, 137, 164, 176, 177, 182, 184, 186, 188
–, reticular 23, 24, 54, 55, 57, 60, 61, 63, **65**, 82, 119, 120, 123, 126, 127, 129, 132, 136, 137, 164, 182, 184
Microfilaments, intermediate, *see* Intermediate microfilaments
–, intracellular (adipose cells) 58, 61
Microglia 68, 144, 147, **156**, 167
Micropinocytotic vesicles, *see* Vesicles, micropinocytotic
Microplicae 22
Microridges, *see* Microplicae
Microvilli 12, 14, 15, 16, 21, 22, 25, 26, 35, 46, 55, 64, 67, 69, 70, 77, 88, 93, 111
Milk 40
Milky spots 77
Mitosis 41, 47, 50, 83, 99, 143
M-line **125**, 130
Mode of action of polypeptide hormones **163**
Modiolus 173
Monocyte 54, **62–63, 68, 71**
Mononuclear phagocyte system 68
Morphological changes in form of epithelial cells **30**
Morula 1
Mossy fibers 152
Motor end plates, *see* End plates, motor
Motor neurons 170
– unit 178
Movement, retinomotor 13
Mucigen granules 35
Mucous cells 39
– droplets 35, 39
– tissue, *see* Gelatinous tissue
Muscle, arrector (hair) 185
– cells, smooth 9, 48, 117, 118, **119, 120**, 180
– –, contraction **121, 122**
– contraction, *see* Myofibrils, contracted
– fascicle **125**, 126
– fibers, *see* Cardiac muscle fibers, Skeletal muscle fibers
– –, extrafusal 182
– –, intrafusal 181, 182, **183**
– relaxation, *see* Myofibrils, relaxed
Muscular tissue 1, 3, **116–142**
Musculature, smooth **117–123**
–, –, innervation **180**
–, –, occurrence **117**

399

Myelinization in the peripheral nervous system 164–166
– in the central nervous system **155**
Myelin lamellae 155, 164–166, 173
– sheath 146, 149, 154, 155, **164**, 166, 175, 179
Myoblasts **124**
Myoepithelial cells 39, 40, **123**
– –, spindle-shaped 123
– –, stellate 123
Myofibrillogenesis **124**
Myofibrils (cardiac musculature) **135–138**
– in the impulse-conducting system 140, **141**
–, transverse section **136**
Myofibrils (skeletal musculature) 125–129, 134
–, contracted **130, 131**
–, leptomeric **183**
–, relaxed **130, 131**
–, transverse section **130**
Myofilaments, see Actin myofilaments, Myosin myofilaments
Myoglobin 127
Myosin myofilaments 119–122, 127, 130, 131, 136
–, molecular structure **131**
Myotendinal junction **132**
Myotome **124**
Myotubes **124**

Neck mucous cells 37
– of glands 34
Necrosis **4**
Nephron 11, 15
Nerve 145, 170, **171, 174,** 175
Nerve cells 2, **146**, 148–150, 152, **157–163**, 190
– –, apolar **158**
– –, arrangement **159**
– –, autonomic 170, **189, 190**
– –, bipolar **157, 158,** 173
– –, form **157**
– –, intraneural 170
– –, motor 170
– –, multipolar **157, 158,** 170, 189, **190**
– –, pseudounipolar **157, 158,** 170, **172**
– –, types **158**
– –, unipolar **158**
Nerve endings, afferent 170, **181–188**
– –, annulospiral **181–183**
– –, around hair root **186**
– –, autonomic 60, **120,** 170, **180**
– –, efferent 170, **178–180**
– –, endoepithelial 185
– – in connective tissue 185
– –, intraepidermal 23, 185
– – of afferent nerve fibers **181–188**
– – of efferent nerve fibers **178–180**
– –, polysynaptic 169
– –, sensory 170, 186
Nerve fascicle 42, 174, 175, **177,** 188
Nerve fibers **164–167**
– –, afferent 146, 170
– –, alpha 182
– –, autonomic 190
– –, efferent 146, 170
– –, "en grappe" **178**
– –, extrafusal 181
– –, gamma 182
– –, intrafusal 181
– –, myelinated 2, 126, **146, 150, 151,** 154, **155, 164, 165,** 172–175, **167,** 177, 178, 181, 184, 187, 189
– –, postganglionic 170
– –, preganglionic 170
– –, regeneration **167**
– –, sensory, Ia 182
– –, – around the hair follicle **186**
– –, unmyelinated 2, 23, 38, 44, 45, 60–63, 74, 77, 79, 109, 126, **146, 150,** 154, **164,** 172, 174, 175, **176,** 177, 181, 184, 187, 189
Nerves, mixed **174**
–, peripheral, see Nerve
–, root, posterior **171, 174**
–, –, anterior **171, 174**
–, structure **175**
Nervous system, peripheral **145, 170**
– –, relationship between neurons and glia cells **146**
Nervous tissue 1, 3, **143–190**
– –, distribution **145**
Neural crest 143
– –, derivative cells **144**
– folds 143
– groove 124, 143
– plate 143
– tube 143, 144
– –, derivative cells **144**
Neuroblasts 143, **144**
Neurocytes, see Nerve cells
Neuroendocrine nerve cells **162, 163**
Neuroepithelium 143
Neurofibrils 157, **161,** 171, 189, 190
Neurofilaments 161, 164, 165, 168
Neuroglia, see Glia
Neurohypophysis 154, 162
Neurokeratin 164, 175
Neurolemma **164**
Neuromuscular spindles 126, **181–183**
– –, capsule **182**
Neurons 143, **146, 157,** see also under Nerve cells
–, cerebrospinal fluid-contacting, see Cerebrospinal fluid-contacting neurons
Neurophysin 162
Neuropil 149, 154, 156, **160**
Neuroplasm 146, 160
Neurosecretory granules 162
– nerve cell, see Neuroendocrine nerve cells
Neurotendinal spindle, see Golgi tendon organ
Neurotubules 162, 164, 165, 168
Nexus **120, 136, 137,** 141, 168, 199
Nissl bodies **160, 162,** 167, 172, 173, 190
Nodes of Ranvier, see Ranvier's nodes
Norepinephrine 58, 61, 180
Notochord **90,** 124, 143
Nuclear bag fibers **181–183**

– chain fibers **181–183**
Nucleus paraventricularis 162
– pulposus 90
– supraopticus 162

Odontoblasts 28, 48, **113**
Oligodendrocytes 144, 146, **147, 155**
Omentum majus 77
Optic stalk 30
– vesicle 30
Orcein staining 82, 86
Organ **2**
Osmium tetroxide 57
Osmoreceptors 147
Ossification, centers of 91, 92, 101
Osteoblasts 48, **91–95**, 98–101, **102, 104**, 105, **109**
Osteoclasts 48, **92, 93, 102, 104**, 105, **109–111**
Osteocytes **91–93, 95, 102, 104**, 105–106, **107–110**
Osteoid **91**, 99, 100, 102, 104, 112
Osteons **105–109**
Osteon, erosion 105
Osteoporosis 111
Osteosynthesis 112
Outer circumferential lamellae **106, 108**
Oviduct 117
Ovum 1
Oxytocin 162

Pacemaker 142
Pancreatic islets **43, 44**
Paraneurium 175
Paratendineum **79**
Parathormone 94, 111
Paraxons **157**
Parietal cells, of gastric glands proper 37
Particle in the nexus 120, 136
Patella, *see* Knee cap
Patellar reflex 182
Periaxial space 181
Pericapillary spaces, *see* Spaces, pericapillary
Perichondrium 83, **84–87**, 95
Pericytes 26, **62–64**, 77, 154, 156
Perikaryon 157, **160**, 161, 167, 172, 189, 190
Perimysium **125**
Perineurial epithelium 171, 174, 175, **176, 177**, 184
Perineurium 171, 174, **175**, 181, 182, 184, 187, 188
–, epithelial lamellae 177
–, structure **176, 177**
Periosteal bony band **95–98, 104**
– bud **97**
Periosteum 92, 95, 97–98, 101, 104, 106, 108, **112**, 125
Peritoneum, visceral 56, 67
Perivascular feet **148, 151, 153, 154**, 156, 160
– space (bony tissue) 105, **109**
Perspiratio insensibilis 8
Phagocytosis (RES) 68
Phagolysosomes (macrophages) 66, 67, 73
– in microglia 156
– in osteoclasts 111
Phosphatase, alkaline 94
Pia mater 153, 154, 174

Pigment cells, *see* Melanocytes
– connective tissue **74**
– epithelium **13**
Pineal body 154
Pinealocyte 144
Pituicytes 144
Pituitary, *see* Hypophysis
Placenta 50
Plaques 19
Plasma cells 54–55, **62, 63, 72**, 115
Plate, epiphyseal, *see* Epiphyseal plate
Polyblast 71
Polyribosomes, helicoidal 124
Precapillaries 178
Precartilaginous tissue **83**
Predentin **113**
Prevertebral sympathetic ganglia 170
Prickle-cell layer, *see* Stratum spinosum
Prickle cells **22, 23**
Primary bundle, *see* Muscle fascicle
Primary ossification center **96**
Primitive organs 56
– streak 48
Processes, cytoplasmic 165, 186
Proelastin molecules 65
Proliferation zone 96, **98–100**
Proliferative phase of menstrual cycle 16, 47
Pronucleus 1
Prostaglandins 69
Proteoglycans 65, 91
Protochondral tissue, *see* Precartilaginous tissue
Proximal tubule (nephron) 15
Pseudoarthrosis 112
Pseudopodia 55
Pulmonary alveoli 31
Pulp cavity, *see* Dental pulp
Purkinje cells (cerebellum) **157, 159**
– – (impulse-conducting system) **140–142**
– fibers **140, 141**
Pyramidal cells 157, **160**

Rami communicantes, gray 170, 174
– –, white 170, 174
Ranvier's cross 165
– nodes 146, 149, 151, 155, 164, **165**, 167, 175, 178
Rays, ultraviolet 31
Regeneration, complete, physiological 47
–, cyclic 47
–, definition **4**
–, of bone **112**
–, reparative 4
Regenerative power 4
Renal corpuscle 15
Replacement cells 17
RES, *see* Reticuloendothelial system
Reserve zone **99**
Residual bodies 66, 156
Resorcin-fuchsin staining 82, 86
Reticular cells 48, **53–55, 68**, 115
– –, phagocytosis **55**
– –, undifferentiated 55

401

Reticular-epithelial cells 27, 29
Reticular fibers 53, 57, 58, 62, 76, 77, 104
– microfibrils, see Microfibrils, reticular
– tissue, see Connective tissue, reticular
Reticuloplasm 72
Reticuloendothelial system 68
Reticulohistiocytic system 68
Reticulum, sarcoplasmic 124, **128, 129**, 137, **138**, 141
Retiform connective tissue, see Connective tissue, retiform
Retinomotor movements 13
Rheotaxis, positive 16
RHS, see Reticulohistiocytic
Ribbons, synaptic, see Synaptic ribbons
Rod cells 13
Ruffled border 111
Russell bodies 72

Sac of gland 41
Sarcolemma 120, **125–127, 132**, 180, 183
Sarcomere 127
Sarcoplasm 125
Sarcoplasmic reticulum, see Reticulum, sarcoplasmic
Satellite cells (neurons) 144, 146, 147, **171, 172**, 189, 190
– – (skeletal musculature) **124**, 126, **127**, 133, 178
Scarlet red 56
Schmidt-Lanterman incisures 155, **164, 166**
Schwann's cells 61, 144, 146, **164**, 166, 167, 172, 173, 184–186, 188, 190
Sclerotome 124
"Scrolls" 69
Sebum 41
"Second messenger" 163
Secretion, apocrine **40**
–, eccrine **35**, 37, 38, **40**
–, heterocrine 33
–, holocrine 33, **41**
–, homocrine 33, **38**
Secretory granules 16, 46
– phase of menstrual cycle 16
Segments, Schmidt-Lanterman 166
Sensory cells 144
Septum, posterior median 147
Serotonin 69, 168
Serous cells 38, 39
– fat cells 58
Sesamoid bone 80
Shaft, see Diaphysis
Sharpey's fibers 80, **108, 113**
Silver staining of the cell boundaries 9, 10
Sinus, endothelial macrophages of the lining 68
Skeletal muscle fibers **125–128**, 178, 179, 182, 184
– – –, myotendinal junction **132**
– – –, red **128**
– – –, white **127**
Skeletal musculature 116, **124–133**, 181, 182
– –, contraction, see Myofibrils (skeletal musculature), contracted

– –, differentiation **124**
– –, histogenesis **124**
– –, innervation 146, **178–179**
– –, regeneration **133**
– –, –, continuous **133**
– –, –, discontinuous **133**
Skin 92
–, injury 47
Smooth muscle cells see Muscle cells, smooth
Soma (nerve cells) 146, 169
Somites 124
Space, subarachnoid 153, 171, 174
–, subdural
– of Virchow-Robin 153
Spaces, intercellular 14, 15, 23, 25, 31
–, pericapillary 162
Spermatozoon 1, 17
Spherulites, see Matrix vesicles
Spicules 91, 102
Spinal cord 148, 170, 174
Spinal ganglia 145, **146**, 170, **171**, 174
– –, capsule **171**, 174
Spinal ganglion cells 144, 146, 157, 170–171, **172**
Spinal nerve, see Nerve
Spindle, neuromuscular 126, **181–183**
Spiral ganglion (of organ of Corti) 26, 173
Spleen 68
Spreading factor 63
Stereocilia 7, 17
Storage (RHS) 68
– adipose tissue **56**
– gland 32, 43, **45**
Stratum basale **22, 23**
– corneum 23
– germinativum, see Germinal layer
– lucidum **23**
– pavimentosum **22, 23**
– spinocellulare, see Stratum spinosum
– spinosum **22, 23**
Striated border 14
Striated musculature, see Cardiac musculature, Skeletal musculature
Stria vascularis 26
Structural adipose tissue **56**
Subarachnoid space, see Space, subarachnoid
Subcapsular sinus (lymph nodes) 53, **54**
Subcommissural organ 154
Subependymal layer 147
Subneural apparatus **179**
Substantia adamantina, see Enamel
– compata (bony tissue) 104, **106–108**
– spongiosa 106, **108**
Succinohydrogenase reaction 127
Sudan III 56
Superficial cells **19**
– epithelia, see Surface epithelia
Supporting tissues 1, **3**, 48, 49, **83–113**
– –, classification **49**
– –, origin **48**
– –, transplantation **114**
Suppressor cells 115

Surface epithelia 7–31
– –, classification 7
– –, functions 31
Sweat gland 123, 185
Sympathetic cells **189, 190**
– trunk 170, 174, **189**
– –, ganglia 170, 174, **189**
Sympathoblasts 144
Synapse 146, 150, 162, 163, **168**
–, axoaxonic 162, **169**, 190
–, –, "en passant" **169**, 180
–, –, inhibitory **169**
–, axodendritic 150, **169**, 190
–, –, branched **169**
–, –, crest **169**
–, –, "en passant" **169**
–, –, interdigitated **169**
–, –, reciprocal **169**
–, axosomatic 150, 162, 163, **169**, 190
–, –, invaginated **169**
–, axospinous **169**
–, chemical **168**
–, classification **168**
–, electrotonic 120, **168**
–, F-type **168**
–, reciprocal **168**
–, S-type **168**
–, type I **168**
–, type II **168**
Synaptic bars **168**
–, clefts **168**, 179
–, ribbons **168**
–, vesicles, see Vesicles, synaptic
Synaptoid vesicles, see Vesicles, synaptoid
Synovial vagina 79

Tactile cells 187
Tangential zone **103**
Tanycytes 147
Tarsal glands 34
Tela submucosa 62, 118
Teloglia cells **178**
Tendon 49, **79**, 125
– cells **79, 80**
– fibers **79**, 132, 184
–, flattened, see Aponeurosis
– microfibrils **80**, 132
–, stretching, see Tendon, Golgi tendon organ
–, transplantation 114
Terminal cisterna 128, **129**
– portions, mucous 39
– –, seromucous **33, 34**
Testosterone 88
Tetraiodothyronine, see Thyroxine
Thermogenesis 60
Thin segment (nephron) 15
Thymus **27, 29**, 115
Thyroid follicles 43, **45**, 46
– follicular cell 43, 45, **46**
– gland 32, 43, 45
– –, kinetics of epithelium **46**

Thyroxine 45, 46, 88, 94
Tigrolysis, see Chromatolysis
Tissue, bradytrophic 78, 85
–, definition **2**
–, general **1–6**, see also Connective tissue, Epithelium, etc.
–, groups **3**
–, lymphatic 53
–, pluripotential, see Mesenchyme
Toluidine blue 69
Tomes' fibers **113**
Tonofibrils 22, 23
Tooth 113
–, neck of 113
–, pulp 113
–, root 113
Trabeculae (bone) 91, 93, 94
Trachea 84, 117
Trajectories 108
Transplantation 6
–, autoplastic **6**
–, heteroplastic **6**
–, homoplastic **6**
–, isoplastic **6**
Triad 124, 128, **129**
Triglycerides 31, 58, 61
Triiodothyronine 45, 46
Trophoblast 1
Tropocollagen molecules 65, 92, 94
Tropomyosin **131**
Troponin-protein complexes **131**
Trypan blue 55
T-system 124, 128, **129, 138**
T-tubule 128, **129**, 137, **138**, 179
Tubuloalveoli 39
Tunica adventitia 62
– albuginea 76
– muscularis 117, 118, 120
Tyrosinase 24

Umbilical cord 25, 52
Union, see Cell union
Unit, motor, see Motor unit
Ureter 5, 117
Uterine glands 47
Uterus 47, 117

Vagina 117
–, fibrous, see Fibrous vagina
Varicosities (autonomic nerve fibers) 180
– (umbilical vessels) 52
Vater-Pacini corpuscles 185, **188**
–, capsule **188**
–, inner core **188**
Vegetative ganglia, see Autonomic ganglia
Vesicles, discoid 19, 20
–, flattened **168**
–, micropinocytotic 58, 61, 66, 119–121, 176, 186
–, spherical **168**
–, synaptic 61, **168**, 179
–, synaptoid 162

Villi, intestinal   14, 118
Villous processes (choroid plexus)   11, 12
Vital staining   55, 68, 154
Vitamins A, C, D_3 94
Vocal cords   82
Volkmann's canals   **106, 108**

Wallerian degeneration   **167**
Wharton's jelly   **52**
White rami communicantes, *see* Rami communicantes, white
Woven bone, *see* Bone, immature

Yolk sac   1

Z-line   127, 129, 130, 136–138
Zone of calcification   **96, 99, 100,** *see also* Primary ossification center
– of cartilage erosion   99
– of endochondral bone formation   99
– of hyaline cartilage   **96, 99**
– of hypertrophic cartilage   96, **98–100**
– of proliferation   **96, 99**
Zones, active   179
Zonula adherens   183
– occludens   9, 35, 154, 176, 177
Zonular fibers   30
Zygote   **1**, 2
Zymogen granules   38